List of Symbols

S^{β}	Beta dual of sequence space	206
(C, α)	α Cesaro mean	233
$\Gamma(f)$	Graph of a function f	169
$D^{+}, D^{-}, D_{+}, D_{-}$	Derivates	127
$[d\nu/d\mu]$	Radon–Nikodym derivative	125
$\lim_{x \to a} f(x) = b$	Limit of function	24
$\lim_{n} x_n = b$	Limit of sequence	24
$\lim_{A}$	Limit with respect to a matrix	230
$\mathbb{N}$	Set of natural numbers $\{1, 2, \ldots\}$	8
$\varnothing$	Empty set	2
$\int$	Integral	91, 118
Π	Product of sets or spaces	69
Q	Rationals	8
$\times$	Product of two sets	3
$[a, b] [a, b)$ $(a, b], (a, b)$	Intervals	14
$<, \leq, >, \geq$		7
∞	Infinity	9
$\setminus$	Set difference	2
$\sim$	Set complement	2
$\cup$	Union of sets	2
$\cap$	Intersection of sets	2
$\subset, \subseteq, \supseteq, \supset$	Set containment	2
$=$	Set equality	2
$\in$	Member of set	2

Modern Analysis

The Prindle, Weber & Schmidt Series in Advanced Mathematics

Brabenec, *Introduction to Real Analysis*
Ehrlich, *Fundamental Concepts of Abstract Algebra*
Eves, *Foundations and Fundamental Concepts of Mathematics, Third Edition*
Keisler, *Elementary Calculus: An Infinitesimal Approach, Second Edition*
Kirkwood, *An Introduction to Real Analysis*
Ruckle, *Modern Analysis: Measure Theory and Functional Analysis with Applications*

The Prindle, Weber & Schmidt Series in Mathematics

Althoen and Bumcrot, *Introduction to Discrete Mathematics*
Brown and Sherbert, *Introductory Linear Algebra with Applications*
Boye, Kavanaugh, and Williams, *Elementary Algebra*
Boye, Kavanaugh, and Williams, *Intermediate Algebra*
Buchthal and Cameron, *Modern Abstract Algebra*
Burden and Faires, *Numerical Analysis, Fourth Edition*
Cass and O'Connor, *Fundamentals with Elements of Algebra*
Cullen, *Linear Algebra and Differential Equations, Second Edition*
Cullen, *Mathematics for the Biosciences*
Dick and Patton, *The Oregon State University Calculus Curriculum Project*
Eves, *In Mathematical Circles*
Eves, *Mathematical Circles Adieu*
Eves, *Mathematical Circles Revisited*
Eves, *Mathematical Circles Squared*
Eves, *Return to Mathematical Circles*
Fletcher and Patty, *Foundations of Higher Mathematics*
Gantner and Gantner, *Trigonometry*
Geltner and Peterson, *Geometry for College Students, Second Edition*
Gilbert and Gilbert, *Elements of Modern Algebra, Second Edition*
Gobran, *Beginning Algebra, Fifth Edition*
Gobran, *Intermediate Algebra, Fourth Edition*
Gordon, *Calculus and the Computer*
Hall, *Algebra for College Students*
Hall and Bennett, *College Algebra with Applications, Second Edition*
Hartfiel and Hobbs, *Elementary Linear Algebra*
Kaufmann, *Algebra for College Students, Third Edition*
Kaufmann, *Algebra with Trigonometry for College Students, Second Edition*
Kaufmann, *College Algebra, Second Edition*
Kaufmann, *College Algebra and Trigonometry, Second Edition*
Kaufmann, *Elementary Algebra for College Students, Third Edition*
Kaufmann, *Intermediate Algebra for College Students, Third Edition*
Kaufmann, *Precalculus, Second Edition*
Kaufmann, *Trigonometry*
Laufer, *Discrete Mathematics and Applied Modern Algebra*
Nicholson, *Elementary Linear Algebra with Applications, Second Edition*
Patty, Fletcher, and Hoyle, *Foundations of Discrete Mathematics*

Pence, *Calculus Activities for Graphic Calculators*
Powers, *Elementary Differential Equations*
Powers, *Elementary Differential Equations with Boundary-Value Problems*
Powers, *Elementary Differential Equations with Linear Algebra*
Proga, *Arithmetic and Algebra, Second Edition*
Proga, *Basic Mathematics, Third Edition*
Rice and Strange, *Plane Trigonometry, Fifth Edition*
Schelin and Bange, *Mathematical Analysis for Business and Economics, Second Edition*
Strnad, *Introductory Algebra*
Swokowski, *Algebra and Trigonometry with Analytic Geometry, Seventh Edition*
Swokowski, *Calculus, Fifth Edition*
Swokowski, *Calculus with Analytic Geometry, Fourth Edition (Late Trigonometry)*
Swokowski, *Calculus of a Single Variable*
Swokowski, *Fundamentals of Algebra and Trigonometry, Seventh Edition*
Swokowski, *Fundamentals of College Algebra, Seventh Edition*
Swokowski, *Fundamentals of Trigonometry, Seventh Edition*
Swokowski, *Precalculus: Functions and Graphs, Sixth Edition*
Tan, *Applied Calculus, Second Edition*
Tan, *Applied Finite Mathematics, Third Edition*
Tan, *Calculus for the Managerial, Life, and Social Sciences, Second Edition*
Tan, *College Mathematics, Second Edition*
Trim, *Applied Partial Differential Equations*
Venit and Bishop, *Elementary Linear Algebra, Third Edition*
Venit and Bishop, *Elementary Linear Algebra, Alternate Second Edition*
Wiggins, *Problem Solver for Finite Mathematics and Calculus*
Willard, *Calculus and Its Applications, Second Edition*
Wood and Capell, *Arithmetic*
Wood, Capell, and Hall, *Developmental Mathematics, Fourth Edition*
Wood and Capell, *Intermediate Algebra*
Zill, *A First Course in Differential Equations with Applications, Fourth Edition*
Zill, *Calculus with Analytic Geometry, Second Edition*
Zill, *Differential Equations with Boundary-Value Problems, Second Edition*

Dedicated to the Memory of
Prof. Dr. Gottfried Köthe
1905–1989

Hier ich immer halten muss,
Dem dort unterm Rasen
Zum getreuen Brudergruss
Sein Leiblied zu blasen!
N. Lenau in *Der Postillion*

Modern Analysis

Measure Theory and Functional Analysis with Applications

WILLIAM H. RUCKLE

CLEMSON UNIVERSITY

PWS-KENT Publishing Company **Boston**

PWS-KENT
Publishing Company

20 Park Plaza
Boston, Massachusetts 02116

PWS-KENT Publishing Company is a division of Wadsworth, Inc.

Library of Congress Cataloging-in-Publication Data
Ruckle, William H.
Modern analysis / William H. Ruckle
p. cm.
Includes index.
ISBN 0-534-92164-7
1. Mathematical analysis. I. Title.
QA300.R79 1991
515—dc20 90-38849
CIP

Printed in the United States of America
91 92 93 94 95 — 10 9 8 7 6 5 4 3 2 1

Managing Editor: Steve Quigley
Editorial Assistant: Beverly Jones
Assistant Editor: Diana Kelly
Production: Helen Walden
Interior Design: S. London
Cover Design: Designworks
Interior Illustration: George Nichols
Manufacturing Coordination: Peter Leatherwood
Typesetting: H. Charlesworth & Co. Ltd
Printing and Binding: Maple-Vail Book Mfr. Group

Contents

Preface

As long as a branch of science offers an abundance of problems, it is alive. D. Hilbert

This book is written to address two contemporary realities in graduate mathematical education: 1. More of the curriculum must be devoted to combinatorial and computational mathematics in order to meet the needs of computer-oriented research. 2. There remains, however, the need to impart to our students a minimal knowledge of topology, measure theory, and functional analysis, without which they cannot effectively function as mathematical scientists. The first reality has fomented lively discussion at numerous meetings and on the pages of several journals and need not be treated here. As for the second reality, consider the following passage from an article entitled "Mathematicians in Operations Research Consulting" by Daniel H. Wagner, which appeared in the November, 1975 issue of the *American Mathematical Monthly.*

> *If I were to pick the single most important graduate course in mathematics to prepare for a career in OR, I would definitely choose real analysis. Among other reasons, real analysis is necessary for a good comprehension of probability, statistics, optimization methods, and numerical analysis. Twenty years ago complex analysis and differential equations typically received the most emphasis in applied mathematics departments, with applications in engineering and physics primarily in view. I would think that today, real analysis is at least as important even for such traditional fields of application.*

In my opinion, Dr. Wagner's remarks also apply to other areas of the mathematical sciences such as statistics and systems theory. Too often, mathematical science faculties have been tempted to create an attractive program by replacing mathematically rigorous courses with courses that are essentially instructions for operating this or that canned program. This is like a music department replacing courses in musical composition with courses in how to play compact discs. The end result of such nonsense would be to produce graduates who cannot be relied upon whenever a situation arises that requires mathematical thought.

This book brings together a collection of topics in modern analysis that should be an essential part of most curricula in mathematics and mathematical sciences.

The book is designed for a first-year graduate course in measure theory, real analysis, topology and functional analysis, sometimes called "baby functional analysis," which is gradually replacing a sequence of several traditional courses for many students. The benefit of this program is that students who concentrate in analysis or topology can begin those courses at a higher level while students with other interests can be exposed to the main results in several branches of analysis without having to study details they may find uninteresting. Of course, I hope that the beauty and power of the mathematics described in this book will arouse the analytic instincts among students who are undecided about their concentration.

In writing this book I have been guided by the following principles.

[1] This book is designed for use with a teacher rather than for self instruction. I have sometimes been frustrated by texts that try to say so much that there is no room for the instructor to add anything let alone cover all the material.

[2] It is better to confront the central issues of a subject than to dwell on many ramifications. Breaking a difficult theorem into dozens of lemmas does not really make it easier for the students to digest. It only makes the student lose sight of the theorem's purpose. Rewriting a major result in several equivalent ways is better left to the exercises than the text.

[3] Wherever possible the main topics should be applied to other branches of the mathematical sciences. These applications are found in the exercises and in the book's final chapter. On the other hand, the applications are not strewn hither and thither in such a way as to obscure the main direction of the theory.

In selecting application topics my first object was to find an area in which the course material could be applied in a clear and appealing way. The importance of the application took second place. After all, if a topic is sufficiently important, it will probably appear later in a required course.

[4] There are three groups of exercises. Group A are routine and intended for the student to do independently in order to reinforce the understanding of the text. Group B are of ordinary difficulty intended for usual daily assignments. Group C are most difficult and intended for occasional assignments and special projects. All of the exercises should be read along with the assignment to help the student become familiar with ideas and terminology found in them.

The material in Chapters Two through Six can be adequately covered in two semesters. If it is necessary to consider Chapter One in any detail then material from Chapters Five and Six can be omitted. The order of presentation reflects my current preferences. However, I have attempted to accommodate instructors with different tastes by constructing a flexible development. The Dependency Table located after this Preface will facilitate rearrangement of topics. For example, if you want to introduce Lebesgue spaces earlier in the program than in the sequence given, the Dependency Table shows that you can introduce them as soon as you complete Sections 1–10, 11, 12 and 17. The model syllabi below describe three independent one semester courses based upon this book.

Course I (Introduction to Modern Analysis). Sections: 1, 2, 3, 4, 5, 6, 7, 8, 9, 11, 12, 13, 17, 18, 19, 24, 27, 28, 34.

Course II (Real Analysis). Sections: 1, 2, 3, 4, 5, 6, 11, 12, 13, 14, 15, 16, 7, 8, 17, 24, 25, 26, 27.

Course III (Elementary Functional Analysis). Sections: 1, 2, 3, 4, 5, 6, 7, 8, 9, 10, 17, 18, 19, 20, 21, 22, 23, 24, 28, 29, 31, 32, 33, 36.

I would like to thank the following reviewers: William Ballard, University of Montana; Bruce A. Barnes, University of Oregon; James W. Carlson, Creighton University; Brad Osgood, Stanford University; Fritz Steinhardt, City College of CUNY; and Daniel C. Weiner, Boston University.

Dependency Table

1:
2: 1
3: 1, 2
4: 1–3
5: 1–4
6: 1–3
7: 1–3
8: 1–3, 7
9: 1–5, 7, 8
10: 1–5, 7–9
11: 1–4
12: 1–5, 11
13: 1–5, 11, 12
14: 1–5, 11, 13
15: 1–5, 11–14
16: 1–5, 11–16
17: 1–10
18: 1–10, 17
19: 1–10, 17, 18
20: 1–10, 17
21: 1–10, 17, 20
22: 1–10, 17, 20, 21
23: 1–10, 17, 20–22
24: 1–10, 17
25: 1–10, 17, 20, 24
26: 1–10, 11, 12
27: 1–10, 11, 12
28: 1–10, 17, 20–22
29: 1–10, 17, 20–22
30: 1–10, 17
31: 1–9
32: 1–10, 17, 24
33: 1–10, 17, 20–22, 28
34: 1–10, 17, 20–22, 29
35: 1–10, 17, 20–22
36: 1–15, 17, 20–22, 26, 27

1

Basic Concepts

... the rational, irrational, and complex numbers should be regarded just as existent as the finite positive whole numbers. G. Cantor

The foundation for modern analysis is (a) the theory of sets, (b) the system $\mathbb{R}$ of real numbers, and (c) linear algebra. Although much more can be written on these subjects, there is enough material in this chapter to form a basis for subsequent discussion. As you finish each section of this chapter, review and use all that you know about that subject, not just what is in the text, to understand subsequent material and to do problems.

In a modern discussion of analysis we use the language and notation of set theory as described in Section 1. However, the bedrock of modern analysis is the system $\mathbb{R}$ of real numbers. Although it is possible to construct the reals from set theoretical concepts, we prefer to give the axiomatic treatment found in Section 2. The most important and probably the most difficult concept in Section 2 is the Completeness Axiom, which is the basis for the topology for the real line.

In the third section we return to set theory, and examine the concepts of cardinality and countability. We also introduce an axiom of set theory known as the Axiom of Choice, which for our purposes we present as a special form of Zorn's Lemma.

The fourth and fifth sections deal with the topology of the real line and with continuous functions from $\mathbb{R}$ into $\mathbb{R}$. Using topological attributes such as openness, compactness, and continuity it is possible to derive rigorously important properties of continuous functions that you first encountered in elementary calculus. The Heine-Borel Theorem, which characterizes the compact subsets of $\mathbb{R}$, will be derived from the Completeness Axiom.

Section 6 is a quick review of important ideas from linear algebra. We shall use Zorn's Lemma to prove the existence of an algebraic basis in every linear space. The distinction between finite dimensional and infinite dimensional linear spaces is crucial here. For studying linear operators between finite dimensional spaces the methods of linear algebra suffice, but for studying linear operators between infinite dimensional spaces, it is usually necessary to introduce some topological ideas.

1

The Language of Set Theory

The notion of a set and the relation of belonging to a set are basic to set theory, that is to say, they are undefined. The statement $x \in A$ means that the *element* or *point* x belongs to the set A. The terms "collection" and "family" are used as synonyms for the word "set" to avoid the expression "set of sets". Using the relation $\in$ we can define other set theoretic relations and operations as follows.

A is equal to B: $A = B$ means $x \in A$ if and only if $x \in B$. The equality symbol "$=$" will always indicate that the quantities on either side are different names for the same entity.

A is contained in B or *B contains A*: $A \subset B$, $B \supset A$, means that $x \in A$ and implies that $x \in B$.

The *union*: If $\mathscr{A}$ is a collection of sets, then $\bigcup \mathscr{A}$, also written $\bigcup_{A \in \mathscr{A}} A$, is the set of all x for which there is $A \in \mathscr{A}$ with $x \in A$.

The *intersection*: If $\mathscr{A}$ is a collection of sets, then $\bigcap \mathscr{A}$, also written $\bigcap_{A \in \mathscr{A}} A$, is the set of all x such that $x \in A$ is true for all $A \in \mathscr{A}$.

The *complement* of B in A: $A \setminus B$ is the set of all x such that $x \in A$ is true but $x \in B$ is false. It is not necessary that $B \subset A$ in order to form $A \setminus B$. If the set C containing A is understood, we write $A^{\sim}$ for $C \setminus A$, which is the complement of A in C.

The *empty set* $\varnothing$ is the set for which $x \in \varnothing$ is always false, that is, the set that contains no element. The set $\varnothing$ can be defined logically as the set of all elements x such that $x \neq x$.

In every discussion involving sets it is understood that every set we encounter is an element of a fixed universe of discourse. This understanding is necessary to avoid Russell's Paradox, which is described in Exercise 9.

Example 1.1:

A finite set can be described by enclosing its members in braces

$$A = \{1.1, 2.3, 3.21\}.$$

Thus the set A consists of three real numbers: 1.1, 2.3, 3.21. The statement $2.3 \in A$ is true, but the statement $1.5 \in A$ is false. If $B = \{1.25, 3, 3.21, 4\}$ then both statements $A \subset B$ and $B \subset A$ are false. The set $A \cup B$ is $\{1.1, 1.25, 2.3, 3, 3.21, 4\}$. Notice that we mention 3.21 only once even though it belongs to both sets. The set $A \cap B$ is $\{3.21\}$. This set is called a *singleton* because it has only one element in it. The set $A \setminus B$ is $\{1.1, 2.3\}$ while the set $B \setminus A$ is $\{1.25, 3, 4\}$.

Example 1.2:

If a set has a large number of members it may not be convenient to list all of them explicitly within braces. Thus if A is the set of all integers from 1 to 100 we might write

$$A = \{1, 2, \ldots, 100\}$$

and depend on the reader's imagination to fill in the other 97 integers. Another possibility is to write

$$A = \{x\colon x \text{ is an integer, } 1 \leqslant x \leqslant 100\}.$$

If we want to describe an infinite set we must use one of the above devices. Thus if B is to consist of all integers greater than 2, we can write

$$B = \{2, 3, \ldots\}$$

or, somewhat more clearly,

$$B = \{x\colon x \text{ is an integer, } x \geqslant 2\}$$

or, equivalently

$$B = \{x\colon x \text{ is an integer, } x > 1\}.$$

If we want C to consist of the interval of real numbers x that are greater than 1 but less than or equal to 2 we can write

$$C = \{x\colon x \text{ is a real number, } 1 < x \leqslant 2\}.$$

For two sets A and B the *product* $A \times B$ is defined as the set of all ordered pairs (a, b) where $a \in A$ and $b \in B$. The ordered pair (a, b) can be cleverly defined as the set $\{\{a, b\}, a\}$, which consists of the set of elements comprising the ordered pair and the element that comes first in it. For the purposes of analysis it is not essential to remember or use this definition. The important thing is that for ordered pairs (a, b) and (a', b') it is true that

$$(a, b) = (a', b') \quad \text{if and only if } a = a' \text{ and } b = b'.$$

Example 1.3:

If $A = \{a, b, c\}$ and $B = \{1, 2\}$ then we have

$$A \times B = \{(a, 1), (a, 2), (b, 1), (b, 2), (c, 1), (c, 2)\}.$$

Any subset O of $A \times B$ is called a *relation*. The set of all a in A such that $(a, b) \in O$ for some b in B is called the *domain* of O, and the set of all b in B such that $(a, b) \in O$ for some a in A is called the *range* of O. If the domain and range of O are both subsets of a set A then O is called a relation *in* A. A relation O is called a *function* if whenever (a, b) and (a, b') are in O it follows that $b = b'$. If F is a function, then each a in the domain of the function determines a unique b in the range. Thus a function is like a machine with a predictable output for a given input. If F is a function it is usual to write $y = F(x)$ instead of $(x, y) \in F$. If $A \subset$ domain (F), then $F(A)$ is the set of all $f(x)$ as x ranges over A. If F is a function from a set S into a set T and $B \subset T$, then $F^{-1}(B)$ is the set of all x in S such that $F(x) \in B$; if $F(S)$ and B are disjoint (i.e., $F(S) \cap B = \varnothing$) then $F^{-1}(B)$ must be the empty set $\varnothing$. [*Caution*: The symbol $F^{-1}(B)$ used in this way does not imply the existence of an inverse function for F.]

Theorem 1.4: Suppose F is a function from S into T, $B \subset T$, and $\mathscr{A}$ is a collection of subsets of T. Then we have

(a)
$$F^{-1}(B^{\sim}) = (F^{-1}(B))^{\sim}.$$

(b)
$$F^{-1}\left(\bigcup_{A \in \mathscr{A}} A\right) = \bigcup_{A \in \mathscr{A}} F^{-1}(A).$$

(c)
$$F^{-1}\left(\bigcap_{A \in \mathscr{A}} A\right) = \bigcap_{A \in \mathscr{A}} F^{-1}(A).$$

Proof:

(a) A point x in S is in $F^{-1}(A^{\sim})$ if and only if $F(x)$ is in $A^{\sim}$, in other words if and only if $F(x)$ is not in A, which is the same as saying x is in $(F^{-1}(A))^{\sim}$.

(b) A point x in S is in $F^{-1}(\bigcup_{A \in \mathscr{A}} A)$ if and only if $F(x)$ is in $\bigcup_{A \in \mathscr{A}} A$, which is true if and only if there is $A \in \mathscr{A}$ such that $F(x) \in A$, that is, $x \in F^{-1}(A)$. Therefore, we conclude x is in $F^{-1}(\bigcup_{A \in \mathscr{A}} A)$ if and only if x is in $\bigcup_{A \in \mathscr{A}} F^{-1}(A)$.

We leave the proof of (c) to you. ■

The following result of set theory is called *DeMorgan's Laws.*

Theorem 1.5: Suppose $\mathscr{A}$ is a collection of subsets of a set C. The following equations are true

(a)
$$\bigcup_{A \in \mathscr{A}} A^{\sim} = \left\{\bigcap_{A \in \mathscr{A}} A\right\}^{\sim}.$$

(b)
$$\bigcap_{A \in \mathscr{A}} A^{\sim} = \left\{\bigcup_{A \in \mathscr{A}} A\right\}^{\sim}.$$

[*Note*: All complements of sets are taken in C.]

Proof: We shall prove (a) by showing that if x is any point of $\bigcup_{A \in \mathscr{A}} A^{\sim}$ it must also be in $\{\bigcap_{A \in \mathscr{A}} A\}^{\sim}$ and that if x is any point of $\{\bigcap_{A \in \mathscr{A}} A\}^{\sim}$ then it must also be in $\bigcup_{A \in \mathscr{A}} A^{\sim}$. The first argument shows that $\bigcup_{A \in \mathscr{A}} A^{\sim} \subset \{\bigcap_{A \in \mathscr{A}} A\}^{\sim}$; the second argument that $\{\bigcap_{A \in \mathscr{A}} A\}^{\sim} \subset \bigcup_{A \in \mathscr{A}} A^{\sim}$.

Suppose $x \in \bigcup_{A \in \mathscr{A}} A^{\sim}$. This implies there is $A \in \mathscr{A}$ such that $x \in A^{\sim}$, the complement of A in C. Thus, x is not in this A so that x is not in $\bigcap_{A \in \mathscr{A}} A$. Hence we conclude $x \in \{\bigcap_{A \in \mathscr{A}} A\}^{\sim}$.

On the other hand, suppose $x \in \{\bigcap_{A \in \mathscr{A}} A\}^{\sim}$. This means that x is not in the intersection $\bigcap_{A \in \mathscr{A}} A$. Thus there must be some A in $\mathscr{A}$ such that x is not in A. That means x is in $A^{\sim}$ for this A so we conclude that $x \in \bigcup_{A \in \mathscr{A}} A^{\sim}$.

We leave the proof of (b) to you. ■

Exercises

A Routine Exercises

1. Suppose $A = \{1.2, 2.3, 2.4\}$ and $B = \{1.1, 2.3, 3.3\}$. Find $A \cup B$, $A \cap B$, and $A \times B$.

2. Let $A=\{x\colon x \text{ is an integer}, x \text{ is even}\}$ and let $B=\{x\colon x \text{ is an integer}, x \text{ is a multiple of } 3\}$. Describe $A \cup B$, $A \cap B$, $A \backslash B$, and $B \backslash A$.
3. If A, B, and C are arbitrary sets prove the *Distributive Laws*:

(a)
$$A \cap (B \cup C) = (A \cap B) \cup (A \cap C).$$

(b)
$$A \cup (B \cap C) = (A \cup B) \cap (A \cup C).$$

4. Let A and B be as in Exercise 1. Suppose we know that F and G are functions with domain in A and range in B. We know F contains the pairs $(1.2, 2.3)$, $(2.4, 3.3)$ and $(x, 1.1)$. What values can x have? We know G contains the pairs $(1.2, 3.3)$, $(2.4, 1.1)$ and (u, v). What are the possible values for u and v?
5. Let $A=\{1, 2\}$ and let $B=\{a, b, c\}$. List all of the functions with domain equal to A and range in B.

B. *Exercises of Intermediate Difficulty*

6. *(a)* If A is a set containing n elements, how many subsets does A have (including $\varnothing$)?
   (b) If A has n elements and B has m elements, how many relations are there with domain in A and range in B?
   (c) How many functions?
   (d) How many functions with domain all of A and range in B?
7. Let $\{A_\alpha\}$ be a collection of sets, and let B be a set. Prove that

(a)
$$B \cup \left(\bigcap_\alpha A_\alpha\right) = \bigcap_\alpha (A_\alpha \cup B).$$

(b)
$$B \cap \left(\bigcup_\alpha A_\alpha\right) = \bigcup_\alpha (A_\alpha \cap B).$$

Formulas (a) and (b) are called the *Distributive Laws.*

8. Suppose A and B are subsets of a set C. Show that $A \subset B$ if and only if $B^\sim \subset A^\sim$.

C. *More Difficult Problems*

9. (*Russell's Paradox*) Given a universe of discourse U let us attempt to form the set Z of all sets A in U for which the statement $A \in A$ is false; that is, A is not an element of itself. (Granted, most sets are not elements of themselves; but some, such as the set of all mathematical ideas, might be.) Prove that Z cannot be in U and, therefore, Z is not a set.
10. A *partition* of a set S is a collection $\{A_\alpha\}$ of nonempty subsets of S such that (1) $\bigcup_\alpha A_\alpha = S$ and (2) if $\alpha \neq \beta$ then $A_\alpha \cap A_\beta = \varnothing$. Given a partition $\{A_\alpha\}$ of a set S define a relation $\mathscr{R}$ in S by $(x, y) \in \mathscr{R}$ if and only if there is α such that both x and y are in A_α.
    (a) Prove that the relation $\mathscr{R}$ has the following properties.

    (E1) For each x in S, $(x, y) \in \mathscr{R}$.
    (E2) If $(x, y) \in \mathscr{R}$ then $(y, x) \in \mathscr{R}$.
    (E3) If $(x, y) \in \mathscr{R}$ and $(y, z) \in \mathscr{R}$ then $(x, z) \in \mathscr{R}$.

Condition (E1) is called the condition of *reflexivity*, condition (E2), the condition of *symmetry* and condition (E3), the condition of *transitivity*. Any relation $\mathscr{R}$ that satisfies these conditions is called an *equivalence relation*.

(b) Suppose $\mathscr{R}$ is an equivalence relation on a set S. Show there is a partition $\{A_\alpha\}$ of S such that $(x, y) \in \mathscr{R}$ if and only if there is α such that x and y both belong to the same A_α.

(c) Criticize the following false proof that (E2) and (E3) imply (E1): Suppose $x \in S$. Let y be any member of S such that $(x, y) \in \mathscr{R}$. Then by (E2) $(y, x) \in \mathscr{R}$. Since (x, y) and (y, x) are in $\mathscr{R}$ it follows by (E3) that $(x, x) \in \mathscr{R}$.

2

The Real Numbers

Although it is possible to construct the system of real numbers from more elementary concepts, we shall consider the set $\mathbb{R}$ of real numbers as one of our basic starting points. The real numbers can be concisely described as "a complete ordered field." The main purpose of this section is to explain this description.

In the first place, a *field* consists of a set F and two functions $+$ and $\cdot$ (called *binary operations*) from $F \times F$ into F. Instead of writing $+(x, y) = z$ or $\cdot(x, y) = z$ it is customary to write $x + y = z$ or $xy = z$ omitting the $\cdot$. The binary operations are required to satisfy the following conditions known as the *field axioms*.

(A1) For all x, y, z in F, $x + (y + z) = (x + y) + z$.

(A2) For all x, y in F, $x + y = y + x$.

(A3) There is 0 in F such that $x + 0 = x$ for all x in F.

(A4) For all x in F there is $-x$ in F such that $x + (-x) = 0$.

(M1) For all x, y, z in F, $x(yz) = (xy)z$.

(M2) For all x, y in F, $xy = yx$.

(M3) There is 1 in F such that $x1 = x$ for all x in F.

(M4) For all x in F with $x \neq 0$ there is x^{-1} in F such that $xx^{-1} = 1$.

(D) For all x, y, z in F, $x(y + z) = xy + xz$.

The analogy between (A1)–(A4) and (M1)–(M4) should be obvious. Conditions (A1)–(A4) assert that F forms an *Abelian group* under addition; (M1)–(M4) imply that nonzero members of F form an *Abelian group* under multiplication. Conditions (A1) and (M1) are *associative laws* while (A2) and (M2) are called *commutative laws* or *Abelian conditions*; (A3) and (M3) are called the *existence of* 0 *and* 1; and (A4) and (M4) are called the *existence of additive and multiplicative inverses.* Statement (D) is the only axiom relating multiplication and addition; it is called the *distributive law*. The algebraic properties of fields can be derived from these fundamental axioms. Except for a few illustrations we shall take all such properties for granted in the case of the field $\mathbb{R}$.

Example 2.1:

Show that the element 0 that satisfies (A3) is unique. Suppose 0 and $0'$ satisfy (A3). Then by (A3), $0 + 0' = 0$ and $0' + 0 = 0'$; but by (A2) we have $0 + 0' = 0' + 0$ so that $0 = 0'$.

We postulate that there exists an order relation $<$ on $\mathbb{R}$ that satisfies statements (O1)–(O4) below. It is customary to write $x < y$ instead of $(x, y) \in <$. The relation $y > x$ is the same as $x < y$.

- ***(O1)*** For all x, y in $\mathbb{R}$ exactly one of the following assertions is valid: $x < y$, $y < x$, $y = x$.
- ***(O2)*** For all x, y, z in $\mathbb{R}$, if $x < y$ and $y < z$, then $x < z$.
- ***(O3)*** For all x, y, z in $\mathbb{R}$, if $x < y$ then $x + z < y + z$.
- ***(O4)*** For all x, y, z in $\mathbb{R}$, if $x < y$ and $z > 0$, then $xz < yz$.

Example 2.2:

Show that if $x < 0$, then $-x > 0$: By (O1) either $-x = 0$, $-x < 0$, or $-x > 0$. If $-x = 0$, then $x = 0 + x = (-x) + x = 0$, which contradicts the assertion that $x < 0$. If $-x < 0$, then by (O2) we have $0 = x + (-x) < x + 0 = x$. This contradicts (O1) since it has already been asserted that $x < 0$.

Example 2.3:

Show that if $x < 0$ and $y < 0$, then $xy > 0$: By Example 2.2 we know that $-x > 0$ so that by (O4), $(-x)y < 0$. By the distributive law we have $(-x)y + xy = (-x + x)y = 0$ so that $(-x)y = -(xy)$. Since $-xy < 0$, it follows from Example 2.2 that $xy > 0$.

For real numbers x and y, the statement $x \leqslant y$ means that $x < y$ or $x = y$. The statement $y \geqslant x$ is equivalent. For a real number x, the *absolute value* $|x|$ of x is defined to be x if $x \geqslant 0$ and $-x$ if $x < 0$. Thus $|x|$ is the distance of the number x from the origin 0, always expressed as a positive number.

The last basic property of the real number system is called the *Completeness Axiom.* This property is the least intuitive and the most difficult to understand, but it is of fundamental importance in describing the topological structure of the real line. In order to state the Completeness Axiom it is first necessary to define some concepts concerning the bounds of a set. A set S of real numbers is said to be *bounded above* if there is a real number M such that $s \leqslant M$ for all s in S; such a number M is called an *upper bound* for S. For example if $Z = \{x\colon x \in \mathbb{R}, x^2 < 2\}$; then 2, 2.2, and 4.7 are all upper bounds for Z. A number L is called a *least upper bound* of a set S if (a) L is an upper bound for S and (b) if $t < L$ then t is not an upper bound for S. That t is not an upper bound for S implies there is $s \in S$ such that $s > t$. The least upper bound of the set Z described above is $\sqrt{2}$. The Completeness Axiom can now be stated as follows.

Completeness Axiom: Every subset of $\mathbb{R}$ that is bounded above has a least upper bound. ■

The set $\mathbb{N} = \{1, 2, \ldots\}$ of *positive integers* (or *natural numbers*) is defined to be the intersection of all sets S such that

(N1) $1 \in S$, and

(N2) if $x \in S$, then $x + 1 \in S$.

A consequence of this definition is the following theorem known as the *Principle of Mathematical Induction.*

Theorem 2.4: If A is a subset of $\mathbb{N}$ such that (a) $1 \in A$ and (b) $x \in A$, implies $x + 1 \in A$, then $A = \mathbb{N}$.

Proof: By hypothesis $A \subset \mathbb{N}$. On the other hand, $\mathbb{N} \subset A$ because by definition $\mathbb{N}$ is the intersection of all subsets of $\mathbb{R}$ that have the properties (a) and (b). ■

The following theorem is known as the *Archimedean Property* of the real numbers.

Theorem 2.5: If x is a real number there is a positive integer n such that $n \geqslant x$.

Proof: We may as well assume that $x > 1$. Let S denote the set of all positive integers m with $m < x$. By the Completeness Axiom, S has a least upper bound u. Since $u - \frac{1}{2}$ is not an upper bound of S there is m in S with $m > u - \frac{1}{2}$. Let $n = m + 1$. Since m is in S, m is a positive integer so that n is also a positive integer. Since $n > u$, n is not in S so that $n \geqslant x$. ■

The set of *integers* is defined as the union of $\{0\}$, $\mathbb{N}$, and the set $\{-n: n \in \mathbb{N}\}$. The set of *rational numbers*, often written Q, is defined to be the set of all ratios m/n $(= mn^{-1})$ where m and n are integers with $n \neq 0$. The set Q forms a field and is thus a subfield of $\mathbb{R}$. The following result, known as the *Density of the Rationals Theorem*, shows that no subinterval of $\mathbb{R}$ can be free of rationals.

Theorem 2.6: If x and y are real numbers with $x < y$ then there is a rational number r with $x < r < y$.

Proof: If x is negative and y is positive we can use 0 for r. If x and y are both negative, then $-y < -x$; if we find r between $-y$ and $-x$, then $-r$ lies between x and y. Thus it suffices to treat the case when x and y are positive. Since $x < y$ it follows that $y - x > 0$. By the Archimedean Property there is an integer p such that $p \geqslant (y - x)^{-1}$. If $n = p + 1$ then $1/n < (y - x)$. Again by the Archimedean Property there is an integer m such that $m > nx$. If m is the smallest such integer then $m > nx \geqslant m - 1$. Dividing the inequality by n we see that $m/n > x \geqslant m/n - 1/n$. The first inequality shows that $x < m/n$, and the second shows that $m/n \leqslant x + 1/n < y$. Thus $r = m/n$ satisfies the conclusions of the theorem. ■

Despite the density of the rationals in $\mathbb{R}$ the following example shows there are real numbers that are not rational.

Example 2.7:

There is no rational number m/n such that $(m/n)^2 = 2$: If there were such a m/n we could reduce the fraction so there would be no common divisor of m and n. Assume m/n is thus reduced. Then $m^2 = 2n^2$ so m^2 is even. Since the square of an odd number is odd it follows that m is even, say $m = 2k$. Thus we have $4k^2 = 2n^2$ so that $n^2 = 2k^2$, which implies n is even. This contradicts the assumption that m and n have no common factor. By the Completeness Axiom the set Z of all real numbers x for which $x^2 < 2$ has a least upper bound b. If we can show that b^2 must equal 2 it follows that b cannot be rational. If $b^2 < 2$ let r be any real number such that $0 < r < \min\{(2 - b^2)/(2b + 1), 1\}$ ("min" stands for minimum; this means r is less than both $(2 - b^2)/(2b + 1)$ and 1). Since $r < 1$ it follows that

$$r < (2 - b^2)/(2b + 1) < (2 - b^2)/(2b + r),$$

so that we have

$$2br + r^2 < 2 - b^2 \quad \text{or} \quad (b + r)^2 < 2.$$

This shows that the number $b + r$ is in Z, and contradicts the fact that b is the least upper bound of Z. Next suppose that $b^2 > 2$. Let r be any real number such that $0 < r < \min\{(b^2 - 2)/(2b), b\}$. Then we have

$$r < (b^2 - 2)/(2b) < (b^2 - 2)/(2b - r),$$

so that $(b - r)^2 > 2$. But if $x^2 < 2$ it follows that x must be less than $(b - r)$ so $(b - r)$ is an upper bound of Z, which is smaller than b. This again contradicts the fact that b is the least upper bound of Z. The only remaining possibility is that $b^2 = 2$.

The *extended* real numbers $\mathbb{R}^*$ consists of $\mathbb{R}$ with two additional objects $+\infty$ (or ∞) and $-\infty$. The objects ∞ and $-\infty$ satisfy the following conventions with respect to addition, multiplication, and order. However, $\mathbb{R}^*$ is not a field since ∞ and $-\infty$ have no inverses with respect to addition or multiplication.

$a + \infty = \infty$	for all a in $\mathbb{R}^*$ except $-\infty$
$a + (-\infty) = -\infty$	for all a in $\mathbb{R}^*$ except ∞
$a_\infty = \infty$	for all a in $\mathbb{R}^*$ with $a > 0$
$a_\infty = -\infty$	for all a in $\mathbb{R}^*$ with $a < 0$
$a(-\infty) = (-a)\infty$	for all a in $\mathbb{R}^*$ with $a \neq o$
$0_\infty = 0(-\infty) = 0$	
$a < \infty$	for all a in $\mathbb{R}$
$a > -\infty$	for all a in $\mathbb{R}$

a/∞ and $a/-\infty$ are 0.

$(-\infty) + \infty$, $\infty + (-\infty)$, ∞/∞, $-\infty/\infty$ and $\infty/-\infty$ are undefined.

The least upper bound of a subset A of real numbers is also called the *supremum* of A and written $\sup(A)$. The greatest lower bound of A is also called the *infimum* of A and written $\inf(A)$. If $\sup(A) \in A$ we sometimes write $\max(A)$ instead of $\sup(A)$. Likewise if $\inf(A) \in A$ we write $\min(A)$. However, we shall not be totally committed to do this as are some authors. If a set A of real numbers is not bounded above then we write $\sup(A) = \infty$, whereas if A is not bounded below we write $\inf(A) = -\infty$. Thus in the setting of $\mathbb{R}^*$ every set has a least upper bound and greatest lower bound.

Exercises

A

1. Show the element 1 of (M3) of the field axioms is unique.
2. Show that if x and y are in R, $-(x+y) = (-x) + (-y)$ and if x and y are nonzero then $(xy)^{-1} = x^{-1}y^{-1}$.
3. Criticize the now-famous false proof that $1 = 2$: Let $x = y$. Then

$$x^2 = xy \quad \text{so} \quad y^2 - x^2 = y^2 - xy.$$

Factor both sides to obtain

$$(y-x)(y+x) = y(y-x).$$

Cancel $(y-x)$ from both sides to obtain $y + x = y$. But since $y = x$, $x + x = 2x = x$. If we let $x = 1$, we obtain $1 = 2$.

4. Prove that the additive and multiplicative inverses in (A3) and (M3) of the field axioms are unique.
5. Show that for all x, y in $\mathbb{R}$,
   (a) $|x+y| \leqslant |x| + |y|$ *(b)* $||x| - |y|| \leqslant |x-y|$
   (c) $|xy| = |x||y|$ *(d)* $|x| = \max(x, -x)$.
6. Show that if $x > 0$ and $y < 0$, then $xy < 0$.
7. Show that $\max(x, y) = (x + y + |x-y|)/2$.
8. What are the least upper bounds of the following sets
   (a) $\{x\colon x \text{ is rational}, x^3 < 3\}$
   (b) $\{x\colon x \text{ is real}, x^3 \leqslant 3\}$
   (c) $\{x\colon x \text{ is an integer}, x^3 < 125\}$.
9. Use mathematical induction to prove the following formulas.
   (a) $1 + 2 + \cdots + n = n(n+1)/2$.
   (b) $1^2 + 2^2 + \cdots + n^2 = n(n+1)(2n+1)/6$.
10. Criticize the following false proof by induction that all horses have the same color: We shall show that if S is a set containing k horses then all of the horses in S have the same color. This is obvious if $n = 1$. Suppose it is true for $k-1$. Let S be a set of k horses where $k > 1$. Remove one of the horses from S. Then by the inductive hypothesis all of the remaining horses are of the same color, say red. Remove another horse from S and put the first one back. Again we have a set of $k-1$ horses, which must all have the same color, namely, red. We know the horse we removed from S the second time was red so if we return him, all of the horses in the original set will have the same color. Therefore, the assertion follows by induction.

11. Adapt the argument at the beginning of Example 2.7 to show there is no rational number m/n such that $(m/n)^2 = 3$.

B

12. *(a)* Prove that every nonzero field F must contain at least 2 elements.

(b) Let $F(2)$ consist of the elements 0, 1 with the binary operations $+$ and $\cdot$ defined as follows: $0+0=1+1=0$, $0+1=1+0=1$, $0\cdot 0=0\cdot 1=1\cdot 0=0$, $1\cdot 1=1$. Show that $F(2)$ with these operations satisfies the field axioms.

(c) If we try to make $F(2)$ into an ordered field by defining $0<1$, which of the conditions (O1)–(O4) will be violated?

13. Prove that the set Q of rational numbers forms an ordered field with the order relation and algebraic operations inherited from $\mathbb{R}$.

14. An *isomorphism* between two fields F and G is a function T with domain all of F and range all of G such that (a) T is one-to-one, that is, if $T(x)=T(y)$, then $x=y$; (b) $T(x+_F y)=T(x)+_G T(y)$; and (c) $T(x\cdot_F y)=T(x)\cdot_G T(y)$. If there is an isomorphism between F and G, then as far as the theory of fields is concerned, F and G are identical. They are said to be *isomorphic* (same structure).

(a) Show that if T is an isomorphism from F to G then the function T^{-1} defined by $T^{-1}(y)=x$ if $T(x)=y$ is an isomorphism from G to F.

(b) Show that if T is an isomorphism from F to G then $T(0_F)=0_G$ and $T(1_F)=1_G$.

(c) Let T be an isomorphism from $\mathbb{R}$ to $\mathbb{R}$. Show first that $T(n)=n$ for each integer n, then show that $T(r)=r$ for each rational r, then show that if $x>0$, then $T(x)>0$ (use the fact that $x=\sqrt{x}\sqrt{x}$), and finally show that $T(x)=x$ for all x in $\mathbb{R}$.

15. If x and y are real numbers with $x<y$, show there is an irrational number z with $x<z<y$.

16. Show that the least upper bound of a set is unique if it exists.

C

17. Let W be a complete ordered field, that is, a set with binary operations $+_W$ and $\cdot_W$, which satisfy conditions (A1)–(A4), (M1)–(M4), (O1)–(O4) and (C). Show that there is an isomorphism T from $\mathbb{R}$ to W. This means that as far as complete ordered fields are concerned, W is identical with $\mathbb{R}$.

3

Functions and Cardinality

If F is a function with domain in the set A and range in the set B then the set $\hat{F}$ of all (b, a) such that $a \in A$, $b \in B$, and $f(a)=b$ is a subset of $B \times A$, and thus a relation. If $\hat{F}$ is also a function, then F is said to be *one-to-one*, and $\hat{F}$ is written F^{-1}.

Example 3.1:

Let A and B both denote the set of positive integers, 1, 2, 3, ... Define F by the formula

$$F(2n-1)=F(2n)=n, \qquad n=1, 2, 3, \ldots.$$

Thus $F(1)=F(2)=1$, $F(3)=F(4)=2$, $F(5)=F(6)=3$, etc. Clearly, the relation $\hat{F}$ is not a function since it contains, for instance, (1, 2) and (1, 1). Thus F is not one-to-one. On the other hand, define G on A by

$$G(n)=2n, \qquad n=1, 2, \ldots$$

so that $G(1)=2$, $G(2)=4$, $G(3)=6$, etc. The domain of G is all of A, but the range of G is the set $\{x\colon x \text{ is an even positive integer}\} \subset B$. The function G is one-to-one because if (b, a) and (b, a') are in $\hat{G}$, it follows that $b=2a$ and $b=2a'$ so that $a=a'$.

Two sets A and B are said to *have the same cardinality* if there is a one-to-one function F that has domain A and range B. Example 3.1 shows that the set of even positive integers has the same cardinality as the set of all positive integers. A set A is called *finite* if it has the same cardinality as a set $\{1, 2, \ldots, n\}$ for some positive integer n. It is easy to see that if A and B are finite sets then A and B have the same cardinality if and only if there are the same number of elements in both A and B. Having the same cardinality for arbitrary sets corresponds to having the same number of members for finite sets. Example 3.1 illustrates that, unlike the finite case, a part (proper subset) of an infinite set can have the same cardinality as the entire set. A set is said to be *countably infinite* if it has the same cardinality as the set of positive integers. A set that is finite or countably infinite is called *countable*. Having the same cardinality is an equivalent relation; in particular it is transitive, that is, if A has the same cardinality as B and B has the same cardinality as C, then A has the same cardinality as C.

The notion of cardinality is due to Georg F. L. P. Cantor (1845–1918) who was the principle founder of set theory. The two diagonal arguments that appear in Theorems 3.2 and 3.3 attest to his ingenuity. There was strong resistance to Cantor's theories during his lifetime, but he also had influential supporters.

Theorem 3.2:

(a) Every subset of a countable set is countable.

(b) If A and B are countable so is $A \times B$.

(c) If $\{A_n\}$ is a countable collection of countable sets then $\bigcup_n A_n$ is countable.

Proof:

(a) Suppose B is a subset of a countable set A. We shall carry out the proof for A, infinite, since the case for A, finite, is easier. Since A is countably infinite there is a one-to-one function F from the set $\mathbb{N}=\{1, 2, \ldots\}$ onto A. If B is empty then we are finished. Otherwise we shall define a function G with domain in $\mathbb{N}$ and range B as follows: Let $G(1)$ be the element b_1 in B for which $F^{-1}(b_1)$ is the smallest. (Here we are using the fact that every non-

empty set of positive integers contains a smallest member. This is equivalent to Theorem 2.4, the Principle of Mathematical Induction.) If $B_2 = B \backslash \{G(1)\}$ is empty, then B has only one element, so we are finished. Otherwise let $G(2)$ be the member b_2 of B_2 for which $F^{-1}(b_2)$ is smallest. If $G(1), G(2), \ldots,$ $G(n-1)$ are defined, either the set $B_n = B \backslash \{G(1), G(2), \ldots G(n-1)\}$ is empty in which case B contains $n-1$ members and is finite; or else we can define $G(n)$ to be element b_n in B_n for which $F^{-1}(b_n)$ is the smallest. Unless B is finite we can continue this process to definite a function G with domain $\mathbb{N}$ and range in B. To see that the range of G is all of B we assume otherwise in order to obtain a contradiction. Suppose the set $B \backslash \{G(n) : n \in \mathbb{N}\}$ is nonempty. Then it must contain a member c for which $F^{-1}(c)$ is a minimum. By the construction of $G(n)$ it follows that $F^{-1}(G(n)) < F^{-1}(c)$ for each n. But it is easy to see that $n \leqslant F^{-1}(G(n))$ for each n so that $n < F^{-1}(c)$ for all n. The number $F^{-1}(c)$ thus contradicts Theorem 2.5, the Archimedean Property of $\mathbb{R}$.

(b) We shall prove that $\mathbb{N} \times \mathbb{N}$ is countable. This is customarily done using the first Cantor diagonal argument shown below.

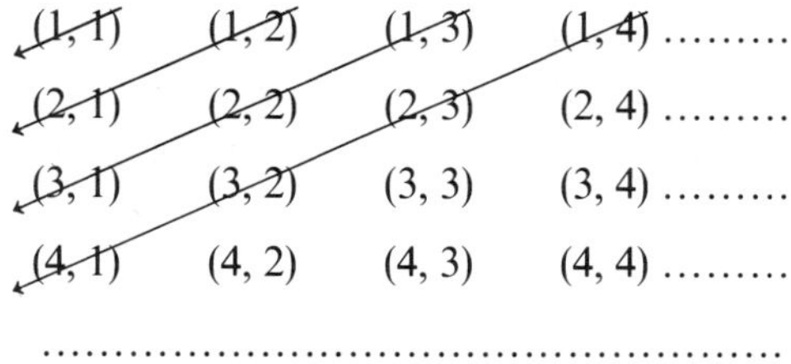

The function described by the arrow is

$$F(m, n) = \frac{(m+n-2)(m+n-1)}{2} + m.$$

We leave to the reader the task of showing that F is one-to-one with domain $\mathbb{N} \times \mathbb{N}$ and range $\mathbb{N}$.

(c) For each $n = 1, 2, \ldots$ let F_n be a one-to-one function from A_n into $\mathbb{N}$. We define a function G from $\bigcup_n A_n$ into $\mathbb{N} \times \mathbb{N}$ as follows:

$G(a) = (n, F_n(a))$ if n is the smallest integer such that $a \in A_n$. If $G(a) = G(b) = (n, m)$ then both a and b are in A_n and $F_n(a) = F_n(b) = m$. Since F_n is one-to-one it follows that $a = b$. Therefore, G is one-to-one. The range of G is a subset of $\mathbb{N} \times \mathbb{N}$, and every such subset is countable. Therefore, $\bigcup_n A_n$ is countable. ■

Theorem 3.3: The set $\mathbb{R}$ of all real numbers is uncountable.

Proof: Using a second diagonal method of Cantor we shall prove that the set I of all real numbers t with $0 \leqslant t \leqslant 1$ is uncountable. Our theorem will then follow since $I \subset \mathbb{R}$. Each t in I has a decimal expansion $t = .r_1 r_2 \ldots$. Some t have two expansions such as $.10000\ldots = .09999\ldots$; if this happens we choose the one that terminates in 9's. Suppose for the sake of obtaining a contradiction that there were a function F from $\mathbb{N}$ onto I. Then we would have

$$F(1) = .r_{11}r_{12}r_{13}\ \cdots$$
$$F(2) = .r_{21}r_{22}r_{23}\ \cdots$$
$$F(3) = .r_{31}r_{32}r_{33}\ \cdots$$
$$\cdots\cdots\cdots\cdots\cdots\cdots\cdots$$

We construct a special t in I as follows: $t = .s_1 s_2 s_3 \ldots$ where $s_i = 5$ if $r_{ii} \neq 5$ and $s_i = 4$ if $r_{ii} = 5$. If it were the case that $F(n) = t = .r_{n1}r_{n2} \ldots$ we would have the contradiction $r_{nn} = s_n \neq r_{nn}$. ∎

Most of the infinite sets encountered in this book will have the same cardinality as $\mathbb{N}$ or the same cardinality as $\mathbb{R}$.

Example 3.4:

Some sets that are countably infinite are:

the set of all integers—positive or negative;
Q, the set of all rational numbers;
the set of all points (x, y) in the Euclidean plane $\mathbb{R}^2$ for which x and y are rational.

Example 3.5:

Some sets that have the same cardinality as $\mathbb{R}$ (sometimes expressed as "have the cardinality of the continuum") are:

the set of irrational numbers;
the set of all real numbers in an interval $[a, b]$, $[a, b)$, $(a, b]$, or (a, b) provided $a < b$;
the Euclidean plane $\mathbb{R}^2$.

If F is a function from a set A with range equal to B, and G is a function from B into a set C we can define the function $G \circ F$ from A into C by the formula $G \circ F(x) = G(F(x))$ for x in A. That $G \circ F$ is a function follows because if $x = y$, then $F(x) = F(y)$ so $G(F(x)) = G(F(y))$. The function $G \circ F$ is called the *composition* of G with F. If both F and G are one-to-one it is not hard to see that $G \circ F$ is one-to-one and that $(G \circ F)^{-1} = F^{-1} \circ G^{-1}$.

In Section 1 we defined the product of a pair of sets. This definition can be extended by induction to products of finitely many sets. However, the definition of the product of an arbitrary collection of sets is quite different. If $\{A_\alpha: \alpha \in \mathscr{A}\}$ is any collection of sets the *product* $\prod_\alpha A_{\alpha \in \mathscr{A}}$ is defined to be the set of all functions F from $\mathscr{A}$ into $\bigcup_\alpha A_\alpha$ such that $F(\alpha) \in A_\alpha$ for all α. If $A_\alpha = \varnothing$ for some β in $\mathscr{A}$, then $\prod_\alpha A_\alpha$ is certainly empty since there is no function F such that $F(\beta) \in A_\beta$. The assertion that $\prod_\alpha A_\alpha$ is nonempty if $\mathscr{A}$ is countable and each A_α is nonempty can be proved using mathematical induction. If $\mathscr{A}$ is uncountable and each A_α is nonempty, then the assertion that $\prod_\alpha A_\alpha$ is nonempty is a basic postulate of Set Theory, called the *Axiom of Choice*. The Axiom of Choice is so named because it asserts the existence of at least one "choice" function f from $\mathscr{A}$ into $\bigcup_{\alpha \in \mathscr{A}} A_\alpha$, which chooses precisely one member of A_α for each α in $\mathscr{A}$. There are many equivalent formulations of the Axiom of Choice. The following version is a special case of *Zorn's Lemma* that will be used in our subsequent study. If $\mathscr{A}$ is a collection of sets, a subcollection $\mathscr{B}$ of $\mathscr{A}$ is called a *chain* if for each pair A, B in $\mathscr{B}$ either $A \subset B$ or $B \subset A$. A set A in $\mathscr{A}$ is called *maximal* if there is no B in $\mathscr{A}$ that properly contains A; that is, if $B \supset A$, then $B = A$. The set A is called *minimal* if there is no B in $\mathscr{A}$ that is properly contained in A.

Postulate 3.6: Suppose $\mathscr{A}$ is a collection of sets.

(a) If for each chain $\mathscr{B}$ in $\mathscr{A}$ the set $\bigcup \mathscr{B}$ is in $\mathscr{A}$ then $\mathscr{A}$ contains a maximal set.

(b) If for each chain $\mathscr{B}$ in $\mathscr{A}$ the set $\bigcap \mathscr{B}$ is in $\mathscr{A}$ then $\mathscr{A}$ contains a minimal set. ■

Exercises

A

1. If A and B are finite sets with n members, how many one-to-one functions are there from A onto B?
2. Prove that the following sets have the same cardinality as $\mathbb{N}$:
   (a) the set $\{0, 1, 2, \ldots\}$
   (b) the set of all integers
   (c) the set of all prime numbers
   (d) the set Q of all rational numbers.
3. *(a)* Prove geometrically that two line segments l and l' of different lengths have the same cardinality by constructing a trapezoid with bases l and l', extending the trapezoid to a triangle and projecting the shorter segment onto the longer. See Figure 3.1.

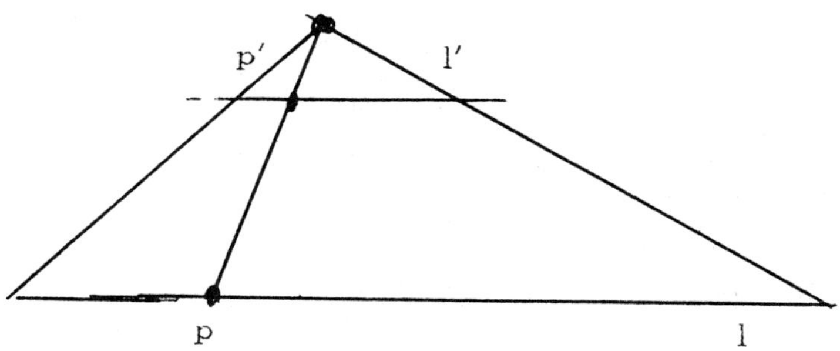

Figure 3.1

(b) Prove the same fact algebraically by constructing a one-to-one linear function with domain $[a, b]$ and range $[c, d]$ for any pair of line segments $[a, b]$ and $[c, d]$.

4. Show that the function $F(x) = 1/x$ is a one-to-one function from the finite segment $(0, 1]$ onto the infinite segment $[1, \infty)$.
5. Prove that if F is one-to-one with domain A and range B, and G is one-to-one with domain B and range C, then $G \circ F$ is one-to-one, and $(G \circ F)^{-1} = F^{-1} \circ G^{-1}$.
6. Prove that the set S is infinite if and only if there is a subset T of S that has the same cardinality as $\mathbb{N}$, the set of positive integers.
7. For each (m, n) in $N \times N$, define $F(m, n) = 2^m 3^n$. Show that F is a one-to-one function with domain $N \times N$ and range contained in $\mathbb{N}$. (Use the result that every integer has a unique decomposition into powers of prime numbers.) Use Theorem 3.2 (a) plus the result above to give another proof of Theorem 3.2 (b).
8. Verify that the function F defined in the proof of Theorem 3.2 (b) is one-to-one. Find another one-to-one polynomial function with domain $\mathbb{N} \times \mathbb{N}$ and range $\mathbb{N}$.
9. Let $\mathscr{B}$ denote a chain of sets. If $A_1, A_2, \ldots, A_n$ are in $\mathscr{B}$ show there is a permutation $\pi(1), \ldots, \pi(n)$ of $\{1, \ldots, n\}$ such that $A_\pi(1) \subset A_\pi(2) \subset \cdots \subset A_\pi(n)$.

B

10. A subset A of a set B is called a *proper* subset if A is not equal to B. Prove that a set B is infinite if and only if it has the same cardinality as a proper subset.
11. Prove that if there is a function F with domain $\mathbb{N}$ and range A then A is countable. (Do not assume F is one-to-one.)
12. Use the following outline to prove the Cantor-Bernstein Theorem: If A has the same cardinality as a subset of B and B has the same cardinality as a subset of A then A and B have the same cardinality. Let F be a one-to-one function with domain A and range a subset of B; let G be a one-to-one function with domain B and range a subset of A.
    (a) For each x in A we say x has origin in A if for a finite number of applications $F^{-1} \circ G^{-1} \circ F^{-1} \ldots F^{-1} \circ G^{-1}(x)$ is not in the range of G. This includes the case when x is not in the range of G. We say x has origin in B if for a finite number of applications $G^{-1} \circ F^{-1} \circ G^{-1} \ldots \circ F^{-1} \circ G^{-1}(x)$ is not in the range of F. Let A_A denote the set of all x in A with origin in A and A_B the set of all x in A with origin in B. Let A_∞ denote the complement of $A_A \cup A_B$ in A. Show how B_A, B_B, and B_∞ can be similarly defined.
    (b) Show that F is a one-to-one function from A_A onto B_A. Show that both F and G^{-1} are one-to-one functions from A_∞ onto B_∞. Show that G^{-1} is a one-to-one function from A_B onto B_B.
    (c) Use the results in (b) to construct a one-to-one function H from A onto B.
13. Prove that if A is a countably infinite set then the family $\mathscr{A}$ of all finite subsets of A is also countable. (Use the fact that the set of all primes is infinite, and proceed as in Exercise 7, or show that $\mathscr{A}$ is a countable union of countable sets.)

14. For x and y in $[0, 1]$ with $x = .s_1 s_2 \ldots$ and $y = .t_1 t_2 \ldots$ define $f(x, y) = .s_1 t_1 s_2 t_2 \ldots$. Prove that f is a one-to-one function from $[0, 1] \times [0, 1]$ onto $[0, 1]$ so that the two sets have the same cardinality.

15. Use mathematical induction to prove that if $A_n \neq \varnothing$ for all $n = 1, 2, \ldots$, then there is a function c from N into $\bigcup_n A_n$ such that $c(n) \in A_n$ for all n. (Begin: A_1 is nonempty so there is an element x in A_1; let $c(1) = x$.)

C

16. A real number x_0 is called *algebraic* if it is the root of a polynomial equation in which all of the coefficients are rational; that is, if there is an integer n and rational numbers $a_0, a_1, \ldots, a_n$ such that $a_0 + a_1 x_0 + \cdots + a_n x_0^n = 0$. The purpose of this exercise is to show that the set $\mathscr{A}$ of algebraic numbers is countable. Since the set R is not countable it follows that there exist real numbers that are not algebraic. This exercise illustrates the use of cardinality theory to establish nonconstructive existence theorems.

(a) Show that if x_0 is algebraic then x_0 is the root of a polynomial equation with integer coefficients.

(b) If

$$p(x) = m_0 + m_1 x + m_2 x^2 + \cdots + m_{n-1} x^{n-1} + m_n x^n$$

is a polynomial with integer coefficients, define the index $i(p)$ of p by the formula

$$i(p) = |m_0| + |m_1| + 2|m_2| + \cdots + (n-1)|m_{n-1}| + n|m_n|.$$

For each k in N let $\mathscr{B}_k$ be the set of all real numbers x_0 such that x_0 is a root of a polynomial with index k. Prove each $\mathscr{B}_k$ is finite. For this you must use the fact that every polynomial of degree n has at most n real roots.

(c) Show that $\mathscr{A} = \bigcup_k \mathscr{B}_k$. Thus $\mathscr{A}$ is countable.

4

Topology of the Real Line

The set $\mathbb{R}$ of real numbers corresponds to the points of a geometric line. The implication of the Completeness Axiom is that $\mathbb{R}$ satisfies our notion of continuity in that it has no gaps. In this section we shall derive the most basic topological properties of the real line.

Let a and b be real numbers with $a < b$. There are four distinct intervals that have greatest lower bound (glb) a and least upper bound (lub) b:

$$(a, b) = \{x: x \text{ is real}, a < x < b\};$$

$$[a, b] = \{x: x \text{ is real}, a \leqslant x \leqslant b\};$$

$$(a, b] = \{x: x \text{ is real}, a < x \leqslant b\}; \quad \text{and}$$

$$[a, b) = \{x: x \text{ is real}, a \leqslant x < b\}.$$

The interval (a, b) contains neither its glb a nor its lub b; it is called an *open* interval. The interval $[a, b]$ contains both its glb and lub; it is called a *closed* interval. The intervals $(a, b]$ and $[a, b)$ are called *half-open*. The half line $\{x: x$ is real, $x > a\}$ is written $[a, \infty)$; the intervals $(-\infty, b)$, $(-\infty, b]$, and $[a, \infty)$ can be defined analogously.

A subset O of $\mathbb{R}$ is called *open* if for each x in O there is an open interval (a, b) such that $x \in (a, b) \subset O$. A subset C of $\mathbb{R}$ is said to be *closed* if $C^\sim$, the complement of C in R, is open. It is easy to see that open intervals are open, closed intervals are closed but half-open intervals are neither closed nor open. The empty set and $\mathbb{R}$ are both open and closed. An open set O containing a point x is called a *nbd* (for "neighborhood") of x.

Theorem 4.1:

(a) If $\mathscr{A}$ is a collection of open sets then $\bigcup \mathscr{A}$ is open.

(b) If $\{A_1, \ldots, A_n\}$ is a finite collection of open sets then $\bigcap_{i=1}^n A_i$ is open.

Proof:

(a) Suppose $x \in S = \bigcup \mathscr{A}$. Then there is $A \in \mathscr{A}$ such that $x \in A$. Since A is open there is an open interval (a, b) with $x \in (a, b) \subset A$. But $A \subset S$ so $x \in (a, b) \subset S$. Since this statement is true for arbitrary x in S, it follows that S is open.

(b) Suppose $x \in S = \bigcap_{i=1}^n A_i$. Then $x \in A_i$ for each i. Since each A_i is open there is an open interval (a_i, b_i) such that $x \in (a_i, b_i) \subset A_i$. Let $a = \max\{a_1, a_2, \ldots, a_n\}$, and let $b = \min\{b_1, b_2, \ldots, b_n\}$. Since each $a_i < x$ it follows that $a < x$. Since each $b_i > x$ it follows that $b > x$. Therefore, for each i we have $x \in (a, b) \subset (a_i, b_i) \subset A_i$ so that $x \in (a, b) \subset S$. Therefore, S is open. ■

Example 4.2:

For each $n = 1, 2, \ldots$ the set $(-1/n, 1/n)$ is open but $\bigcap_n (-1/n, 1/n) = \{0\}$, which is not an open set. Thus in Theorem 4.1 (b) it is necessary to assume that the collection of sets is finite.

Corollary 4.3:

(a) If $\mathscr{G}$ is a collection of closed sets then $\bigcap \mathscr{G}$ is closed.

(b) If $\{C_1, C_2, \ldots, C_n\}$ is a finite collection of closed sets then $\bigcup_{i=1}^n C_i$ is closed.

Proof: In order to establish this proposition we use Theorem 4.1 and the DeMorgan Laws, Theorem 1.4. For example, to prove (a) we note that the complement of $\bigcap \mathscr{G}$ is equal to $\bigcup_{C \in \mathscr{G}} C^\sim$ because of Theorem 1.3 (a), and the latter set is open by Theorem 4.1 (a). ■

A subset K of $\mathbb{R}$ is called *compact* if whenever $\mathscr{A}$ is a collection of open sets such that $\bigcup \mathscr{A} \supset K$ there is a finite subcollection $\{A_1, A_2, \ldots, A_n\} \subset \mathscr{A}$ such that $K \subset \bigcup_{i=1}^n A_i$. We can say that every open covering of K contains a finite subcovering. An arbitrary collection $\mathscr{A}$ of sets is said to have the *finite intersection property* (fip) if whenever $A_1, \ldots, A_n$ is a finite subcollection of $\mathscr{A}$, the intersection $\bigcap_{i=1}^n A_i$ is nonempty.

Theorem 4.4:

(a) Every compact subset K of $\mathbb{R}$ is closed.

(b) If K is a compact subset of $\mathbb{R}$ and C is a subset of K, which is closed, then C is also compact.

(c) A subset K of $\mathbb{R}$ is compact if and only if the following condition holds:

(*) If $\mathscr{G}$ is a collection of closed subsets of K, which has the finite intersection property, then $\bigcap \mathscr{G}$ is nonempty.

Proof:

(a) If S is a subset of $\mathbb{R}$, which is not closed, then there is a point x in $S^{\sim}$ such that $(x - 1/n, x + 1/n)$ is not contained in $S^{\sim}$ for any n. This means that $(x - 1/n, x + 1/n)$ contains points of S for each n. For each n let $O_n = (-\infty, x - 1/n) \cup (x + 1/n, \infty)$. Then $\cup O_n = R\backslash\{x\} \supset S$. However $\{O_1, \ldots, O_{n}\}$ does not cover S for any n since $\bigcup_{i=1}^n O_i$ contains no points of $(x - 1/(n+1), x + 1/(n+1))$.

(b) Suppose C is a closed subset of a compact set K. Let $\mathscr{A}$ be an open covering of C. Then $\{C^{\sim}\} \bigcup \mathscr{A}$ is an open covering of K so it contains a finite subcovering $\mathscr{B}$. If $\mathscr{B}$ does not contain $C^{\sim}$ then it is the desired finite subcovering of C. If $\mathscr{B}$ does contain $C^{\sim}$ then $\mathscr{B}\backslash\{C^{\sim}\}$ is the desired subcovering of C.

(c) It is easiest to verify (c) by proving the contrapositive statement: A subset K is compact if and only if whenever $\mathscr{G}$ is a collection of closed subsets of K such that $\bigcap \mathscr{G} = \varnothing$, then there is a finite subcollection $C_1, C_2, \ldots, C_n$ of K such that $\bigcap_{i=1}^n C_i = \varnothing$. If K is compact and $\bigcap \mathscr{G} = \varnothing$ then the collection $\{C^{\sim} : C \in \mathscr{G}\}$ is an open covering of K since by the DeMorgan Laws $K\backslash\bigcup_{C\in\mathscr{G}} C^{\sim} = K \cap (\bigcap \mathscr{G}) = \varnothing$. Thus there is a finite subcovering $C_1^{\sim}$, $C_2^{\sim}, \ldots, C_n^{\sim}$. Since $K \subset \bigcup_{i=1}^n C_i^{\sim}$ and $C_i \subset K$ for each i it follows that $\bigcap_{i=1}^n C_i = \varnothing$.

On the other hand suppose K has the property (*). Let $\mathscr{A}$ be an open covering of K. For each $A \in \mathscr{A}$ let $C(A) = A^{\sim} \cap K$. Then each $C(A)$ is a closed subset of K and by the DeMorgan Laws, $\bigcap_{A\in\mathscr{A}} C(A) = \varnothing$. Thus there are $A_1, A_2, \ldots, A_n$ with $\bigcap_{i=1}^n C(A_i) = \varnothing$. It follows from the DeMorgan Laws that $\bigcup_{i=1}^n A_i \supset K$. ∎

A subset B of $\mathbb{R}$ is called *bounded* if it has both an upper and lower bound, in other words, if there is a real number b such that $B \subset [-b, b]$. If K is a compact set then K is bounded because the collection $\{(-n, n) : n = 1, 2, \ldots\}$ being an open covering of the real line covers K as well. Thus there is n such that $K \subset \bigcup_{j=1}^n (-j, j) \subset (-n, n)$. The following theorem known as the *Heine-Borel Theorem for R* characterizes the compact subsets of R.

Theorem 4.5: A subset K of $\mathbb{R}$ is compact if and only if it is closed and bounded.

Proof: We have already noted that compact sets are closed (Theorem 4.4) and bounded.

In order to prove the converse it suffices to show that every closed interval $[a, b]$ is compact. For every closed subset of a compact set is compact (Theorem 4.4), and every bounded set is a subset of a closed interval. The proof relies strongly upon the Completeness Axiom. Suppose $\mathscr{A}$ is an open covering of $[a, b]$. Let C be the set of all real numbers r such that (a) $a \leqslant r \leqslant b+1$ and (b) there is a finite subcollection $A_1, A_2, \ldots, A_n$ of $\mathscr{A}$ such that $\bigcup_{j=1}^{n} A_j \supset [a, r]$. Because of (a), $b+1$ is an upper bound of C, and since there is $A \in \mathscr{A}$ such that $a \in A$ it follows that $a \in C$. By the Completeness Axiom, C has a least upper bound M. We shall prove that $M > b$. Suppose otherwise for the sake of reaching a contradiction. Then $M \geqslant a$ since a is in C so that $M \in [a, b]$. Since $\mathscr{A}$ is an open covering of $[a, b]$ there is $A \in \mathscr{A}$ such that $M \in A$. Since A is open there is an open interval (u, v) such that $M \in (u, v) \subset A$. If $M = a$ then let $c = a$; otherwise let c be any real number such that $\sup(a, u) < c < M$. Let d be any real number such that $M < d < \inf(v, b+1)$. Since $c < M$ there is c' in C such that $c < c' < M$. This means there is a finite subcollection $A_1, A_2, \ldots, A_n$ of $\mathscr{A}$, which covers $[a, c']$. But adding the set A to this subcollection we cover $[a, d]$ because $A_1 \cup A_2 \cup \ldots \cup A_n \cup A \supset [a, c'] \cup (u, v) \supset [a, d]$. This proves that d is in the set C contradicting the assertion that M is an upper bound of C. Therefore, we conclude that $M > b$; and this implies that b is in C so that there is a finite subcovering of $[a, b]$. ■

A subset B of $\mathbb{R}$ is called *connected* if there is no pair of disjoint open sets O_1, O_2 such that (a) both $O_1 \cap B$ and $O_2 \cap B$ are nonempty and (b) $B = \{O_1 \cap B\} \cup \{O_2 \cap B\}$. Such a pair of sets is called a *separation* for B. Singletons are certainly connected. If the definition of connectedness is to fit our intuitive notion, intervals should be connected as well. Theorem 4.7 establishes this fact and characterizes all connected subsets of $\mathbb{R}$.

Lemma 4.6: If $\mathscr{A}$ is a collection of connected subsets of $\mathbb{R}$ and $\bigcap \mathscr{A}$ is nonempty then $\bigcup \mathscr{A}$ is connected.

Proof: Suppose O_1, O_2 is an attempted separation for $\bigcup \mathscr{A}$. Let $x \in \bigcap \mathscr{A}$. Then x is in O_1 or O_2 but not both. If x is in O_1 and $A \in \mathscr{A}$ then $A = \{A \cap O_1\} \cup \{A \cap O_2\}$ so one of the intersections must be empty since A is connected. But $x \in A \cap O_1$ so $A \cap O_2$ is empty. Since this is true for all A in $\mathscr{A}$ it follows that $O_2 \cap (\bigcup \mathscr{A}) = \varnothing$. This shows that every attempted separation for $\bigcup \mathscr{A}$ must fail so $\bigcup \mathscr{A}$ is connected. ■

Theorem 4.7: Every interval open, closed, half closed, bounded, and unbounded is connected. The only connected subsets of $\mathbb{R}$ are singletons and intervals.

Proof: We first show that every closed bounded interval $[a, b]$ is connected. Let O_1, O_2 be an attempted separation of $[a, b]$. One of the sets, say O_1, must contain a. Let C be the set of all r such that $a \leqslant r \leqslant b+1$ and $[a, r] \subset O_1$. Then C is bounded above so it has a least upper bound M. If we prove that $M > b$

it follows that $[a, b] \subset O_1$ so that the attempted separation will fail. If $M \leqslant b$ then either $M \in O_1$ or $M \in O_2$. If $M \in O_2$, then M is strictly greater than a; also since O_2 is open there are u, v with $M \in (u, v) \subset O_2$. Since $u < M$ there is x in C with $\sup(a, u) < x < M$. This means that $[a, x] \subset O_1$ so $x \in O_1$. Thus we have $x \in O_1 \cap O_2 \cap [a, b]$ so the attempted separation fails. Thus we must assume $M \in O_1$. If $M \in O_1$, then $[a, M] \subset O_1$ because $[a, t] \subset O_1$ for all t with $a \leqslant t < M$. Since O_1 is open there are u, v such that $M \in (u, v) \subset O_1$. If $M < y < \inf(b+1, v)$, then we have $[a, y] \subset O_1$ contradicting the fact that M is an upper bound of C. Thus, we have $M > b$.

Since $(a, b] = \bigcup_n [a - 1/n, b]$, it follows from Lemma 4.6 that $(a, b]$ is connected. Similarly $[a, b)$ and (a, b) are connected. Since $[a, \infty) = \bigcup_n [a, a+n]$ it follows that $[a, \infty)$ is connected. Similarly, the other infinite intervals are connected.

Next, suppose C is a connected subset of $\mathbb{R}$. If x and y are in C with $x < y$ and t is any number in (x, y) then t is in C, otherwise $(-\infty, t)$, (t, ∞) would be a separation of C. Thus if $m = \inf(C)$ and $M = \sup(C)$ where m can be $-\infty$ and M can be $+\infty$ it follows that $(m, M) \subset C$. Therefore, if C has two or more points it must be an interval. ■

Theorem 4.8: If O is an open subset of $\mathbb{R}$ then O is a countable union of disjoint open intervals, that is,

$$O = \bigcup_n (a_n, b_n) \quad \text{with } (a_i, b_i) \cap (a_j, b_j) = \varnothing \quad \text{when } i \neq j.$$

Proof: For each x in O we define $C(x)$, the component of x, to be the union of all intervals (u, v) such that $x \in (u, v) \subset O$. Each $C(x)$ is open by Theorem 4.1 and connected by Lemma 4.6. If $C(x) \cap C(y) \neq \varnothing$, then $C(x) \cup C(y)$ is connected by Lemma 4.6 so that $C(x)$ must equal $C(y)$. Since each $C(x)$ is open we can choose a single rational number $\rho(C(x))$ from each $C(x)$. The correspondence from $C(x)$ to $\rho(C(x))$ is a one-to-one function from the collection of all components onto a subset of Q the rationals. Thus $\{C(x) : x \in O\}$ contains countably many distinct sets. It is clear that $\bigcup_{x \in O} C(x) = O$. Since each component is open and connected it is an open interval. If $\{(a_i, b_i) : i = 1, 2, \ldots\}$ is any listing of the distinct components, then $\bigcup_n (a_n, b_n) = O$. ■

A real number x is called a *limit point* of a subset S of $\mathbb{R}$ if each open set O that contains x contains a point in S different from x. The following theorem about the existence of limit points is called the *Bolzano-Weierstrass Theorem.*

Theorem 4.9: If B is a bounded infinite subset of $\mathbb{R}$ then B has a limit point.

Proof: Since B is bounded it is contained in some closed interval $[a, b]$. If B has no limit point, then no point x of $[a, b]$ is a limit point of B so there is an open set $O(x)$ containing x such that $O(x) \cap B$ contains no other point of B except possibly x. The collection $\{O(x) : x \in [a, b]\}$ is an open covering of $[a, b]$ so it contains a finite subcovering $\{O(x_1), O(x_2), \ldots, O(x_n)\}$ by Theorem 4.5. Since $O(x_i)$ contains no point of B except possibly for x_i and $B \subset [a, b] \subset \bigcup_i O(x_i)$ it follows that $B \subset \{x_1, \ldots, x_n\}$ so that B must be finite. Therefore, if B is infinite it must have a limit point. ■

Exercises

A

1. Find the inf(S) and sup(S) for the following sets S and show that your answer is correct.
   (a) $S = \{x: x = r\sqrt{2},\ r \text{ is rational with } -\sqrt{2} < r < \sqrt{2}\}$.
   (b) $S = \{n + 1/n: n = 1, 2, \ldots\}$.
   (c) $S = (-1, 2] \cup \{3, 3\frac{1}{2}, 3\frac{3}{4}, 3\frac{7}{8}, \ldots\}$.
2. If S is a nonempty set show that $\inf(S) \leqslant \sup(S)$. Show that according to the definition $\inf(\varnothing) = \infty$ while $\sup(\varnothing) = -\infty$.
3. Show that the glb and lub of a set are unique.
4. Show that a subset O of $\mathbb{R}$ is open if and only if for each x in O there is $\varepsilon > 0$ such that $(x - \varepsilon, x + \varepsilon) \subset O$.
5. Show that the collection $\mathscr{A} = \{(0, 1/n): n = 1, 2, \ldots\}$ has the finite intersection property but that $\bigcap \mathscr{A}$ is empty.
6. Show that for each irrational number z, $Q \cap (-\infty, z)$, $Q \cap (z, \infty)$ is a separation for the set Q of rational numbers. Find a similar separation for the set $Q^{\sim}$ of irrational numbers.
7. Verify that $(a, b] = \bigcup_n [a + 1/n, b]$ and that $[a, \infty) = \bigcup_n [a, a + n)$.
8. Show that a subset S of $\mathbb{R}$ is both open and closed if and only if $S = \mathbb{R}$ or $S = \varnothing$.

B

9. Show that every set S of real numbers that is bounded below has a greatest lower bound.
10. The length of an interval (a, b), $[a, b)$, $(a, b]$, or $[a, b]$ is defined to be $b - a$. Show that for each $\varepsilon > 0$ there is a union of open intervals $\bigcup_n (a_n, b_n)$ that contains the set Q of rational numbers and for which the total length, $\sum_n (b_n - a_n)$, is less than ε.
11. Show that a subset O of $\mathbb{R}$ is open if and only if for each $x \in O$ there are rational numbers $a < x < b$ such that $(a, b) \subset O$. Show that if C is a closed subset of R, which contains the set Q of rational numbers, then $C = \mathbb{R}$.
12. Show that a subset S of $\mathbb{R}$ is connected if and only if for each real number r not in S one of the intervals $(-\infty, r)$, (r, ∞) has an empty intersection with S.
13. (a) Show that a is a limit point of the intervals $[a, b]$ and $(a, b]$.
   (b) Show that if M is the least upper bound of a set S either $M \in S$ or M is a limit point of S. Is it possible for M to be both a member of S and a limit point of S?
   (c) Show that each point in $\mathbb{R}$ is a limit point of the set Q of rationals.
   (d) Show that x is a limit point of S if and only if every open set O that contains x contains infinitely many points of S. Thus every set that has a limit point must be infinite.
   (e) Give an example of an infinite set with no limit point.

14. The set S' of all limit points of a set is called the *derived set* of S.
  (a) Show that S' is always closed.
  (b) Show that $S \cup S'$ is closed and that if C is a closed set that contains S, then C contains $S \cup S'$. The set $S \cup S'$ is called the *closure* of S and written $\bar{S}$.
  (c) Show that if $S \cap S' = \varnothing$ then S must be countable.

15. Show that the collection of all intervals of the form (a, b) with a and b in the set Q of rationals is countable.

16. If $\mathscr{A}$ is a collection of open sets show that there is a countable subcollection $\{O_i: i = 1, 2, \ldots\}$ such that $\bigcup_{\mathscr{A}} = \bigcup_i O_i$.

17. A real number x is called a *condensation point* of a set S if every open set O containing x contains uncountably many points of S. Show that if S is uncountable then the set of noncondensation points of S is countable so S must contain uncountably many condensation points. (Use Exercise 16.)

18. If S is a nonempty subset of $\mathbb{R}$ and x is in $\mathbb{R}$ then the *distance* from x to S is defined by the formula

$$d(x, S) = \inf\{|x - y| : y \in S\}.$$

  (a) If C is a closed subset of $\mathbb{R}$ and $d(x, C) = 0$ then $x \in C$.
  (b) If x is a limit point of S then $d(x, C) = 0$.

19. If S and T are two subsets of $\mathbb{R}$ then the distance between S and T is given by the formula

$$d(S, T) = \inf\{|s - t| : s \in S,\ t \in T\}.$$

  (a) What is $d(S, T)$ if $S = (0, 1)$ and $T = (2, 3)$?
  (b) If C is a closed subset of $\mathbb{R}$ and x is a point not in C then there is y in C such that $|x - y| = d(x, C)$.
  (c) If K is a compact subset of R and C is a closed subset of $\mathbb{R}$, then there are points x in K and y in C such that $|x - y| = d(K, C)$.
  (d) Give an example of two closed subsets C and D of $\mathbb{R}$ such that $C \cap D = \varnothing$ but $d(C, D) = 0$.

5

Real Functions and Sequences

In this section we shall study functions with domain and range in $\mathbb{R}$. In the special case when the domain of the function x is the set $\mathbb{N}$ of positive integers we usually write x_n instead of $x(n)$ for the value of x at the integer n. Such a function is called a *sequence*. It is possible to consider sequences whose domains are countable sets other than $\mathbb{N}$. For example, we might consider the sequence

$$x = \{\ldots\ x_{-2}, x_{-1}, x_0, x_1, x_2, \ldots\}$$

whose domain is the set of all integers. Differences in such cases from the standard treatment are mostly technical. Let a be a limit point of a set S of real numbers. A function f defined on S is said to *converge* to the *limit* b at a if for each nbd O

of b there is a nbd U of a such that whenever x is in $(U \cap S)\backslash\{a\}$, it follows that $f(x) \in O$. This is symbolized by writing

$$\lim_{x \to a} f(x) = b.$$

A sequence $\{x_n\}$ is said to *converge* to a number b if for each nbd O of b there is an integer M such that $x_n \in O$ for each $n > M$. This is expressed symbolically by writing

$$\lim_{n} x_n = b.$$

The following theorem asserts that the topological definitions of convergence stated above are equivalent to the "ε-definitions" usually encountered in calculus.

Theorem 5.1: Suppose f is a function defined on a set S and a is a limit point of S.

(a) The function f converges to b at a if and only if the following condition holds

for each $\varepsilon > 0$ there is $\delta > 0$ such that $|b - f(x)| < \varepsilon$ when $(x \in S$ and$)$ $0 < |a - x| < \delta$.

(b) A sequence (x_n) converges to x if and only if the following condition holds

for each $\varepsilon > 0$ there is M such that $|x - x_n| < \varepsilon$ when $n > M$.

Proof:

(a) Suppose f converges to b at a and $\varepsilon > 0$ is given. The open interval $(b - \varepsilon, b + \varepsilon)$ is a nbd of b so there is a nbd U of a such that $f(x)$ is in $(b - \varepsilon, b + \varepsilon)$, that is,

$$|b - f(x)| < \varepsilon \tag{5–1}$$

when $x \in (U \cap S)\backslash\{a\}$. Since U is a nbd of a there is an open interval (p, q) such that $a \in (p, q) \subset U$. Let $\delta = \inf\{|p - a|, |q - a|\}$. Then if $0 < |a - x| < \delta$, x is in U so (5–1) holds.

Suppose f satisfies the $\varepsilon - \delta$ condition with respect to a, b. Let O be a nbd of b. Then there is an open interval (p, q) with $b \in (p, q) \subset O$. If $\varepsilon = \inf\{|p - b|, \ |q - b|\}$ then $y \in O$ if $|b - y| < \varepsilon$. Let δ be such that $0 < |b - f(x)| < \varepsilon$ when $|a - x| < \delta$. Let U be the nbd $(a - \delta, a + \delta)$ of a. Then $f(x) \in O$ if $x \in (U \cap S)/\{a\}$. ■

The next theorem, stated without proof, relates convergence to the algebraic properties of $\mathbb{R}$.

Theorem 5.2: Suppose f and g are functions defined on $S \subset \mathbb{R}$, with limit point a such that f converges to b at a and g converges to c at a. Then we have the following

(a) $f + g$ converges to $b + c$ at a;

(b) fg converges to bc at a;

(c) f/g converges to b/c at a if $c \neq 0$;

(d) if $f(x) \leqslant g(x)$ for all x in $U\backslash\{a\}$ for U a nbd of a, then $b \leqslant c$. ■

Example 5.3:

Define $f(x)$ to be $1/n$ if x is a rational number of the form m/n where m and n have no common divisor and 0 otherwise. If r is any number, then $\lim_{x\to r} f(x) = 0$. To verify this let O be any nbd of 0. There is then $\varepsilon > 0$ such that $(-\varepsilon, \varepsilon) \subset O$. Let M be such that $M > 1/\varepsilon$. For each $k = 1, 2, \ldots, M$, let A_k consist of all rational numbers of the form $\{m/k: m \text{ is an integer, and } |m/k - r| < 1\}$. Then each A_k is finite so $A = \bigcup_{k=1}^{M} A_k$ is finite. Let δ be the distance from r to the set $A\backslash\{r\}$. If m/n is in $(r - \delta, r + \delta)\backslash\{r\}$, n must be larger than M so that $1/n < \varepsilon$. It follows that $f(m/n) \in O$ for all rational numbers in $(r - \delta, r + \delta)\backslash\{r\}$. If x is an irrational number $(r - \delta, r + \delta)$ then $f(x) = 0$, which is also in O.

A sequence (x_n) is called *increasing* if $x_n \leqslant x_{n+1}$ for each n. If $x_n < x_{n+1}$ for all n then (x_n) is called *strictly increasing*. The terms *decreasing* and *strictly decreasing* have analogous definitions. According to this definition a constant sequence is increasing. In order to avoid ambiguity, some authors reserve the term "increasing" for strictly increasing sequences and "nondecreasing" for increasing sequences. A sequence is *bounded above* if its range is bounded above. A sequence being a function from $\mathbb{N}$ into $\mathbb{R}$, its range is the set $\{x_1, x_2, \ldots\}$ where a value in the range may be named more than once if the sequence takes that value more than once.

Theorem 5.4: If an increasing sequence (x_n) is bounded above then it converges to the least upper bound of its range.

Proof: Denote $\sup(x_n)$ by x. Let O be any nbd of x. Then there are a and b with $a < x < b$ such that $(a, b) \subset O$. Since $a < x$, a is not an upper bound of (x_n) so there is M such that $a < x_M$. If $n > M$ we have $a < x_M \leqslant x_n$ so that $x_n \in (a, b) \subset O$. Therefore, by definition we have $\lim_n x_n = x$. ■

An analogous proof shows that every bounded decreasing sequence converges to the greatest lower bound of its range.

Example 5.5:

Define a sequence (a_n) as follows:

$$a_1 = 1, \qquad a_{n+1} = a_n + (2 - a_n^2)/(2a_n + 1).$$

If $\sqrt{2} - 1 < a_k < \sqrt{2}$ for $k = n$, then the inequality is true for $k = n + 1$. To verify this, note first that

$$\begin{aligned}\sqrt{2} - a_{n+1} &= \sqrt{2} - a_n - (2 - a_n^2)/(2a_n + 1)\\ &= (\sqrt{2} - a_n)(1 - (\sqrt{2} + a_n)/(2a_n + 1))\\ &= (\sqrt{2} - a_n)((a_n + 1 - \sqrt{2})/(2a_n + 1)) > 0,\end{aligned}$$

which gives the inequality $a_{n+1} < \sqrt{2}$. The inequality $\sqrt{2} - 1 < a_{n+1}$ results from the inductive hypothesis plus the fact that $a_{n+1} > a_n$, which is a consequence of $(2-a_n^2)/(2a_n+1)>0$. It follows from these calculations that (a_n) is a strictly increasing sequence bounded above by $\sqrt{2}$. By Theorem 5.4 this sequence converges to a limit a. The value of a can be calculated by using the following device:

$$a = \lim_n a_n = \lim_n a_{n+1} = \lim_n (a_n + (2-a_n^2)/(2a_n+1))$$
$$= a + (2-a^2)/(2a+1)$$

so that $2-a^2=0$ or $a=\sqrt{2}$.

A sequence (y_n) is called a *subsequence* of a sequence (x_n) if there is an increasing sequence $m(1)<m(2)<\ldots$ of integers such that $y_n = x_{m(n)}$ for all $n=1,2,\ldots$. For example, the sequence $(\frac{1}{2},\frac{1}{4},\frac{1}{8},\ldots)$ is a subsequence of the sequence $(1,\frac{1}{2},1,\frac{1}{4},1,\frac{1}{8},1,\ldots)$; the increasing sequence of integers is easily seen to be $2,4,6,\ldots$. The sequence $(\frac{1}{2},\frac{1}{4},\frac{1}{8},\ldots)$ is also a subsequence of $(1,\frac{1}{2},\frac{1}{3},\frac{1}{4},\frac{1}{5},\ldots)$ but now the increasing sequence of integers is not as easy to calculate. For a sequence (x_n) in $\mathbb{R}$ we define a sequence of extended real numbers (u_n) by setting

$$u_n = \inf\{x_j : j \geqslant n\}.$$

Since u_{n+1} is the greatest lower bound of a set that is a subset of $\{x_n, x_{n+1},\ldots\}$ we know that $u_{n+1}\geqslant u_n$. We define

$$\liminf_n x_n = \sup_n u_n.$$

For example, if $x_n=-n$ for each n, then $u_n=-\infty$ for each n so that

$$\liminf_n x_n = -\infty.$$

The quantity $\liminf_n x_n$ is called the *inferior limit* of the sequence (x_n). In an analogous way we can define the *superior limit* $\limsup_n x_n$.

Theorem 5.6: If (x_n) is a bounded sequence then there is a subsequence $(x_{n(k)})$ of (x_n) that converges to $\liminf_n x_n$.

Proof: For each $k\in N$ let $u_k=\inf\{x_n : n\geqslant k\}$. Then by definition we have

$$\sup_k u_k = \liminf_n x_n.$$

We define a subsequence $(x_{n(k)})$ of (x_n) as follows: Let $x_{n(1)}$ be any member of (x_n) such that

$$u_1 + \tfrac{1}{2} > x_{n(1)} \geqslant u_1.$$

If $x_{n(k)}$ has been determined, let $x_{n(k+1)}$ be any member of the sequence $(x_{n(k)+1}\, x_{n(k)+2}\,\ldots)$ for which we have

$$u_{n(k)+1} + 1/k > x_{n(k+1)} \geqslant u_{n(k)+1}.$$

Since the sequence $(u_{n(k)+1})$ is an increasing sequence in R, which is bounded above, it converges to its least upper bound, $\liminf_n x_n$, by Theorem 5.4. Since $\liminf_n x_n$ is equal to $\lim_k u_{n(k)+1}$ and $\lim_k 1/k=0$ it follows that $\liminf_n x_n$ is equal to $\lim_k x_{n(k)}$. ■

A function f with domain $S \subset \mathbb{R}$ and range in $\mathbb{R}$ is defined to be *continuous* at a limit point a in S if

$$\lim_{x \to a} f(x) = f(a).$$

If the point a in S is *isolated*, that is, not a limit point of S, then f is said to be continuous at a without satisfying any condition. The function f is said to be *continuous* on the set S if it is continuous at each point $a \in S$.

Theorem 5.7:

(a) A function f with domain $S \subset \mathbb{R}$ and range in $\mathbb{R}$ is continuous at a point $a \in S$ if and only if the following condition holds

for each nbd O of $f(a)$ there is a nbd U of a such that $U \cap S \subset f^{-1}(O)$.

(b) The function f is continuous on S if and only if the following condition holds

(*) for each open subset O of $\mathbb{R}$, there is an open set U such that $f^{-1}(O) \cap S = U \cap S$.

Proof:

(a) We have $\lim_{x \to a} f(x) = f(a)$ if and only if for each nbd O of $f(a)$ there is a nbd U of a such that $f(x) \in O$ for all x in $U \cap S$, which is true if and only if $U \cap S \subset f^{-1}(O)$.

(b) If the function f is continuous on S then (a) is true for every a in S. If O is open then for all b in $f(S) \cap O$ there is an open set U_b such that $U_b \cap S \subset f^{-1}(O)$. Let U be defined by

$$U = \cup \{U_b \colon b \in f(S) \cap O\}.$$

Then U is open and $U \cap S \subset f^{-1}(O) \cap S$. If $x \in f^{-1}(O) \cap S$ then $f(x)$ is in $O \cap f(S)$ so $x \in U_x \cap S \subset U \cap S$. Therefore, we have $U \cap S = f^{-1}(O) \cap S$.

If (*) is true and $a \in S$ then f is continuous at a because if O is a nbd of $f(a)$ then we can take the set U given in (b) to verify that (a) holds. ■

The notion of a compact set is the topological analog of the notion of a finite set. Thus we may suspect that a property that is true for an arbitrary function defined on a finite set will carry over to a continuous function defined on a compact set. The following result is one of the most prominent examples of this idea.

Theorem 5.8: If f is a continuous real valued function whose domain is a compact subset K of $\mathbb{R}$ there are points x_m and x_M in K such that

$$f(x_m) \leqslant f(x) \leqslant f(x_M) \quad \text{for all } x \text{ in } K.$$

In other words, f is bounded on K and assumes values that are maximum and minimum.

Proof: We first show that the set $f(K)$ is compact. Let $\{U_\alpha\}$ be any covering of $f(K)$ by open sets. By Theorem 5.7 (b) there are open sets V_α such that $V_\alpha \cap K =$

$f^{-1}(U_\alpha) \cap K$. Then $\{V_\alpha\}$ is an open covering of $Kx \in K$, which implies $f(x) \in f(K)$ so $f(x) \in U_\alpha$ for some α so $x \in f^{-1}(U_\alpha) \cap K \subset V_\alpha$. Since K is compact we can extract a finite subcovering $V_1, \ldots, V_n$. Then the corresponding $U_1, \ldots, U_n$ covers $f(K)$ because if $y \in f(K)$, $y = f(x)$ for some x in K so there is j such that $x \in V_j \cap K \subset f^{-1}(U_j)$, which implies that $f(x) \in U_j$. It follows that $f(K)$ is compact.

Since $f(K)$ is compact it is bounded. Let $C = \sup(f(K))$. For each $n \in N$ the set $O_n = \{y\colon y < C - 1/n\}$ is an open set of $\mathbb{R}$ so by Theorem 5.7 (b) there is an open set U_n such that

$$U_n \cap K = f^{-1}(O_n) \cap K = \{x\colon x \in K,\ f(x) < C - 1/n\}.$$

If a finite subcollection $\{U_1, U_2, \ldots, U_k\}$ of $\{U_n\}$ covered K it would follow that $K \subset f^{-1}(O_k)$ because $O_k \supset O_i$ for $i < k$. If K is a subset of $f^{-1}(O_k)$ it follows that $f(x) < C - 1/k$ for all x in K, contradicting the assertion that C is the least upper bound for $f(K)$. Therefore, no finite subcollection of $\{U_n\}$ covers K. Since K is compact $K \setminus \bigcup_n f^{-1}(O_n)$ is nonempty. If x_M is a point in $K \setminus \cup f^{-1}(O_n)$ we have $f(x_M) \geqslant C - 1/n$ for all n so $f(x_M) = C$.

The existence of x_m is established in an analogous way. ■

Theorem 5.9: A function f on S is continuous at $a \in S$ if and only if, for every sequence (x_n) in S, which converges to a, the sequence $(f(x_n))$ converges to $f(a)$.

Proof: Suppose f is continuous at a. Let O be a nbd of $f(a)$. By Theorem 5.7 (a) there is a nbd U of a such that $f(x) \in O$ whenever $x \in U \cap S$. If $\lim_n x_n = a$ and for each $x_n \in S$ there is M such that $x_n \in U$ when $n > M$. Thus when $n > M$, $f(x_n) \in O$ so that we conclude $\lim_n f(x_n) = f(a)$.

Suppose f is not continuous at a. Then there is a nbd O of $f(a)$ such that $U \cap S$ is not contained in $f^{-1}(O)$ for any nbd U of a. For $n = 1, 2, \ldots$ let x_n be a point in $(a - 1/n, a + 1/n) \cap S$ but not in $f^{-1}(O)$. Then (x_n) is a sequence in S that converges to a, but $f(x_n)$ is never in O so that $(f(x_n))$ does not converge to $f(a)$. ■

To conclude this section we shall introduce some further concepts of convergence for functions on $\mathbb{R}$ that will be of use later. If f is a real valued function that is defined on a subset S of $\mathbb{R}$, which contains the half-open interval $[a, c)$ for some $c > a$, then we say that f has the *limit b from the right* at a, if the following condition holds for each $\varepsilon > 0$ there is $\delta > 0$ such that

$$|f(x) - b| < \varepsilon \quad \text{if } a < x < a + \varepsilon.$$

To describe this symbolically we write

$$\lim_{x \downarrow a} f(x) = b.$$

Another common way to write this is $f(a+) = b$. We leave to the reader the construction of an analogous definition for *limit from the left*, which is written $\lim_{x\uparrow a} f(x) = b$ or $f(a-) = b$. A function f is said to be *continuous from the right at a* if $\lim_{x\downarrow a} f(x) = f(a)$. A similar definition for *continuous from the left* can be made using the limit from the left at a.

Suppose f is a real valued function defined on a set S, which contains a nbd $(a - \varepsilon, a + \varepsilon)$ of a. For $\delta \leqslant \varepsilon$ let $\sup(f, \delta)$ denote the extended real number $\sup\{f(x)\colon$

$a-\delta<x<a+\delta\}$. If $\delta_1<\delta_2$ then we have $\sup(f,\delta_1)\le\sup(f,\delta_2)$. The *limit superior* of f as x goes to a can now be defined by the formula

$$\limsup_{x\to a} f(x)=\inf\{\sup(f,\delta)\colon 0<\delta\leqslant\varepsilon\}.$$

An analogous definition can be given for $\liminf_{x\to a} f(x)$, the *limit inferior* of f as x goes to a.

Exercises

A

1. Prove Theorem 5.1 (b).
2. Prove Theorem 5.2.
3. Show by example that if $f(x)<g(x)$ for each x, $\lim_{x\to a} f(x)$ need not be strictly less than $\lim_{x\to a} g(x)$.
4. State and prove a theorem for sequences analogous to Theorem 5.2.
5. Find $\lim_n a_{k(n)}$ for those subsequences $(a_{k(n)})$ that converge when the sequence (a_n) is given by the following.
   (a) $\frac{1}{2}, 0, \frac{3}{4}, 0, \frac{7}{8}, 0, \ldots$.
   (b) $((-1)^n+(-1)^n/n)_{n=1}^{\infty}$.
   (c) (r_n) where r_n is any ordering of the rational numbers with $0<r_n<1$.
6. Find the inferior and superior limits for the sequences in Exercise 5.
7. An infinite series $\sum_{n=1}^{\infty} a_n$ is said to *converge* to the number a if the sequence, $\sum_{n=1}^{k} a_n$, of partial sums converges to a. Use Theorem 5.4 to prove that a series of positive terms converges if and only if the sequence of partial sums is bounded above.
8. Suppose f and g are continuous at a point a. Show that $f+g$ and fg are continuous at a. Show that f/g is continuous at a if $g(a)\neq 0$. Show that $\sup(f,g)$ defined by $\sup(f,g)(x)=\sup\{f(x),g(x)\}$ is continuous at a.
9. Show that the functions $f(x)=b$ and $g(x)=x$ are continuous on all of $\mathbb{R}$. Use this and the rules for multiplication and division to prove that all polynomials, $p(x)=a_0+a_1x+\cdots+a_nx^n$ are continuous on all of $\mathbb{R}$. Show that all rational functions, that is, functions $r(x)=p(x)/q(x)$ where p and q are polynomials are continuous wherever the denominator $q(x)$ is different from 0.
10. Give examples of a continuous function on a nonclosed interval that is bounded but does not assume its maximum on the interval, and a noncontinuous function on an interval $[a,b]$ that does not assume its maximum on the interval.
11. Write out the definition for the inferior limit of a sequence. Show that if (x_n) is a sequence that is bounded below, there is a subsequence $(x_{n(k)})$ that converges to $\liminf_n x_n$.
12. State definitions for
    (a) limit from the left;
    (b) continuous from the left;
    (c) lim inf of a function at a.

(d) Give an example of a function with different limit superior and inferior at a point a.

B

13. Show that a sequence (x_n) of real numbers converges if and only if it satisfies the following condition (the *Cauchy Criterion*): For each $\varepsilon > 0$, there is M such that $|x_m - x_n| < \varepsilon$ whenever $m, n > M$. (To show that this condition implies that a sequence converges, you show the sequence is bounded, the use Theorem 5.6, and finally show the entire sequence converges to the superior limit.)

14. Prove that the function defined in Example 5.3 is continuous at every irrational and discontinuous at every rational.

15. Show that if (a_n) is any sequence with $a_{n+1} = a_n + (2 - a_n^2)/(2a_n + 1)$ for all n, then $\lim_n a_n = \sqrt{2}$ if $a_1 > -\frac{1}{2}$ and $\lim_n a_n = -\sqrt{2}$ if $a_1 < -\frac{1}{2}$.

16. Define a sequence (a_n) by $a_1 = 1$, $a_{n+1} = 1/(1 + a_n)$. Show that this sequence converges and find its limit.

17. Let f be a continuous function with domain $[0, 1]$ and range contained in $[0, 1]$. Show there is $x \in [0, 1]$ such that $f(x) = x$.

18. Let χ_Q be the characteristic function of the set of rationals, that is, $\chi_Q(x) = 1$ for $x \in Q$, $\chi_Q(x) = 0$ for $x \in Q^{\sim}$. Show that χ_Q is not continuous at any point but that f assumes its maximum and minimum on every interval.

19. If x is a limit point of the range of the sequence (x_n) prove there is a subsequence of (x_n) that converges to x.

20. If (x_n) is a bounded sequence show that $\limsup_n x_n$ is the largest limit point of the range of (x_n) and that $\liminf x_n$ is the smallest limit point.

21. Show that $\lim_n x_n = x$ if and only if $\limsup_n x_n = \liminf x_n = x$.

22. If (x_n) is bounded above or (y_n) is bounded below show that

$$\limsup_n x_n + \liminf_n y_n \leqslant \limsup_n (x_n + y_n)$$
$$\leqslant \limsup_n x_n + \limsup_n y_n.$$

23. Let M be an integer greater than 1. If (a_n) is any sequence of integers with $0 \leqslant a_n < M$ for each n show that the series $\sum_{n=1}^{\infty} a_n M^{-n}$ converges to a number in $[0, 1]$. Show that every number x in $[0, 1]$ has an expansion of this kind that is unique unless x has the form b/M^n where b and n are integers, in which case there are exactly two expansions.

C

24. A function f is called *lower semicontinuous* (lsc) at a point x of its domain S if for each open interval of the form (c, ∞) that contains $f(x)$ there is a nbd U of x such that $S \cap \subset f^{-1}(c, \infty)$.

(a) Construct a corresponding definition for a function being upper semicontinuous (usc) at x.

(b) Show that a function is continuous at x if and only if it is both lsc and usc at x.

(c) Suppose $f(t)=0$ for $0 \leqslant t < \frac{1}{2}$ and $f(t)=1$ for $\frac{1}{2} < t \leqslant 1$. How must $f(\frac{1}{2})$ be defined to make f lsc? usc?

(d) Show that f is lsc at x if and only if for each sequence (x_n) converging to x we have $f(x) \leqslant \liminf_n f(x_n)$.

(e) If (f_n) is a sequence of lower continuous functions then the function defined by $f(x)=\sup_n f_n(x)$ is lower semicontinuous at those points where it is not infinite.

(f) Show that if f is lsc on a compact set K then its range is bounded below and it assumes its greatest lower bound on K.

6

Linear Spaces and Operators

A *linear space* over a field F is a set E and two functions one from $E \times E$ into E denoted by $+$ and one from $F \times E$ into E denoted by $\cdot$ or by juxtaposition. These operations are required to satisfy the following conditions known as the *linear space axioms.*

(A1) For all x, y, z in E, $x+(y+z)=(x+y)+z$.

(A2) For all x, y in E, $x+y=y+x$.

(A3) There is 0 in E such that $x+0=x$ for x in E.

(A4) For all x in E there is $-x$ in E such that $x+(-x)=0$.

(P1) For all a, b in F and x in E, $a(bx)=(ab)x$.

(P2) For all a, b in F and x in E, $(a+b)x=ax+bx$.

(P3) For all a in F and x, y in E, $a(x+y)=ax+ay$.

(P4) For all x in E, $1x=x$.

Conditions (A1)–(A4) assert that E forms a group under addition. Conditions (P1)–(P2) describe the action of the field F upon the group E. We are primarily interested in the case when the field is either $\mathbb{R}$, the field of real numbers, or $\mathbb{C}$, the field of complex numbers. These two fields are collectively known as the *scalar fields.* A few of the many important examples of linear spaces are listed below.

Example 6.1:

$\mathbb{R}^n$, the product of n copies of $\mathbb{R}$, is often called *Euclidean n-space.* A point in $\mathbb{R}^n$ is designated as an *n-tuple*

$$x=(x_1, x_2, \ldots, x_n)$$

where each $x_i \in \mathbb{R}$. Addition in $\mathbb{R}^n$ is given by the formula

$$(x_1, x_2, \ldots, x_n)+(y_1, y_2, \ldots, y_n)=(x_1+y_1, x_2+y_2, \ldots, x_n+y_n).$$

Example 6.2:

If S is any set, let $\Omega(S)$ consist of the set of all functions f from S into $\mathbb{R}$. The sum $f+g$ of two functions f and g is the function whose value at $s \in S$ is $f(s)+g(s)$. For a in $\mathbb{R}$ and f in $\Omega(S)$, af is the function whose value at s is $af(s)$. We shall refer to these definitions of sum and scalar multiple as the *usual* sum and scalar multiple of scalar valued functions defined on a set. The space $\Omega(\mathbb{N})$ where $\mathbb{N}$ denotes the set of integers represents the space of all real sequences; it is denoted by ω.

Example 6.3:

Suppose S is a subset of the real line. Let $C(S)$ denote the set of all real valued functions that are continuous on S. With the usual definitions for sum and scalar multiple $C(S)$ is a linear space. Special cases of $C(S)$ are $C(\mathbb{R})$, the space of all continuous functions on $\mathbb{R}$, $C[a, b]$ where $[a, b]$ is a closed interval, and $C(a, b)$ where (a, b) is an open interval. Notice that we have dropped a set of parentheses for aesthetic reasons, writing $C(a, b)$ instead of $C((a, b))$, which would be strictly correct.

Example 6.4:

The space $P(\mathbb{R})$ of all real valued polynomials of one real variable with the usual definitions for sum and scalar multiple is a linear space.

Example 6.5:

Let φ denote the set of all sequences (x_n) in ω that are eventually 0. This means that there is an integer M that depends on the sequence (x_n) such that $x_n = 0$ for $n \geqslant M$. With the same definition for sum and scalar multiple as in ω, φ is also a linear space.

If F is a subspace of a linear space E, and F is a linear space with respect to the operations defined on E, then F is called a *subspace* (or *linear subspace*) of E.

Example 6.6:

Let c denote the set of all convergent real sequences, c_0 the set of all real sequences that converge to 0, and l^1 the set of all sequences (x_n) for which $\sum_n |x_n| < \infty$. Then it is not hard to see that

$$\varphi \subset l^1 \subset c_0 \subset c \subset \omega$$

and that each of these sets forms a linear subspace of each larger one.

A subset S of a linear space E is called *linearly independent* (li) if whenever $x_1, x_2, \ldots, x_n$ are distinct members of S and $a_1, a_2, \ldots, a_n$ are scalars (that is, members of the field F) for which we have

$$\sum_{i=1}^{n} a_i x_i = 0,$$

it must follow that $a_i = 0$ for each i. For example, the three points $(1, 0, 0, 0)$, $(0, 1, 0, 0)$, and $(0, 0, 1, 0)$ of the space $\mathbb{R}^4$ form a li set. A subset S of a linear space E is said to *span* E if for each x in E there are $x_1, x_2, \ldots, x_n$ in S and scalars $a_1, a_2, \ldots, a_n$ such that

$$x = \sum_{i=1}^{n} a_i x_i.$$

It is easy to see that S spans a subspace F of E if and only if F is the smallest subspace of E that contains S; we denote this by writing

$$F = sp(S) \quad \text{or} \quad F = [S].$$

A linearly independent set S that spans a space E is called an *algebraic basis* (a-basis) for E. If a linear space has a finite a-basis then every a-basis for E has the same number of elements. In this case we say E has *finite dimension*; otherwise we say E has *infinite dimension.*

Example 6.7:

The space $\mathbb{R}^n$ has dimension n. An a-basis for $\mathbb{R}^n$ is the set of vectors

$$\{e_1 = (1, 0, \ldots, 0), e_2 = (0, 1, \ldots, 0), \ldots, e_n = (0, 0, \ldots, 1)\}.$$

This set is called the *usual* basis.

Theorem 6.8: Suppose S is a subset of a linear space E. The following statements are equivalent.

(a) S is an a-basis for E.

(b) Every x in E has a unique representation

$$x = \sum_{i=1}^{n} a_i x_i$$

where $x_1, x_2, \ldots, x_n$ are distinct members of S and each a_i is a scalar.

(c) S is a maximal linearly independent set, that is, S is li and if T properly contains S then T is not li.

(d) S is a minimal set that spans E, that is, $sp(S) = E$, and if T is a proper subset of S, then $sp(T)$ is a proper subset of E. ■

The next theorem is our first application of Postulate 3.6 on page 00. This postulate is not necessary if the space E is spanned by a countable set, but it is necessary if E has uncountable dimension.

Theorem 6.9: Every linear space E has an a-basis.

Proof: Let $\mathscr{A}$ denote the collection of all li subsets S of E. If $\mathscr{B}$ is a chain in $\mathscr{A}$ we show that $T = \bigcup \mathscr{B}$ is also in $\mathscr{A}$. Suppose $x_1, x_2, \ldots, x_n$ are distinct members of T and $a_1, a_2, \ldots, a_n$ are scalars such that $\sum_{i=1}^n a_i x_i = 0$. For each i there is S_i in $\mathscr{B}$ such that $x_i \in S_i$. Since $\mathscr{B}$ is a chain there is a largest S_i, say S_k. This means, $S_i \subset S_k$ for each i so that each $x_i \in S_k$. But $S_k \in \mathscr{A}$ so S_k is li. This implies that each $a_i = 0$. We have shown that $\mathscr{A}$ satisfies the hypotheses of Postulate 3.6. Therefore, $\mathscr{A}$ has a maximal member S that is an a-basis of E by Theorem 6.8. ■

A *linear operator* or *linear transformation* from a linear space E into a linear space F over the same field is a function T from E into F, which satisfies the following condition of linearity for all scalars a and b and all x, y in E

$$T(ax + by) = aT(x) + bT(y).$$

We shall also denote the value $T(x)$ of a linear transformation T by Tx if no difficulty arises from this notation.

Example 6.10:

Define D on $P(\mathbb{R})$ by $Df = f'$ where f' denotes the derivative of f. It is known from elementary calculus that the derivative of a polynomial is a polynomial and that differentiation satisfies the conditions of linearity. Therefore, D is a linear operator from $P(\mathbb{R})$ into $P(\mathbb{R})$. It is proved in linear algebra that a linear operator from a space E with dimension n into a space F with dimension m can be represented by an $m \times n$ matrix after fixing a-bases $x_1, x_2, \ldots, x_n$ for E and $y_1, y_2, \ldots, y_m$ for F. If we take the functions $1, x, x^2, \ldots, x^n, \ldots$ as the a-basis for $P(\mathbb{R})$ the operator D is represented by the infinite matrix.

$$\begin{bmatrix} 0 & 1 & 0 & 0 & \cdots \\ 0 & 0 & 2 & 0 & \cdots \\ 0 & 0 & 0 & 3 & \cdots \\ \cdots & \cdots & \cdots & \cdots & \cdots \end{bmatrix}$$

with respect to this a-basis.

Two linear spaces E and F are said to be *isomorphic* if there is a one-to-one linear operator with domain E and range F. This linear operator is called an *isomorphism*. The following theorem relates a-basis to isomorphism.

Theorem 6.11: Two linear spaces E and F are isomorphic if and only if a-bases for E and F have the same cardinality. ■

A subset C of a real or complex linear space E is called *convex* if for each pair of points x and y in C the line segment that joins x to y, namely the set

$$L(x, y) = \{tx + (1 - t)y : 0 \leqslant t \leqslant 1\}$$

is also in C.

Theorem 6.12:

(a) If $\mathscr{A}$ is any collection of convex sets, then $\bigcap \mathscr{A}$ is convex.

(b) If $\mathscr{A}$ is a chain of convex sets, then $\bigcup \mathscr{A}$ is convex.

Proof:

(a) Suppose x and y are in $\bigcap \mathscr{A}$ and $0 \leqslant t \leqslant 1$. Since x and y are in C for each C in $\mathscr{A}$ so is $tx + (1-t)y$. Hence, this point is also in the set $\bigcap \mathscr{A}$.

(b) That $\mathscr{A}$ is a chain means that if C and D are in $\mathscr{A}$, then either $C \subset D$ or $D \subset C$. Suppose x and y are in $\bigcup \mathscr{A}$. Suppose $x \in C$ and $y \in D$. If $C \subset D$ then x is also in D so $tx + (1-t)y \in D \subset \bigcup \mathscr{A}$ for all t in $[0, 1]$. A similar assertion can be made if $D \subset C$. ■

Exercises

A

1. Determine which of the following subsets of $\mathbb{R}^3$ are linear spaces and justify your conclusions.
   (a) S_1 consists of all (x_1, x_2, x_3) with $x_1 = 0$.
   (b) S_2 consists of all (x_1, x_2, x_3) with $x_1 = 1$.
   (c) S_3 consists of all (x_1, x_2, x_3) with $2x_1 - x_2 + x_3 = 0$.
   (d) S_4 consists of all (x_1, x_2, x_3) with $x_1 + x_2 = 0$ and $x_2 + x_3 = 0$.
   (e) S_5 consists of all (x_1, x_2, x_3) with $x_1^2 = x_2^2$.
   Which of these subsets are convex?
2. Determine on a-basis for each of the linear spaces in Exercise 1.
3. Show that the set of powers $\{1, x, x^2, \ldots\}$ is li in $P(\mathbb{R})$.
4. Show that in a linear space E over a field F, $0x = 0$ for all x in E and $(-1)x = -x$.
5. State the facts needed to show that the sets $C[a, b]$, φ, l^1, c, and c_0 are linear spaces.
6. For each n let e_n denote the sequence that has 1 in the nth place and 0's elsewhere. Show that $\{e_n : n = 1, 2, \ldots\}$ forms an a-basis for φ. Show that this set does not form an a-basis for l^1 since there is no finite linear combination of these vectors that equals the vector $(1, \frac{1}{2}, \frac{1}{4}, \frac{1}{8}, \ldots)$.
7. Find an isomorphism from φ onto $P(\mathbb{R})$.
8. Which of the following functions from $\mathbb{R}^3$ into $\mathbb{R}^3$ are linear operators? Justify your answer.
   (a) $T_1(x_1, x_2, x_3) = (x_2, x_3, x_1)$.
   (b) $T_2(x_1, x_2, x_3) = (2x_2 + x_3, x_3^2, x_1)$.
   (c) $T_3(x_1, x_2, x_3) = (x_1 - 2x_2 + 2x_3, x_3 - x_1, 2x_1 + 3x_2 - x_3)$.
   (d) $T_4(x_1, x_2, x_3) = (2x_2 + x_3, x_1, 1)$.
9. If T is a linear operator from a linear space E into a linear space D, show that $T0 = 0$ where the first 0 is the 0 of E and the second 0 is the 0 of F.
10. What value must x take so that the set

$$\{(1, -2, 1), (2, 3, 0), (x, 1, 1)\}$$

    is not li?
11. For f in $C[0, 1]$ define Sf to be the function on $[0, 1]$ whose value at t is

given by the formula

$$Sf(t) = \int_0^t f(s)\,ds.$$

Show that S is a linear transformation from $C[0, 1]$ into $C[0, 1]$.

12 Is every subspace of a real or complex linear space convex?

13. Suppose T is a linear operator from a real linear space E into a real linear space F. Show that if C is a convex subset of E, $T(C)$ is convex in F; and if D is a convex subset of F, $T^{-1}(D)$ is convex in E.

B

14. Let $C^\infty(\mathbb{R})$ denote the space of all functions f on $\mathbb{R}$ such that $f^{(n)}$, the nth derivative of f, exists for all $n = 1, 2, \ldots$. Show that $C^\infty(\mathbb{R})$ is a linear space and $Df = f'$ is a linear operator from $C^\infty(\mathbb{R})$ into itself.

15. Show that every sequence x in c can be written in the form

$$x = \left(\lim_n x_n\right) e + u$$

where e is the sequence $(1, 1, 1, \ldots)$ and u is in c_0. Thus $sp(\{e\} \cup c_0) = c$. This is expressed by saying that c_0 has codimension one in c.

16. Let $F(2)$ be the two-point field of Exercise 12 in Section 2. Let $F(2)^n$ consist of all n-tuples $(a_1, a_2, \ldots, a_n)$ with $a_i \in F(2)$ for each i.

(a) Show that $F(2)^n$ satisfies the linear space axioms with the usual definition of sum and scalar multiplication.

(b) How many points are there in the space $F(2)^n$?

(c) How many linear transformations from $F(2)^n$ into $F(2)^m$? (Use the bases $e_1, e_2, \ldots, e_n$ for $F(2)^n$ and $e_1, e_2, \ldots, e_m$ for $F(2)^m$ and count the number of matrices.)

17. If T is a linear operator from a linear space E into a linear space D, show that $T(E)$ is a linear subspace of D. Show that the set of all x in E such that $Tx = 0$ is a linear subspace of E. Show that if $T(S)$ is li, then S is li.

18. Show that if C and D are convex subsets of $\mathbb{R}^n$ then the set $C + D = \{x + y\colon x \in C,\ y \in D\}$ is also convex.

C

19. *(a)* Show that $\mathbb{R}$ forms a linear space over the field Q of rationals.

(b) If x is not an algebraic number, show that the set $\{1, x, x^2, \ldots\}$ is linearly independent over Q. (See Exercise 16 of Section 3.)

(c) Show that the dimension of $\mathbb{R}$ with respect to Q must be uncountable. (Assume there were a countable a-basis and from this derive the false assertion that $\mathbb{R}$ is countable.)

(d) Any a-basis H of $\mathbb{R}$ over Q is called a *Hamel* basis. If H is any such basis then each r in $\mathbb{R}$ has a unique representation $r = \sum_{h\in H} f_h(r)h$ where $f_h(r)$ is a function from $\mathbb{R}$ into Q such that $f_h(x + y) = f_h(x) + f_h(y)$ for all x, y in $\mathbb{R}$, $f_h(sx) = sf_h(x)$ for s in Q, and x in $\mathbb{R}$ and $f_h(r) = 0$ for all but finitely many h.

(e) If α is any function from a Hamel basis H into $\mathbb{R}$ and the linear transformations f_h are given as above, show that the function T defined by $Tx = \sum_{h \in H} \alpha(h) f_h(x)$ is a linear transformation from $\mathbb{R}$ into $\mathbb{R}$ with respect to Q. That is, if s and t are in Q and x and y are in $\mathbb{R}$, $T(sx + ty) = sT(x) + tT(y)$.

(f) If T is a function from $\mathbb{R}$ into $\mathbb{R}$ such that $T(x + y)$ for all x and y in $\mathbb{R}$, show that for all rational numbers m/n, $T(m/n) = (m/n)T(1)$. Show that if T is continuous $T(x) = xT(1)$ for all x in $\mathbb{R}$. But show there are uncountably many discontinuous S such that $S(1) = T(1)$.

2

Metric Spaces

I tell them that if they study mathematics
they will find in it the best remedy
against the lusts of the flesh.
Thomas Mann in *The Magic Mountain*

This chapter is an introduction to some basic ideas of point set topology, in particular to the concept of metric space. Topology may be described as the mathematical study of those properties of a space that are preserved by homeomorphisms, that is, continuous functions that have continuous inverses. A metric space is a set on which a topology is induced by a distance-like function. Although metric topological spaces are the most common in analysis, we shall encounter a few topologies that are not induced by a metric.

In Section 7 we give the abstract definition of a topology and a topological space, and then we define fundamental topological concepts such as open, closed, separable, interior point, and limit point. Equally fundamental to topology is the concept of continuous function. In Section 7 we define continuity and show that continuous functions preserve compactness and connectedness. These concepts are relevant to all topologies whether they are induced by a metric or not.

The class of metric spaces introduced in Section 8 is of primary importance in analysis. The usual Euclidean distance on $\mathbb{R}^n$ is the prototypical example of a metric. A metric on a space, provides a topology for the space just as the absolute value function on $\mathbb{R}$ that we discussed in Section 4. In Section 8 we describe explicitly how this is accomplished; and we discuss how basic topological properties, particularly compactness, can be characterized in a metric setting. Some very strong results, in particular the Baire Category Theorem, hold for metric spaces that are complete. Complete metric spaces and their properties are discussed in Section 9.

The final section of this chapter concerns the product of topological spaces. We first define a topology on a finite product of topological spaces and we determine that if the factor spaces are metric spaces—complete metric spaces or compact metric spaces—then the product space is also of that type. We next extend the definition of product space to arbitrary products. The main theorem in this setting is the Tychonoff Theorem, which asserts that the product of compact spaces is compact. We also verify that connectedness is preserved by products and metric space is preserved by countable products.

7

Topological Spaces

A *topology* on a nonempty set S is a collection $\mathscr{T}$ of subsets of S that satisfies the following three conditions.

($\mathscr{T}$1) The entire set S and the empty set $\varnothing$ belong to $\mathscr{T}$.

($\mathscr{T}$2) If $\mathscr{B}$ is a subcollection of $\mathscr{T}$ then $\bigcup \mathscr{B}$ is in $\mathscr{T}$.

($\mathscr{T}$3) If $\{O_1, O_2, \ldots, O_n\}$ is a finite subcollection of $\mathscr{T}$ then $\bigcap_{i=1}^{n} O_i$ is in $\mathscr{T}$.

If you compare assertions ($\mathscr{T}$2) and ($\mathscr{T}$3) to Theorem 4.1 on page 18, you will see that the axioms for a topology are satisfied by the collection of open subsets of $\mathbb{R}$ defined in Section 4. This topology on $\mathbb{R}$ is known as the *usual topology*. The pair $(S, \mathscr{T})$ is called a *topological space* and the sets in $\mathscr{T}$ are called the *open* (or $\mathscr{T}$-open) *subsets* of S.

An open set containing the point x is called a *nbd* (for neighborhood) of x. A subset C of S is called *closed* if $C^{\sim}$, the complement of C in S, is open. The *interior* of a set A, written $\operatorname{int}(A)$, is the union of all open subsets contained in A. The *closure* of a set A, written $\bar{A}$, is the intersection of all closed sets that contain A. A set A is said to be *dense* in a set B if $\bar{A} = B$. A set A is called *separable* if A contains a countable dense subset B.

We specifically warn the reader that contrary to the ordinary usage of the words, *the terms "open" and "closed" are not opposites with respect to topology.* (Therefore, do not assume that: (1) If O is not closed, then O is open; (2) Since O is open it is not closed; (3) Every set is either open or closed.) For example, the empty set and the whole space are always both open and closed. In the usual topology on $\mathbb{R}$, the half-open interval $[0, 1)$ is neither open nor closed.

The following example describes the two most obvious topologies on a set. The main importance of the example is that it shows that the definition of topology allows structures on a set that have little potential use.

Example 7.1:

If S is any set, the *discrete* topology $\mathscr{T}_{\max}$ consists of all subsets of S. The collection $\mathscr{T}_{\min}$, which consists only of S and $\varnothing$, is called the *indiscrete* topology.

If $\mathscr{T}_1$ and $\mathscr{T}_2$ are two topologies on a set S with $\mathscr{T}_1 \subset \mathscr{T}_2$ we say that $\mathscr{T}_1$ is *weaker* than $\mathscr{T}_2$ and $\mathscr{T}_2$ is *stronger* than $\mathscr{T}_1$. It is easy to see that $\mathscr{T}_1$ is weaker than $\mathscr{T}_2$ if and only if every $\mathscr{T}_1$ open set is $\mathscr{T}_2$ open if and only if every $\mathscr{T}_1$ closed set is $\mathscr{T}_2$ closed.

If $(S, \mathscr{T})$ is a topological space and T is a nonempty subset of S, then the collection $\mathscr{T}_T$ of all sets of the form $O \cap T$ as O ranges over $\mathscr{T}$ satisfies axioms ($\mathscr{T}$1)–($\mathscr{T}$3). The topological space $(T, \mathscr{T}_T)$ is called a *subspace* of $(S, \mathscr{T})$, and the topology $\mathscr{T}_T$ is called the *relative topology* on T. The notion of relative topology is important because it allows us to discuss topological concepts such as closure and separability with respect to a subset of a space. This is illustrated in the following example.

Example 7.2:

Let I denote the half-open interval $[a, b)$. If $\mathscr{T}$ denotes the usual topology on R, then the relative toplogy $\mathscr{T}_I$ consists of all subset of the form $O \cap I$ where O is open in $\mathbb{R}$. Since $I = \mathbb{R} \cap I$ it follows that I itself is open in I even though it is not an open subset of $\mathbb{R}$. The point a is an interior point of I in the relative topology $\mathscr{T}_I$ even though it is not an interior point of I in the usual topology of R.

Let $(S, \mathscr{T})$ be a topological space. A point x in S is called a *limit point* of a subset A of S if every nbd of x contains a point in A that is not equal to x. The definition does not require x to be in A. If S is any set with more than two points, then every point is a limit point of every set with respect to the indiscrete topology $\mathscr{T}_{\min}$; however, the discrete topology $\mathscr{T}_{\max}$ allows no limit points.

Theorem 7.3: A set A is closed if and only if every limit point x of A is in A.

Proof: Sufficiency Suppose A is a set in a topological space $(S, \mathscr{T})$ that contains all of its limit points. We shall prove that $A^\sim$ is open. If $x \in A^\sim$, then x is not a limit point of A so there is a nbd O_x of x such that $O_x \cap A = \varnothing$; that is, $O_x \subset A^\sim$. Since $A^\sim$ is equal to the union of all such O_x, and the union of open sets is open, it follows that $A^\sim$ is open.

Necessity Suppose A is closed and x is in $A^\sim$. Since $A^\sim$ is open, it is a nbd of x that contains no point of A. Hence x is not a limit point of A. Therefore, A contains all of its limit points. ■

A topological space $(S, \mathscr{T})$ is called a *Hausdorff space* and the topology $\mathscr{T}$ is called a *Hausdorff topology* if for each pair of points x and y in S there are nbds O_x of x and O_y of y such that $O_x \cap O_y = \varnothing$. The Hausdorff property is the most important of several separation axioms. It describes a minimal degree of separation, which a topological space needs in order to be of much use in analysis. The topology $\mathscr{T}_{\min}$ does not have the Hausdorff property unless the set S contains only one point; on the other hand, the topology $\mathscr{T}_{\max}$ is always Hausdorff. The usual topology on R is also Hausdorff.

A *separation* for a subset B of a topological space $(S, \mathscr{T})$ is a pair of disjoint open sets O_1, O_2 such that (a) both $O_1 \cap B$ and $O_2 \cap B$ are nonempty and (b) $B = (O_1 \cap B) \cup (O_2 \cap B)$. The set B is *connected* if there is no separation for B. It is clear that singletons are always connected. In $\mathscr{T}_{\max}$ these are the only connected sets. In $\mathscr{T}_{\min}$ every set is connected. Theorem 4.7 on page 20 asserts that the only connected subsets of R with the usual topology are singletons and intervals.

The following theorem is an improvement of Lemma 4.6 on page 20.

Theorem 7.4: If $\mathscr{A}$ is a collection of connected subsets of a topological space $(S, \mathscr{T})$ and $A \cap B$ is nonempty for all A and B in $\mathscr{A}$, then $\bigcup \mathscr{A}$ is connected.

Proof: Let (O_1, O_2) be an attempted separation of $\bigcup \mathscr{A}$. For each $A \in \mathscr{A}$ either $A \subset O_1$ or $A \subset O_2$ because if both $A \cap O_1$ and $A \cap O_2$ were nonempty, then

(O_1, O_2) would be a separation for A. If A is a fixed set in $\mathscr{A}$ and $A \subset O_1$, then for each $B \in \mathscr{A}$, $A \cap B \neq \varnothing$ so that $B \cap O_1 \neq \varnothing$. This implies $B \subset O_1$. Hence we have $\bigcup \mathscr{A} \subset O_1$ so the separation fails. ■

Continuity is a key notion in topology. The term "continuous" connotes the transformation of an object A into an object B in such a way that A can be stretched or squeezed but not broken. A function f from a topological space $(S, \mathscr{T}_1)$ to a topological space $(T, \mathscr{T}_2)$ is said to be *continuous* at the point x in S if for each nbd U of $f(x)$, there is a nbd V of x such that $V \subset f^{-1}(U)$. If the domain of f is a subset S_1 of $(S, \mathscr{T}_1)$ then f is said to be continuous at x in S_1 if it is continuous with respect to the relative topology on S_1. See Theorem 5.7 on page 27. The function f is said to be continuous on S if it is continuous at each x in S. The following theorem illustrates the connection between the concepts of continuous function, open set and closed set.

Theorem 7.5: For a function f from the topological space $(S, \mathscr{T}_1)$ into the topological space $(T, \mathscr{T}_2)$, the following statements are equivalent.

(a) f is continuous on S.

(b) $f^{-1}(U)$ is open for each open subset U of T.

(c) $f^{-1}(C)$ is closed for each closed subset C of T.

Proof: (The arrow $\Rightarrow$ means *implies.*)

(a) $\Rightarrow$ *(b).* If U is open in T then for each $y = f(x)$ in U there is a nbd V_x of x such that $V_x \subset f^{-1}(U)$. It follows that we have

$$f^{-1}(U) = \bigcup \{V_x \colon x \in V_x \subset f^{-1}(U)\}$$

so that $f^{-1}(U)$ is open.

(b) $\Rightarrow$ *(c).* If C is closed in T then $C^{\sim}$ is open in T so $f^{-1}(C^{\sim})$ is open in S. But $f^{-1}(C) = \{f^{-1}(C^{\sim})\}^{\sim}$ by Theorem 1.4 (a) on page 4, so $f^{-1}(C)$ is closed.

(c) $\Rightarrow$ *(b).* The proof proceeds as that to (b) $\Rightarrow$ (c).

(b) $\Rightarrow$ *(a).* Suppose x is any point of S, and U is a nbd of $f(x)$. Then $f^{-1}(U)$ is open so it is a nbd of x, which satisfies the definition of continuity. ■

The fields $\mathbb{R}$ of real numbers and $\mathbb{C}$ of complex numbers are called the *scalar fields.* A function from an arbitrary set into a scalar field is called a *scalar function.* The algebraic properties of the collection of scalar functions defined on a set S depends on the range rather than the set S. Specifically, let S be any set and let f and g be functions from S into the field $\mathbb{R}$. The functions af (for $a \in \mathbb{R}$), $f+g$, fg, f/g, $|f|$, $\sup(f, g)$, $\inf(f, g)$ can be defined from S into $\mathbb{R}$ in pointwise. Thus we have

$(af)(x) = af(x)$	for all $x \in S$, $a \in \mathbb{R}$;
$(f+g)(x) = f(x) + g(x)$	for all $x \in S$;
$fg(x) = f(x)g(x)$	for all $x \in S$;
$(f/g)(x) = f(x)/g(x)$	for all $x \in S$ with $g(x) \neq 0$;
$\sup(f, g)(x) = \sup\{f(x), g(x)\}$	for all $x \in S$;
$\inf(f, g)(x) = \inf\{f(x), g(x)\}$	for all $x \in S$;
$\lvert f\rvert(x) = \lvert f(x)\rvert$	for all $x \in S$.

The following theorem shows that if S is a topological space, then the set of continuous real valued functions on S is also closed under the algebraic and lattice operations we just described. You can state corresponding definitions for complex valued functions on S (but now inf, and sup cannot be defined) and prove a corresponding theorem.

Theorem 7.6: Suppose f and g are functions from a topological space $(S, \mathcal{T})$ into $\mathbb{R}$ with the usual topology. If f and g are continuous at the point x in S so are the functions: af, $f+g$, fg, f/g (provided $g(x) \neq 0$), $|f|$, $\sup(f, g)$, and $\inf(f, g)$.

Proof: af The constant function C_a defined by $C_a(x) = a$ is continuous on S because if V is any nbd of a, then $f^{-1}(V) = \mathbb{R}$, which we know is open. Thus af is continuous because the product of continuous functions is continuous, which we prove below.

$\boldsymbol{f+g}$ Let U be a nbd of $c = f(x) + g(x)$. Since U is an open set in $\mathbb{R}$ there is $\varepsilon > 0$ such that $(c - \varepsilon, c + \varepsilon) \subset U$. Define sets U_1 and U_2 by

$$U_1 = \left(f(x) - \frac{\varepsilon}{2}, f(x) + \frac{\varepsilon}{2}\right) \quad \text{and} \quad U_2 = \left(g(x) - \frac{\varepsilon}{2}, g(x) + \frac{\varepsilon}{2}\right).$$

By the definition of continuity there are nbds V_1 and V_2 of x such that $V_1 \subset f^{-1}(U_1)$ and $V_2 \subset f^{-1}(U_2)$, If $y \in V_1 \cap V_2$, then we have

$$|(f(x) + g(x)) - (f(y) + g(y))| \leqslant |f(x) - f(y)| + |g(x) - g(y)| < \frac{\varepsilon}{2} + \frac{\varepsilon}{2} = \varepsilon$$

so that $f(y) + g(y) \in U$. Therefore, $V_1 \cap V_2$ is a nbd of x contained in $(f+g)^{-1}(U)$. By definition, $f+g$ is continuous at x.

$\boldsymbol{fg}$ If V is a nbd of $f(x)g(x)$, then there is ε with $0 < \varepsilon < 1$ such that we have

$$(f(x)g(x) - \varepsilon, f(x)g(x) + \varepsilon) \subset V.$$

Let $\delta > 0$ be defined by the formula

$$\delta = \frac{\varepsilon}{2}(|f(x)| + |g(x)| + 1)$$

so that $|f(x)|\delta$ and $(|g(x)| + 1)\delta$ are both less than $\varepsilon/2$. Since f and g are continuous, the set U defined by

$$U = f^{-1}(f(x) - \delta, f(x) + \delta) \cap g^{-1}(g(x) - \delta, g(x) + \delta)$$

is a nbd of x. If u is in U we have

$$\begin{aligned} |f(u)g(u) - f(x)g(x)| &= |f(u)g(u) - f(x)g(u) + f(x)g(u) - f(x)g(x)| \\ &\leqslant |g(u)||f(u) - f(x)| + |f(x)||g(u) - g(x)| \\ &< |g(u)|\delta + |f(x)|\delta \\ &\leqslant (|g(x)| + 1)\delta + |f(x)|\delta < \varepsilon. \end{aligned}$$

$\boldsymbol{f/g}$ We assume $g(x) \neq 0$ and show $1/g$ is continuous at x. Then we apply the previous assertion to $f/g = f(1/g)$. Let U be a nbd of $c = 1/g(x)$. Let $\varepsilon > 0$ be such that $U_1 = (c - \varepsilon, c + \varepsilon) \subset U$ and $\varepsilon < 1$. Let $\delta = \varepsilon|g(x)|^2/2$; let $U_2 = (g(x) - \delta, g(x) + \delta)$; and let V be a nbd of x such that $V \subset g^{-1}(U_2)$. If $y \in V$ then

$|g(y)-g(x)|<\delta$ so that we have

$$|g(x)|<|g(y)|+\delta<\frac{|g(y)|+|g(x)|}{2},$$

which implies that $|g(y)|>|g(x)|/2$ or that $1/|g(y)|<2/|g(x)|$. Therefore, we conclude that

$$\left|\frac{1}{g(y)}-\frac{1}{g(x)}\right|=\left|\frac{(g(x)-g(y))}{g(x)g(y)}\right|<\left\{\left(\frac{\varepsilon}{2}\right)|g(x)|^2\right\}\left\{\frac{1}{|g(x)|}\right\}\left\{\frac{2}{|g(x)|}\right\}=\varepsilon.$$

$|f|$ If U is a nbd of $c=|f(x)|$ there is $\varepsilon>0$ such that $(c-\varepsilon, c+\varepsilon)\subset U$. If V is a nbd of x such that $V\subset f^{-1}(c-\varepsilon, c+\varepsilon)$ then V is also a subset of $|f|^{-1}(c-\varepsilon, c+\varepsilon)$ because if y is in V we have

$$\|f(x)|-|f(y)\|\leqslant|f(x)-f(y)|<\varepsilon.$$

The continuity of $\sup(f, g)$ and $\inf(f, g)$ follows from the preceding argument plus the identities

$$\textbf{\textit{sup}}\,(f, g)=\tfrac{1}{2}\{(f+g)+|f-g|\}\quad\text{and}\quad \textbf{\textit{inf}}\,(f, g)=\tfrac{1}{2}\{(f+g)-|f-g|\}.\ \blacksquare$$

The following theorem shows that the formal definitions for connected set and continuous function conform to our intuitive ideas. That is to say, a continuous function will not "break" a connected set.

Theorem 7.7: Suppose f is a continuous function from a topological space $(S, \mathscr{T}_1)$ into a topological space $(T, \mathscr{T}_2)$. If B is a connected subset of S then $f(B)$ is a connected subset of T.

Proof: If U, V is a separation of $f(B)$, we show $f^{-1}(U)$, $f^{-1}(V)$ is a separation of B. The sets $f^{-1}(U)$ and $f^{-1}(V)$ are open because U and V are open and f is continuous. The intersection $f^{-1}(U)\cap B$ is nonempty because if $y\in U\cap f(B)$, there is x in B with $f(x)=y$, and this x is in $f^{-1}(U)\cap B$. Similarly the intersection $f^{-1}(V)\cap B$ is nonempty. The sets $f^{-1}(U)$ and $f^{-1}(V)$ are disjoint because if u were in their intersection, $f(u)$ would be in $U\cap V$, which we know is empty. Finally, if x is in B, then $f(x)$ is in the union of $U\cap f(B)$ and $V\cap f(B)$ so x must be in the union of $f^{-1}(U)\cap B$ and $f^{-1}(V)\cap B$. ■

A subset C of a topological space $(S, \mathscr{T})$ is called *compact* if from every collection $\mathscr{B}$ of open sets that covers C, that is, $C\subset\bigcup\mathscr{B}$, we can extract a finite subcovering. It is easy to see that every finite set in any topological space is compact. The notion of a set being compact is the topological analog of a set being finite. This means, for example, that a continuous real valued function on a compact set behaves similarly to an arbitrary real valued function on a finite set. The proof of the following theorem resembles that of Theorem 4.4 (b) and (c) on page 19.

Theorem 7.8:

(a) If C is compact and K is a closed subset of C, then K is compact.

(b) A topological space $(S, \mathscr{T})$ is compact if and only if the following condition holds: (*) If $\mathscr{G}$ is a collection of closed subsets of S that has the fip, then $\bigcap\mathscr{G}$ is nonempty. ■

The statement that corresponds to (a) of Theorem 4.4 on page 19 is the following.

Theorem 7.9: Every compact subset C of a Hausdorff space $(S, \mathscr{T})$ is closed.

Proof: We shall prove that no point of $C^{\sim}$ is a limit point of C. Then C will be closed by Theorem 7.3. Suppose x is a given point in $C^{\sim}$. For each point y in C there are disjoint nbds U_y of x and V_y of y. The collection $\{V_y: y \in C\}$ is an open covering of C so it must contain a finite subcovering $\{V_{y,1}, V_{y,2}, \ldots, V_{y,n}\}$. The intersection $U = \bigcap_{i=1}^{n} U_{y,i}$ of the corresponding sets $U_{y,i}$ is a nbd of x, which is disjoint from each $V_{y,i}$ and hence from $\bigcup_{i=1}^{n} V_{y,i}$, which contains C. Therefore, x is not a limit point of C. ■

The following example shows that we must use the Hausdorff condition in Theorem 7.9.

Example 7.10:

Let S be an infinite set. Let $\mathscr{T}$ consist of all sets O such that $O^{\sim}$ is finite or $O = \varnothing$. Thus a nonempty set is open if and only if it has a finite complement, and it is closed if and only if it is finite or else it is all of S. It is not hard to see that $\mathscr{T}$ is a topology. Every subset A of S is compact because if $\{O_\alpha\}$ is a covering of A, any nonempty O_γ in this covering must contain all but finitely many points of A. Therefore, only finitely many additional sets from $\{O_\alpha\}$ are required to cover A.

In proving Theorem 5.8 we showed that the image of a compact subset of $\mathbb{R}$ under a continuous function from $\mathbb{R}$ into $\mathbb{R}$ is also compact. The following theorem shows this fact is true in a more general setting.

Theorem 7.11: Suppose f is a continuous function from a topological space $(S, \mathscr{T})$ into a topological space $(T, \mathscr{T}_1)$. If C is a compact subset of S then $f(C)$ is a compact subset of T.

Proof: Let $\{O_\alpha\}$ be any covering of $f(C)$ by open sets. Since f is continuous, $f^{-1}(O_\alpha)$ is open for each α. The collection $\{f^{-1}(O_\alpha)\}$ is an open covering of C because if $x \in C$, $f(x) \in f(C)$, so $f(x) \in O_\alpha$ for some α, which implies $x \in f^{-1}(O_\alpha)$. Since C is compact there is a finite subcovering $\{f^{-1}(O_1), \ldots, f^{-1}(O_n)\}$. It is not hard to see that the corresponding collection $\{O_1, \ldots, O_n\}$ is a finite subcovering of C. ■

Two topological spaces $(S_1, \mathscr{T}_1)$ and $(S_2, \mathscr{T}_2)$ are said to be *homeomorphic* if there is a one-to-one continuous function f with domain equal to S_1 and range equal to S_2 such that the inverse function f^{-1} is also continuous. If f is a homeomorphism, then the function T defined on $\mathscr{T}_1$ by $T(O) = f(O)$ is a function from $\mathscr{T}_1$ onto $\mathscr{T}_2$ that preserves unions and intersections. From the point of view of topology, the space $(S_1, \mathscr{T}_1)$ is indistinguishable from the space $(S_2, \mathscr{T}_2)$.

Theorem 7.12: Suppose f is a continuous one-to-one function from a compact topological space $(S_1, \mathscr{T}_1)$ onto a Hausdorff topological space $(S_2, \mathscr{T}_2)$. Then f is a homeomorphism.

Proof: Since f is one-to-one, f^{-1} is a function from S_2 onto S_1. It suffices to prove that f^{-1} is continuous, that is, that $(f^{-1})^{-1}(O) = f(O)$ is open for each open subset O of S_1. If O is an open subset of S_1, then its complement $O^{\sim}$ is closed. Since S_1 is compact, $O^{\sim}$ is compact by Theorem 7.8 (a). Thus $f(O^{\sim})$ is compact in S_2, hence closed by Theorem 7.9. This means that $f(O^{\sim})^{\sim}$ is open, but as the reader may show, $f(O^{\sim})^{\sim} = f(O) = f^{-1}(f^{-1}(O))$. This proves that f^{-1} is continuous. ■

Exercises

A

1. Make a list of all topologies on sets of 1, 2, and 3 points.
2. If $\mathscr{T}_1$ and $\mathscr{T}_2$ are topologies on a set S, show $\mathscr{T}_1 \cap \mathscr{T}_2$ is a topology on S but $\mathscr{T}_1 \cup \mathscr{T}_2$ need not be.
3. For a set A let A' denote the set of limit points of A. Let A be a subset of a Hausdorff topological space.
   (a) If x is a limit point of A', then x is a limit point of A.
   (b) A' is closed.
   (c) Show that $\bar{A} = A \cup A'$.
   (d) Let $S = \{a, b, c\}$, $\mathscr{T} = \{\varnothing, S, \{a\}, \{b, c\}\}$. If $A = \{a, c\}$ show that $A' = \{b\}$, $(A')' = \{c\}$.
4. *(a)* Show $\operatorname{int}(A)$ is the largest open set contained in A and $\bar{A}$ is the smallest closed set that contains A.
   (b) Show that $(\bar{A})^{\sim} = \operatorname{int}(A^{\sim})$.
   (c) Prove that $(\operatorname{int}(A))^{\sim} = (\overline{A^{\sim}})$.
5. *(a)* Show that if $(S, \mathscr{T})$ is a Hausdorff space, then each singleton subset $\{x\}$ of S is closed.
   (b) Show that the only Hausdorff topology on a finite S is $\mathscr{T}_{\max}$.
6. *(a)* Show that a set B is connected in $(S, \mathscr{T})$ if B is connected in the relative topology $\mathscr{T}_B$.
   (b) Prove that $S = \{a, b, c\}$, $\mathscr{T} = \{S, \varnothing, \{c\}, \{a, c\}, \{b, c\}\}$. The set $B = \{a, b\}$ is connected in $\mathscr{T}$ but not in $\mathscr{T}_B$.
7. Let $\{B_\alpha\}$ be a collection of connected sets and let B be a connected set such that $B \cap B_\alpha$ is nonempty for each α. Show that the set $B \cup (\bigcup_\alpha B_\alpha)$ is connected.
8. Show that every function f from a topological space $(S, \mathscr{T}_{\max})$ into another topological space is continuous.
9. If B is a connected subset of a topological space S and f is a continuous function from S into $\mathbb{R}$ show that $f(B)$ is an interval or a singleton. If B is compact and connected, show that $f(B)$ is a closed interval $[a, b]$ or a singleton.

10. Let (x_n) be a sequence of real numbers with $\lim_n x_n = x_0$.
(a) Show that $\{x_n\}$ forms a compact set of R if and only if x_0 is in $\{x_n\}$.
(b) Show that $\{x_0\} \cup \{x_n\}$ is compact.

11. If f is a continuous function from $(S_1, \mathcal{T}_1)$ into $(S_2, \mathcal{T}_2)$, and g is a continuous function from $(S_2, \mathcal{T}_2)$ into $(S_3, \mathcal{T}_3)$, show that the composite function $g \circ f$ defined by $g \circ f(x) = g(f(x))$ is continuous from $(S_1, \mathcal{T}_1)$ into $(S_3, \mathcal{T}_3)$. Use this to show that if f is continuous from S into R then $|f|$ is also continuous.

B

12. *(a)* Prove that a topological space $(S, \mathcal{T})$ is connected if and only if every continuous function f from S into the space $\{0, 1\}$ with the discrete topology must be constant.
(b) Use this fact plus Exercise 11 to show that if S is connected and f is continuous, then $f(S)$ is connected.

13. Let $\mathbb{N}^*$ denote the set $\{\infty, 1, 2, \ldots\}$ where ∞ is any point not in $\mathbb{N} = \{1, 2, \ldots\}$. Call every subset U of $\mathbb{N}$ open; call a set U containing ∞ open if and only if U contains all but finitely many points of $\mathbb{N}$.
(a) Show that the collection $\mathcal{T}$ thus defined forms a topology on $\mathbb{N}^*$.
(b) Show that a function f on $\mathbb{N}^*$ into $\mathbb{R}$ is continuous if and only if $\lim_n f(n) = f(\infty)$.
(c) Restate Theorem 7.5 as a theorem about sequences and their limits, given (a) and (b) of this exercise.

14. Suppose f is a continuous function from a topological space $(S, \mathcal{T})$ into $\mathbb{R}$. Show that if A is a connected subset of S and there are points u and v in A with $f(u) = a$ and $f(v) = b$ where $a < b$, then for each c with $a < c < b$ there is x in A with $f(x) = c$.

15. Suppose A is a connected subset of $(S, \mathcal{T})$ and $A \subset B \subset \bar{A}$. Show that B is connected.

C

16. A topological space $(S, \mathcal{T})$ is said to satisfy the first axiom of countability if for each point $s \in S$ there is a countable collection $\mathcal{U}(s)$ of open sets containing s such that whenever O is an open set that contains s there is $U \in \mathcal{U}(s)$ such that $U \subset O$. The space $(S, \mathcal{T})$ is said to satisfy the second axiom of countability if there is a countable collection of open sets $\mathcal{U}$ such that for each x in S and each open set O containing x there is U in $\mathcal{U}$ such that $x \in U \subset O$.
(a) Show that if a space satisfies the second axiom of countability then it also satisfies the first axiom.
(b) Show $\mathbb{R}$ with the usual topology satisfies the second axiom of countability.
(c) Show that in a space $(S, \mathcal{T})$ that satisfies the first axiom of countability a point x is a limit point of a set A if and only if there is a sequence (u_n) in A that converges to x.
(d) Let $S = \mathbb{N} \times \mathbb{N}$ the set of all pairs (m, n) of positive integers. Define $\mathcal{T}_*$ in this way: every set (including $\varnothing$) that does not contain $(1, 1)$ is in $\mathcal{T}_*$; if U contains $(1, 1)$ then U is in $\mathcal{T}_*$ if and only if U contains all but finitely

many of the "columns" $\{(j, 1), (j, 2), (j, 3), \ldots\}$ for all but finitely many j. Show that $\mathscr{T}_A$ is a topology.

(e) Show that there is no countable collection $\mathscr{U}(1, 1)$ in $\mathscr{T}_A$ such that if O is an open set containing $(1, 1)$, there is U in $\mathscr{U}(1, 1)$ with $(1, 1) \in U \subset O$.

(f) Let A be the set $\{(k, k): k = 2, 3, \ldots\}$. Show that $(1, 1)$ is a limit point of A, but no sequence in A converges to anything.

17. Suppose $(S, \mathscr{T})$ is a noncompact Hausdorff topological space. Let p_∞ be a point that is not in S. Let $S^* = S \cup \{p_\infty\}$. Define a collection $\mathscr{T}^*$ of subsets of S^* as follows: if $U \in \mathscr{T}$, then $U \in T^*$; if $p_\infty \in U$, then $U \in \mathscr{T}^*$ if and only if $S^* \backslash U$ is compact in $(S, \mathscr{T})$.

(a) Show that $(S^*, \mathscr{T}^*)$ is a compact Hausdorff topological space. It is called the *one point compactification* of $(S, \mathscr{T})$.

(b) Show that the relative $\mathscr{T}^*$-topology on S coincides with the topology $\mathscr{T}$.

(c) Show that S is dense in S^*.

8

Metric Spaces

A *metric* on a set S is a function ρ from $S \times S$ into $\mathbb{R}$ that satisfies the following conditions.

(M1) $\rho(x, y) \geqslant 0$ for all x, y in S, $\rho(x, y) = 0$ if and only if $x = y$.

(M2) $\rho(x, y) = \rho(y, x)$ for all x, y in S.

(M3) $\rho(x, z) \leqslant \rho(x, y) + \rho(y, z)$ for all x, y, z in S.

The first condition is called the *positivity* condition; the second, the *symmetry* condition; the third, *the triangle inequality*. The pair (S, ρ) is called a *metric space*; we shall also denote a metric space by the single character S when the metric ρ is implicitly understood. The three conditions (M1)–(M3) are attributes of our notion of distance. For example the metric defined on $\mathbb{R}$ by $\rho(x, y) = |x - y|$ certainly satisfies these three conditions. However, the following two examples show that a metric may not have all the intuitive properties of distance.

Example 8.1:

Define ρ on $\mathbb{R} \times \mathbb{R}$ by $\rho(x, y) = \inf(1, |x - y|)$. It is clear that ρ satisfies (M1) and (M2). To see that ρ satisfies (M3) note that if either $\rho(x, y)$ or $\rho(y, z)$ are greater than or equal to one, then (M3) is true since $\rho(x, z)$ is never greater than one. If $\rho(x, y)$ and $\rho(y, z)$ are both less than one then we have

$$|x - z| \leqslant |x - y| + |y - z| = \rho(x, y) + \rho(y, z)$$

so that (M3) holds since $\rho(x, z) \leqslant |x - z|$. This example shows that a metric need not become arbitrarily large.

Example 8.2:

For any set S define a metric ρ on $S \times S$ by $\rho(x, y) = 1$ whenever $x \neq y$ and $\rho(x, y) = 0$ if $x = y$. If x, y, and z are in S and both $\rho(x, y)$ and $\rho(y, z)$ are 0 it follows that $x = z$ so that $\rho(x, z) = 0$. Thus (M3) holds. It is equally clear that (M1) and (M2) also hold.

If (S, ρ) is a metric space, the *metric topology* $\mathscr{T}_\rho$ determined by ρ is defined by designating a subset O of S as open if for each x in O there is $\varepsilon > 0$ such that O contains the set $B(x, \varepsilon)$ of all y such that $\rho(x, y) < \varepsilon$. The set $B(x, \varepsilon)$ is called the *open ball* with center x and radius ε. We leave to the reader the task of showing that arbitrary unions and finite intersections of open sets are open. To verify that $B(x, \varepsilon)$ is open, suppose y is any point in $B(x, \varepsilon)$. Then $\varepsilon' = \varepsilon - \rho(x, y) > 0$ and $B(y, \varepsilon') \subset B(x, \varepsilon)$ because if $z \in B(y, \varepsilon')$ we have

$$\rho(x, z) \leqslant \rho(x, y) + \rho(y, z) < \rho(x, y) + \varepsilon' = \rho(x, y) + (\varepsilon - \rho(x, y)) = \varepsilon.$$

Most of the metrics that you will encounter in this course are derived from norms. Suppose E is a real or complex linear space. A *norm* on E is a function $\| \ \|$ from E into the set of nonnegative real numbers that has the following properties.

(N1) For $x \in E$, $\|x\| = 0$ if and only if $x = 0$.

(N2) For $x \in E$ and a in the scalar field $\|ax\| = |a| \, \|x\|$.

(N3) For $x, y \in E$, $\|x + y\| \leqslant \|x\| + \|y\|$.

Property (N2) is known as *scalar homogeneity* and (N3) is called the *Triangle Inequality*. If $\| \ \|$ is a norm on the linear space E the function $\rho(x, y) = \|x - y\|$ is a metric on E. The space E with the topology defined by this metric is called a *normed* space.

Example 8.3:

One of the most important examples of a metric space is $\mathbb{R}^n$ with the Euclidean distance defined by

$$\rho(x, y) = \left\{ \sum_{i=1}^{n} (x_i - y_i)^2 \right\}^{1/2}.$$

This distance is derived from the norm

$$\|x\| = \left\{ \sum_{i=1}^{n} x_i^2 \right\}^{1/2}.$$

We shall usually write $\|x - y\|$ rather than $\rho(x, y)$ when referring to the Euclidean metric. It is easy to see that the norm on $\mathbb{R}^n$ satisfies (N1) and (N2). That the norm satisfies (N3) as well is a consequence of the following theorem. The inequality in Theorem 8.4 (a) is called *Cauchy's Inequality* and the inequality in Theorem 8.4 (b) is called the *Triangle Inequality*. Both of these inequalities are important, and we shall study them in greater detail in the future.

Theorem 8.4: For $x=(x_1, \ldots, x_n)$ and $y=(y_1, \ldots, y_n)$ in $\mathbb{R}^n$ we have

(a) $|\sum_{i=1}^n x_i y_i| \leqslant \|x\| \, \|y\|$, and
(b) $\|x+y\| \leqslant \|x\| + \|y\|$.

Proof:
(a) Since $0 \leqslant (|x_i| - |y_i|)^2 = x_i^2 + y_i^2 - 2|x_i|\,|y_i|$ we have

$$|x_i y_i| \leqslant \tfrac{1}{2}(x_i^2 + y_i^2) \quad \text{for each } i.$$

Thus if $\|x\| = \|y\| = 1$ it follows that

$$\left|\sum_{i=1}^n x_i y_i\right| \leqslant \sum_{i=1}^n |x_i y_i| \leqslant \frac{1}{2}\left\{\sum_{i=1}^n x_i^2 + y_i^2\right\}$$

$$= \tfrac{1}{2}(\|x\|^2 + \|y\|^2) = 1.$$

If both $\|x\|$ and $\|y\|$ are nonzero then $\|(x/\|x\|)\| = \|(y/\|y\|)\| = 1$ so that we have

$$\left|\sum_{i=1}^n \left(\frac{x_i}{\|x\|}\right)\left(\frac{y_i}{\|y\|}\right)\right| \leqslant 1.$$

Statement (a) now follows if we multiply both sides of this inequality by $\|x\| \, \|y\|$. If either $\|x\|$ or $\|y\|$ are 0 then both sides of (a) are 0.

(b) This inequality follows from (a) since we have

$$\|x+y\|^2 = \sum_{i=1}^n (x_i + y_i)^2 \leqslant \sum_{i=1}^n x_i^2 + y_i^2 + 2|x_i y_i|$$

$$\leqslant \|x\|^2 + \|y\|^2 + 2\|x\| \, \|y\| = (\|x\| + \|y\|)^2. \blacksquare$$

Example 8.5:

Let $B(S)$ denote the linear space of all bounded real valued functions on a set S. The function defined on $B(S)$ by

$$\|f\| = \sup\{|f(s)|: s \in S\}$$

is a norm. To verify (N3) suppose f and g are in $B(S)$. For each s in S we have

$$|f(s) + g(s)| \leqslant |f(s)| + |g(s)| \leqslant \|f\| + \|g\|.$$

Thus if we take the supremum over all s in S on the left-hand side of the inequality above we obtain

$$\|f+g\| \leqslant \|f\| + \|g\|.$$

The topology derived from this metric is called the *topology of uniform convergence* on S. In this course you will find this norm in numerous different settings.

In a metric space the topology, that is the collection of open sets, is completely determined by convergent sequences. This is not true in an arbitrary topological space as Exercise 16 of Section 7 indicates. A sequence (x_n) in a topological space

$(S, \mathcal{T})$ is said to *converge* to a point x if for every nbd O of x there is $M \in \mathbb{N}$ such that $x_n \in O$ when $n > M$. For a metric space (S, ρ) it is not hard to show that (x_n) converges to x if for each $\varepsilon > 0$ there is $M > 0$ such that $\rho(x, x_n) < \varepsilon$ for all $n > M$.

Theorem 8.6: A subset C of a metric space S is closed if and only if every convergent sequence (x_n) in C converges to a point in C.

Proof: Necessity Suppose (x_n) is a sequence in C that converges to a point x. If $x = x_j$ for some j then x is certainly in C. Suppose x is not in $\{x_n\}$. If O is any nbd of x there are points of $\{x_n\}$ in O since $\lim_n x_n = x$. These points are in C and different from x so x is a limit point of C. Since C is closed x must be in C by Theorem 7.2.

Sufficiency Suppose C is a subset of S, which contains all limits of convergent sequences in C. We shall show that C contains all of its limit points so that it is closed by Theorem 7.2. Suppose x is a limit point of C. For $n = 1, 2, \ldots$ let x_n be a point in $B(x, 1/n) \cap C$ different from x. To show that (x_n) converges to x let U be any nbd of x. By the characterization of open sets in S there is $\varepsilon > 0$ such that $B(x, \varepsilon) \subset U$. If M is an integer with $1/M < \varepsilon$ it follows that $B(x, 1/n) \subset U$ when $n > M$. Therefore, $x_n \in U$ when $n > M$ so that $x \in C$. ■

Since the convergent sequences in a metric space S determine the closed subsets, they also determine the open subsets. Therefore, the following theorem is true.

Theorem 8.7: Suppose $\mathcal{T}_1$ and $\mathcal{T}_2$ are two metric topologies on a set S. The following statements are equivalent.

(a) $\mathcal{T}_1$ is weaker than $\mathcal{T}_2$.

(b) For each x in S and $\mathcal{T}_1$ open ball $B_1(x, \varepsilon)$ of x there is a $\mathcal{T}_2$ open ball $B_2(x, \delta)$ of x such that $B_2(x, \delta) \subset B_1(x, \varepsilon)$.

(c) If the sequence (x_n) converges to x in $\mathcal{T}_2$ it also converges to x in $\mathcal{T}_1$.

Proof: We shall prove the chain of implications (a) $\Rightarrow$ (b) $\Rightarrow$ (c) $\Rightarrow$ (a). Thus, if one of the statements above is valid then all of the statements are valid.

(a) $\Rightarrow$ (b). This is true since each $\mathcal{T}_1$ open ball $B_1(x, \varepsilon)$ is also a $\mathcal{T}_2$ nbd of x.

(b) $\Rightarrow$ (c). Suppose (x_n) converges to x in $\mathcal{T}_2$ and U is a $\mathcal{T}_1$ nbd of x. Then there is a $\mathcal{T}_1$ open ball $B_1(x, \varepsilon)$ contained in U. By (b) there is a $\mathcal{T}_2$ open ball $B_2(x, \delta)$ contained in $B_1(x, \varepsilon)$. Since (x_n) converges to x in $\mathcal{T}_2$ there is M such that $x_n \in B_2(x, \delta) \subset U$ when $n > M$.

(c) $\Rightarrow$ (a). Suppose C is closed in $\mathcal{T}_1$ and (x_n) is a sequence in C that converges to x in $\mathcal{T}_2$. Because of (c), (x_n) converges to x in $\mathcal{T}_1$ so x is in C. Hence C is closed in $\mathcal{T}_2$ by Theorem 8.6. Therefore, we have $\mathcal{T}_1 \subset \mathcal{T}_2$. ■

In the following theorem condition (b) is called *countable compactness*, condition (c) is called the *Bolzano-Weierstrass Property*, and condition (d) is called *sequential compactness*.

Theorem 8.8: For a subset C of a metric space S the following statements are equivalent.

(a) C is compact.

(b) Every countable covering of C by open sets contains a finite subcovering.

(c) Every infinite subset B of C has a limit point in C.

(d) Every sequence in C contains a subsequence that converges to a point of C.

Proof:

(a) ⇒ (b). Obvious.

(b) ⇒ (c). Suppose there were an infinite subset of C that has no limit point in C. Let $B = \{x_1, x_2, \ldots\}$ be a countable subset of this set. Then B has no limit point in C either. For each i let V_i be a nbd of x_i that contains no other point of C and let U be the complement of B. Then $\{U, V_1, V_2 \ldots\}$ is a countable open covering of C. However, this covering contains no finite subcovering since for each i, x_i is contained only in V_i.

(c) ⇒ (d). Let (x_n) be a sequence in C. If the range $\{x_n\}$ of the sequence is finite, then at least one number x must be repeated infinitely often so there is a constant subsequence that converges to x. Otherwise, $\{x_n\}$ has a limit point x. For each $k = 1, 2, \ldots$, let $x_{n(k)}$ be a point in $B(x, 1/k) \cap \{x_n\}$ not equal to x. Then $(x_{n(k)})$ converges to x by an argument similar to that in the sufficiency proof of Theorem 8.6.

(d) ⇒ (b). Suppose $\{U_i : i = 1, 2, \ldots\}$ is a countable open covering of C with no finite subcovering. For each n let x_n be in the set $C \setminus \bigcup_{i=1}^{n} U_i$. Then (x_n) has no convergent subsequence with limit in C. For suppose x in C were the limit of a subsequence $(x_{n(k)})$ of (x_n). Since (U_i) covers C there is M such that $x \in U_M$. Since U_M is a nbd of x there is H such that $x_{n(k)} \in U_M$ when $n(k) > H$. But this contradicts the fact that no x_n is in U_M for $n > M$.

(b) ⇒ (a). We first show that if (b) is true then for each $\varepsilon > 0$ there is a finite set $F(\varepsilon)$ such that for each x in C there is u in $F(\varepsilon)$ such that $\rho(u, x) < \varepsilon$. Suppose to the contrary that there were no such $F(\varepsilon)$ for some $\varepsilon > 0$. Then we could find a sequence (x_n) such that $\rho(x_n, x_m) > \varepsilon$ for each m and each n. This implies that (x_n) has no limit point, contradicting (b) ⇒ (c).

Next we form the set $D = \bigcup_n F(1/n)$, which is countable since each $F(1/n)$ is finite. Thus the collection $\mathscr{B}$ of open balls $\{b(u, 1/n): u \in D \text{ and } n = 1, 2, \ldots\}$ is countable. If x is any point C, and U is a nbd of x, there is m such that $B(x, 1/m) \subset U$. There is $u \in D$ such that $\rho(u, x) < \frac{1}{2}m$ so that $x \in B(u, \frac{1}{2}m) \subset B(x, 1/m) \subset U$. In summary, for each nbd U of a point x in C there is B in the countable collection $\mathscr{B}$ such that $x \in B \subset U$.

Let $\mathscr{U}$ be an arbitrary open covering of C. Let $\mathscr{B}'$ consist of all B in $\mathscr{B}$ for which there is U in $\mathscr{U}$ with $B \subset U$. For each B in $\mathscr{B}'$ let $U(B)$ be any chosen U in $\mathscr{U}$ with $B \subset U(B)$. The collection $\{U(B): B \in \mathscr{B}'\}$ is countable. To see that this subcollection covers C let x in C be given. Then there is U in $\mathscr{U}$ with $x \in U$. But there is also B in $\mathscr{B}$ with $x \in B \subset U$. Since $B \subset U(B)$ it follows that $x \in U(B)$. ■

At this point we note that if an A is a subset of a metric space (S, ρ) then the relative topology on A is the metric topology induced by ρ restricted to A.

A function f from a metric space (S, ρ) into a metric space (T, d) is said to be *uniformly continuous* if for each $\varepsilon > 0$ there is $\delta > 0$ such that $d(f(x), f(y)) < \varepsilon$ whenever $\rho(x, y) < \delta$. The important difference from continuity is that the number δ depends only on ε and not on x.

Theorem 8.9: If f is a continuous function from a compact metric space (S, ρ) into a metric space (T, d) then f is uniformly continuous.

Proof: Let $\varepsilon > 0$ be given. For each x in S there is $\delta_x > 0$ such that $d(f(x), f(y)) < \varepsilon/2$ when $\rho(x, y) < \delta_x$. Since the collection $\{B(x, \delta_x/2: x \in S\}$ is an open covering S so we can extract from it a finite subcovering that corresponds to points $x_1, x_2, \ldots, x_n$. Let δ be the minimum of the numbers $\{\delta_{x_i}/2: i = 1, 2, \ldots, n\}$. Suppose $\rho(x, y) < \delta$. Let x_k be such that $x \in B(x_k, \delta_{x_k}/2)$. By the triangle inequality we then have

$$\rho(y, x_k) \leqslant \rho(y, x) + \rho(x, x_k) < \delta + \frac{\delta_{x_k}}{2} \leqslant \delta_{x_k}.$$

Therefore, both $d(f(x), f(x_k))$ and $d(f(y), f(x_k))$ are less than $\varepsilon/2$ so that we have

$$d(f(x), f(y)) \leqslant d(f(x), f(x_k)) + d(f(x_k), f(y)) < \varepsilon. \blacksquare$$

We conclude this section with a brief introduction to many of the metric spaces considered in this book. This introduction will provide you with a reference for the metric spaces definition as well as a source of examples. Additional spaces of integrable functions are defined in Section 26, and some spaces of differentiable functions are defined in Section 28. Properties of some of the spaces are treated more fully in Chapter 5.

Example 8.10:

Normed and Metric Spaces.

Finite Dimensional Spaces

1. The Euclidean metric. The Euclidean norm $\|x\|$ of an n-tuple of scalars $x = (x_1, x_2, \ldots, x_n)$ is given by the formula
$$\|x\| = \left\{\sum_{i=1}^{n} |x_i|^2\right\}^{1/2}.$$
In Example 8.3 it was shown that $\|x\|$ is a norm on $\mathbb{R}^n$. A similar argument shows that it is also a norm on $\mathbb{C}^n$, the space of complex n-tuples.
2. Minkowski norms. If $1 \leqslant p < \infty$ then the function $\|x\|_p$ on $\mathbb{R}^n$ or $\mathbb{C}^n$ defined by
$$\|x\|_p = \left\{\sum_{i=1}^{n} |x_i|^p\right\}^{1/p}$$
is a norm. That $\|x\|_p$ satisfies (N1) and (N2) is easy to show; that it satisfies (N3) results from a finite dimensional version of the Minkowski inequality, which is derived below. The function $\|x\|_\infty$ defined by the formula
$$\|x\|_\infty = \max\{|x_i|: 1 \leqslant i \leqslant n\}$$
is clearly a norm on $\mathbb{R}^n$ or $\mathbb{C}^n$. In fact it is a special case of Example 8.5 with $S = \{1, 2, \ldots, n\}$. The norms $\|x\|_p$ for $1 \leqslant p \leqslant \infty$ are known as *Minkowski norms.* It can be shown that all of the Minkowski norms determine

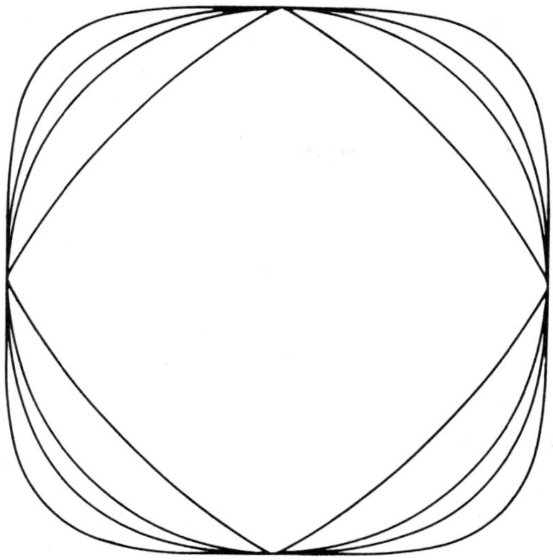

Figure 8.1

the same topology on $\mathbb{R}^n$. See Exercise 16 on page 58. Figure 8.1 depicts the sets $\{x: x \in \mathbb{R}^2, \|x\|_p = 1\}$ for $p = \frac{5}{4}$, 2, $\frac{5}{2}$, and 3. These sets are called *unit spheres* of the Minkowski space.

Spaces of Sequences

1. ω: The space of all sequences is denoted by ω. It is a linear space with coordinatewise addition and scalar multiplication. There is no useful norm topology on ω, but a useful metric is the *Fréchet metric* given by the formula

$$\rho(x, y) = \sum_{n=1}^{\infty} 2^{-n}|x_n - y_n|/(1 + |x_n - y_n|).$$

It can be shown that if (x_n) is a sequence of sequences, that is, $x_n = (x_{nj})$ for each n, then $\lim_n x_n = x$ with respect to the Fréchet metric if and only if $\lim_n x_{ni} = x_i$ for each i. See Exercise 20 on page 58.

2. The space 1^∞ and the uniform norm on $\mathbb{N}$. The space of bounded sequences is denoted by 1^∞. The uniform norm $\|x\|_\infty$ defined by the formula

$$\|x\| = \sup_j |x_j|$$

clearly satisfies (N1)–(N3). It is easy to see that 1^∞ is a special case of $B(S)$ when $S = \mathbb{N}$.

3. The spaces c_0 and c. The space of all convergent sequences is denoted by c, and the space of all sequences that converge to 0 is denoted by c_0. The main norm used on these spaces is the uniform norm defined in 2 (above) that makes them topological subspaces of 1^∞.

4. The spaces 1^p for $1 \leqslant p < \infty$. The space 1^p consists of all scalar sequences

$x = (x_j)$ such that

$$\|x\|_p = \left\{\sum_{j=1}^{\infty} |x_j|^p\right\}^{1/p} < \infty.$$

The proof that $\|x\|_p$ is a norm on 1^p uses a version of the Hölder and Minkowski inequalities given below.

5. The spaces *bs* and *cs*. The space *bs* consists of all scalar sequences $x = (x_j)$ for which

$$\|x\|_{bs} = \sup_n \left|\sum_{j=1}^{n} x_j\right| < \infty.$$

The space *cs* consists of all sequences x such that $\sum_{j=1}^{\infty} x_j$ converges. Since the partial sums of convergent series are bounded it follows that $cs \subset bs$. The norm $\|x\|_{bs}$ is the norm usually used on *cs*.

6. The spaces *bv* and bv_0. The space *bv* consists of all sequences x such that

$$\|x\|_{bv} = |x_1| + \sum_{j=1}^{\infty} |x_{j+1} - x_j| < \infty.$$

Since $\sum_j |x_{j+1} - x_j| < \infty$ for x in *bv*, x is a Cauchy sequence and is thus in c; hence, $bv \subset c$. Another norm that is useful on *bv* is

$$\|x\|_{bv1} = \lim_n |x_n| + \sum_{j=1}^{\infty} |x_{j+1} - x_j|.$$

Function Spaces

1. $B(S)$ with the uniform norm was already mentioned in Example 8.5.
2. $C(S)$. If S is a metric space and (f_n) is a sequence of scalar valued functions on S that converges to f in the uniform norm, then certainly $(f_n(s))$ converges to $f(s)$ for each s in S. Moreover it can be shown that f is continuous. Thus the space $C(S)$ of all bounded continuous functions on S is a closed subspace of $B(S)$ in the uniform norm. If S is compact, it is not necessary to assume that the members of $C(S)$ are bounded because every continuous scalar valued function is bounded on S.
3. Other norms on $C[a, b]$. It is sometimes useful to consider norms other than the uniform norm on the space $C[a, b]$ of continuous scalar valued functions on the interval $[a, b]$. A more natural setting for these norms is described in Section 27. For $1 \leqslant p < \infty$ define the function $\|f\|_p$ on $C[a,b]$ by the formula

$$\|f\|_p = \left\{\int_a^b |f(t)|^p \, dt\right\}^{1/p}.$$

The analogy between these norms, the Minkowski norms and the norms defined on 1^p, is evident. It is obvious that $\|f\|_1$ is a norm and the other norms $\|f\|_p$ satisfy (N1) and (N2). That these norms also satisfy (N3) is a result of the Minkowski inequality, which is proved below.

The following theorem is known as *Hölder's Inequality*.

Theorem 8.11: If p and q are nonnegative reals with $1/p + 1/q = 1$, then for f and g in $C[a, b]$ we have

$$\int_a^b |fg| \, dt \leqslant \|f\|_p \|g\|_q.$$

When both $\|f\|_p$ and $\|g\|_q$ are nonzero then equality holds if and only if there is $a > 0$ such that $|g|^q = a|f|^p$ for each t.

Proof: The first step is to prove the inequality

$$a^{1/p} b^{1/q} \leqslant \frac{a}{p} + \frac{b}{q} \tag{8–1}$$

for all nonnegative a, b. To do this, let $0 < \alpha < 1$, and consider the function

$$f(t) = t^\alpha - \alpha t + \alpha - 1.$$

Then $f'(t) = \alpha(t^{\alpha-1} - 1)$, so that $f(1) = f'(1) = 0$; $f'(t)$ is positive for $0 < t < 1$; and $f'(t)$ is negative for $t > 1$. This implies that $f(t) \leqslant 0$ for $t \geqslant 0$. For $b = 0$, inequality (8–1) is obvious. If $b > 0$ let $t = a/b$ and $\alpha = 1/p$. Then we have

$$f\left(\frac{a}{b}\right) = \left(\frac{a}{b}\right)^{1/p} - \left(\frac{1}{p}\right)\left(\frac{a}{b}\right) + \frac{1}{p} - 1 \leqslant 0.$$

If we multiply this through by b we obtain

$$a^{1/p} b^{(1-1/p)} \leqslant \frac{a}{p} + b\left(1 - \frac{1}{p}\right),$$

which is clearly equivalent to inequality (8–1). The function $f(t) = 0$ only when $t = 1$ so that equality holds in (8–1) only when $a = b$.

Now suppose $\|f\|_p = \|g\|_q = 1$. For each t in S we have

$$|f(t)g(t)| \leqslant \frac{|f(t)|^p}{p} + \frac{|g(t)|^q}{q} \tag{8–2}$$

by applying inequality (8–1) with $a = |f(t)|^p$ and $b = |g(t)|^q$. If we integrate both sides of (8–2) over $[a, b]$ we obtain

$$\int_a^b |fg| \, dt \leqslant \int_a^b \frac{|f|^p}{p} \, dt + \int_a^b \frac{|g|^q}{q} \, dt = \frac{1}{p} + \frac{1}{q} = 1. \tag{8–3}$$

If $\|f\|_p$ or $\|g\|_q$ are 0 then both sides of the Hölder inequality are 0 so we may assume that both are nonzero. Then we may apply the preceding paragraph to the functions $f/\|f\|_p$ and $g/\|g\|_q$, both of which have norm one, to obtain the inequality

$$\int_a^b \left|\frac{f}{\|f\|_p}\right| \left|\frac{g}{\|g\|_q}\right| dt \leqslant 1$$

from which the Hölder inequality immediately follows by multiplying both sides by $\|f\|_p \|g\|_q$.

Equality can hold in (8–2) if and only if $|f(t)|^p = |g(t)|^q$ and equality can hold in (8–3) if and only if this equality holds for all t in S. Thus equality holds in the Hölder relation if and only if $\|g\|^q |f|^p = \|f\|^p |g^q$ for all t. ■

We apply the Hölder inequality to derive the next result, which is known as *Minkowski's Inequality.*

Theorem 8.12: If f and g are $C[a, b]$, then we have

$$\|f+g\|_p \leqslant \|f\|_p + \|g\|_p \quad \text{for all } 1 \leqslant p < \infty.$$

Proof: For each t in $[a, b]$ we have

$$|f(t)+g(t)|^p \leqslant (|f(t)+g(t)|)^{p-1}|f(t)| + (|f(t)+g(t)|)^{p-1}|g(t)|$$

so that if we integrate both sides over S we obtain

$$\|f+g\|_p^p \leqslant \int_a^b |f+g|^{p-1}|f|\, dt + \int_a^b |f+g|^{p-1}|g|\, dt.$$

We use the Hölder inequality on the two integrals on the right-hand side to obtain

$$\int_a^b |f+g|^{p-1}|f|\, dt \leqslant \|f\|_p \||f+g|^{p-1}\|_q$$

and

$$\int_a^b |f+g|^{p-1}|g|\, dt \leqslant \|g\|_p \||f+g|^{p-1}\|_q.$$

By definition we have

$$\||f+g|^{p-1}\|_q = \left\{\int_a^b |f+g|^{(p-1)q}\, dt\right\}^{1/q} = \|f+g\|_p$$

because $(p-1)q = (p-1)(1-1/p)^{-1} = (p-1)p(p-1)^{-1} = p$. Combining these inequalities we find

$$\|f+g\|_p^p \leqslant \|f+g\|_p^{p/q}\{\|f\|_p + \|g\|_p\},$$

which gives the Minkowski inequality after we divide both sides by $\|f+g\|_p^{p/q} = \|f+g\|^{p-1}$. ■

Exercises

A

1. *(a)* Prove that a metric space has the Hausdorff property.
   (b) Show that if C is a closed subset of a metric space and u is a point not in C, then there is $\varepsilon > 0$ such that $\rho(u, x) > \varepsilon$ for all x in C.
   (c) Show that with C and u as in (b) there are disjoint open sets U, V with $u \in U$, and $C \subset V$.
   (d) Prove that if C and D are disjoint closed subsets of a metric space then there are disjoint open sets U containing C and V containing D.
2. *(a)* If ρ is the metric on $\mathbb{R}$ given by $\rho(x, y) = |x - y|$, show that for each x in $\mathbb{R}$ and each $\varepsilon > 0$ the closure of $B(x, \varepsilon)$ is the set of all y in $\mathbb{R}$ such that $|x - y| \leqslant \varepsilon$.

(b) If ρ is the metric of Example 8.2 show that the closure of $B(x, \varepsilon)$ need not be the set of all y in S such that $\rho(x, y) \leqslant \varepsilon$. For all ε determine $B(x, \varepsilon)$, $\overline{B(x, \varepsilon)}$, and $\{y: \rho(x, y) \leqslant \varepsilon\}$.

3. *(a)* Prove that the metric on a metric space (S, ρ) is jointly continuous in the following sense: For each pair x, y in S and each $\varepsilon > 0$ there is $\delta > 0$ such that $|\rho(x, y) - \rho(u, v)| < \varepsilon$ for u, v in S with both $\rho(x, u)$ and $\rho(y, v) < \delta$.

(b) Show that for each x in S the function ρ_x defined from S into R by $\rho_x(y) = \rho(x, y)$ is continuous.

4. Use Exercise 3 (b) above to show that for each x in S and each $\varepsilon > 0$ the set $\{y: y \in S, \rho(x, y) \leqslant \varepsilon\}$ is always closed.

5. The distance between a point x and a set B in a metric space (S, ρ) is given by the formula $\rho_B(x) = \inf\{\rho(x, y): y \in B\}$. Show that for $B \subset S$, $\bar{B} = \{x: x \in S, \rho_B(x) = 0\}$. Conclude that if C is a closed subset of a metric space and x is not in C then $\inf\{\rho(x, y): y \in C\} > 0$.

6. Show that a function f from a metric space (S_1, ρ) to a metric space (S_2, σ) is continuous at $x \in S_1$ if $\lim_n f(x_n) = f(x)$ whenever $\lim_n x_n = x$. (Imitate the proof of Theorem 5.9 on page 28.)

7. State a condition derived from Theorem 8.7 (b) for the metric topology $\mathscr{T}_1$ on a set S to be equal to the metric topology $\mathscr{T}_2$.

8. Show that for the conditions in Theorem 8.8, (a) $\Rightarrow$ (b) $\Rightarrow$ (c) $\Rightarrow$ (d) for an arbitrary topological space.

9. Show that if (S, ρ) is a metric space, the function $\sigma(x, y) = \inf\{1, \rho(x, y)\}$ is a metric on S that gives the same metric topology as the original metric ρ.

10. *(a)* Show that (S, ρ) and (T, σ) are metric spaces, the function τ defined by $\tau((x_1, y_1), (x_2, y_2)) = \rho(x_1, x_2) + \sigma(y_1, y_2)$ is a metric on $S \times T$.

(b) Prove that the function P_1 from $(S \times T, \sigma)$ onto (S, ρ) defined by $P_1(x, y) = x$ is continuous.

11. Draw the sets $\{y: y \in \mathbb{R}^2, \|y\|_\infty \leqslant 1\}$, $\{y: y \in \mathbb{R}^2, \|y\|_1 \leqslant 1\}$, and $\{y: y \in \mathbb{R}^2, \|y\|_{3/2} \leqslant 1\}$. Describe the corresponding sets in $\mathbb{R}^3$.

12. *(a)* If $x_m = (x_{m1}, x_{m2}, \ldots, x_{mn})$ is in $\mathbb{C}^n$ for $m = 1, 2, \ldots$ show that the sequence (x_m) converges in the Euclidean metric to $x = (x_1, \ldots, x_n)$ if and only if for each $i = 1, 2, \ldots n$ we have $\lim_m x_{mi} = x_i$.

(b) For each $m = 1, 2, \ldots$ let e_m be the sequence $(\delta_{m1}, \delta_{m2}, \ldots)$ where $\delta_{mn} = 1$ if $m = n$ and $\delta_{mn} = 0$ if $m \neq n$. Show that $\lim_{mn} \delta_m = 0$ for each n but that (e_m) does not converge to 0 in 1^∞ because $\|e_m\| = 1$ for each m.

13. Determine whether the following functions are defined and continuous (a) from 1^2 into 1^∞; (b) from 1^2 into 1^2; (c) from ω into ω.

(a) $F(x_j) = (jx_j)$ *(b)* $F(x_j) = (x_j/j)$

(c) $F(x_j) = (x_j^2)$ *(d)* $F(x_j) = (x_j^{1/2})$.

B

14. *(a)* If C_1 and C_2 are nonempty compact subsets of a metric space (S, ρ) show there are points $x_1 \in C_1$ and $x_2 \in C_2$ such that $\rho(x_1, x_2) \leqslant \rho(y_1, y_2)$ whenever $y_1 \in C_1$ and $y_2 \in C_2$.

(b) Show that if S is the real line $\mathbb{R}$ it suffices for one of the sets to be compact and one closed.

(c) Give an example of two closed subsets C_1 and C_2 of $\mathbb{R}$ for which the conclusion of (a) does not hold.

15. *(a)* Suppose $\mathcal{T}_1$ and $\mathcal{T}_2$ are two norm topologies on a linear space E determined by the norms $\| \ \|_1$ and $\| \ \|_2$ respectively. Show $\mathcal{T}_1 \subset \mathcal{T}_2$ if and only if there is $M > 0$ such that $\|x\|_1 \leqslant M\|x\|_2$ for all x in E. (Use Theorem 8.7.)

(b) State a condition derived from (a) that implies $\mathcal{T}_1 = \mathcal{T}_2$.

16. *(a)* Prove that if x and y are in $\mathbb{R}^n$ or $\mathbb{C}^n$ and if p and q are positive real numbers such that $1/p + 1/q = 1$, then

$$\sum_{j=1}^{n} |x_j y_j| \leqslant \|x\|_p \|y\|_q.$$

(Imitate the proof of Theorem 8.10.)

(b) Show $\|x + y\| \leqslant \|x\| + \|y\|$.

(c) Show that for all x in $\mathbb{R}^n$ (or $\mathbb{C}^n$), $\|x\|_\infty \leqslant \|x\|_p \leqslant \|x\|_1 \leqslant n\|x\|_\infty$ for all p with $1 < p < \infty$.

(d) Use (c) to show that the metrics on $\mathbb{R}^n$ or $\mathbb{C}^n$ determined by the Minkowski norms all induce the same topology. (Use Exercise 15.)

17. *(a)* Show that if (S, ρ) is a separable metric space, there is a countable collection $\mathcal{B}$ of open sets with the property that if U is any nbd of a point x in S there is V in $\mathcal{B}$ with $x \in V \subset U$. This is known as the *Lindelöf Property*. (Imitate part of the proof of (b) $\Rightarrow$ (a) of Theorem 8.8.)

(b) Show that every open covering $\mathcal{U}$ of a separable metric space (S, ρ) contains a countable subcovering.

(c) Show that $\mathbb{R}^n$ is separable by showing the set of all x in $\mathbb{R}^n$ with rational coordinates forms a dense subset.

18. A function f from a metric space (S, ρ) onto a metric space (T, σ) is called an *isometry* if $\sigma(f(x), f(y)) = \rho(x, y)$ for all x, y in S.

(a) Prove that an isometry is a one-to-one continuous function and the inverse function f^{-1} is also continuous.

(b) Prove that the function $f(x) = 1/x$ is a one-to-one continuous mapping from $(0, 1]$ onto $[1, \infty)$, which has a continuous inverse, but it is not an isometry.

19. *(a)* Suppose f is a uniformly continuous function defined from a dense subspace D of a metric space (S, ρ) into a metric space (T, d). Show that there is a unique continuous extension of f to S.

(b) Show that (a) need not hold if f is continuous but not uniformly continuous.

20. Show that a sequence $x_n = (x_{n1}, x_{n2}, \ldots)$ in ω converges to $x = (x_1, x_2, \ldots)$ in the Fréchet metric if and only if $\lim_n x_{nj} = x_j$ for all $j = 1, 2, \ldots$.

21. *(a)* Prove that if $(x_j) \in 1^p$ and $(y_j) \in 1^q$ where $1/p + 1/q = 1$, then $(x_j y_j) \in 1^1$ and $\sum_{j=1}^{\infty} |x_j y_j| \leqslant \|(x_j)\|_p \|(y_j)\|_q$.

(b) Prove that if x and y are in 1^p then $x + y$ is in 1^p and $\|x + y\|_p \leqslant \|x\|_p + \|y\|_p$.

22. *(a)* Show that the function S defined from $C[0, 1]$ into $C[0, 1]$ by

$$Sf(t) = \int_0^t f(s)\, ds$$

is continuous from $C[0, 1]$ into $C[0, 1]$, but does not have range equal to all of $C[0, 1]$.

(b) The function defined by $Df = f'$ on the subspace of $C[0, 1]$ consisting of functions having continuous derivatives at each point, is not continuous from its domain into $C[0, 1]$. (Show that $t^n/n \to 0$ but $D(t^n/n)$ does not converge to 0.)

23. *(a)* Prove that the topology determined by the uniform norm is stronger than the topology determined by each norm $\| \ \|_p$ on $C[a, b]$.

(b) Show that if $f_n(t) = t^n$ for $0 \leqslant t \leqslant 1$ we have $\lim_n \|f_n\|_p = 0$ for each $p \geqslant 1$, but $\|f_n\| = 1$ for each n for the uniform norm.

24. *(a)* Show that for $1 \leqslant p < q \leqslant \infty$, $1^p \subset 1^q$.

(b) Prove the inclusion is proper by exhibiting a sequence that is in 1^q but not 1^p.

25. For a sequence $x = (x_1, x_2, \ldots)$ let $Q_n x$ denote the sequence $y = (y_1, y_2, \ldots)$ where $y_i = 0$ for $i < n$ and $y_i = x_i$ for $i \geqslant n$.

(a) Show $\lim_n Q_n x = 0$ for all x in the spaces 1^p for $1 \leqslant p < \infty$, c_0, cs, bv_0, and ω.

(b) Prove that there are x in 1^∞, c, bs, and bv for which $\lim_n Q_n x$ does not exist.

C

26. If S is a finite set of points in a linear space then the smallest convex set C, which contains S is called the *convex hull* of S, and the points in S are called the *vertices* of C.

(a) Prove that the *unit ball* $B_1 = \{x: x \in \mathbb{R}^n: \|x\|_1 \leqslant 1\}$ is the convex hull of the set $\{de_j: d = 1 \text{ or } -1, j = 1, 2, \ldots, n\}$ where e_j is the n-vector that has a 1 in the jth coordinate and 0's elsewhere.

(b) Prove that B_1 has $2n$ vertices.

(c) Show that the set $B_\infty = \{x: x \in R^n, \|x\|_\infty \leqslant 1\}$ is the closed convex hull of the set $\{(r_1, r_2, \ldots, r_n): r_i = 1, -1\}$.

(d) Prove that B_∞ has 2^n vertices.

(e) Use (a)–(d) to construct an isometry from $\mathbb{R}^n$ with the $\| \ \|_\infty$ norm onto a subspace of $\mathbb{R}^m$ with the $\| \ \|_1$ norm if $m \geqslant 2^{n-1}$.

(f) Prove that no unit ball of a real normed space of dimension n can be the convex hull of fewer than $2n$ points.

(g) Can the unit ball of a real normed space of dimension n be the convex hull of more than 2^n points?

9

Complete Metric Spaces

The idea of a complete metric space is related to the completeness of the real line $\mathbb{R}$. A sequence (x_n) in a metric space (S, ρ) is called a *Cauchy Sequence* if for each $\varepsilon > 0$ there is $M \in \mathbb{N}$ such that $\rho(x_m, x_n) < \varepsilon$ when m and n are greater than M.

From an intuitive viewpoint a Cauchy sequence is bunching together, trying to converge. A sequence that converges is Cauchy because if x is its limit and M is such that $\rho(x, x_n) < \varepsilon/2$ when $n > M$, it follows that

$$\rho(x_m, x_n) \leqslant \rho(x_m, x) + \rho(x, x_n) < \varepsilon$$

when m and n are greater than M. A metric space (S, ρ) is called *complete* if for each Cauchy sequence (x_n) in S, there is a point x in S to which (x_n) converges. In other words, a metric space is complete if it contains enough points to provide a limit for every Cauchy sequence in the space. Being complete is not a topological property; that is, a metric space S that is complete may be homeomorphic to a metric space T that is not complete. In fact, in Example 9.1 (b) below we illustrate two metrics ρ and σ that induce the same topology on a set S; but one metric space (S, ρ) is complete while (S, σ) is not complete. Example 9.1 also shows how two essentially different metrics can determine the same topology on a set.

Example 9.1:

(a) With its usual metric $\mathbb{R}$ is complete. If (x_n) is a Cauchy sequence in $\mathbb{R}$, there is $M > 0$ such that $|x_m - x_n| < 1$ for $m, n > M$. If $b = \sup\{|x_1|, |x_2|, \dots |x_M|, |x_{M+1}| + 1\}$ then $|x_n| \leqslant b$ for each n. Thus (x_n) is a sequence in the compact set $[-b, b]$. By Theorem 8.8 on page 50 there is a subsequence $(x_{k(n)})$ of (x_n) that converges, say to x. But x is also the limit of (x_n). For suppose $\varepsilon > 0$ is given. Let M_1 be such that $k(n) > M_1$ implies $|x - x_{k(n)}| < \varepsilon/2$. Let M_2 be such that $m, n > M_2$ implies $|x_m - x_n| < \varepsilon/2$. If $n > \sup(M_1, M_2)$, let $k(j)$ be any subscript in $\{k(i)\}$ such that $k(j) > n$. Then we shall have

$$|x - x_n| \leqslant |x - x_{k(j)}| + |x_{k(j)} - x_n| < \frac{\varepsilon}{2} + \frac{\varepsilon}{2} = \varepsilon.$$

(b) Let S be the half-open interval $(0, 1]$. If σ denotes the usual metric on $\mathbb{R}$ then (S, σ) is not complete. For example, $(1/n)$ is a Cauchy sequence in S that does not have a limit in S (because the limit must be 0). If we define ρ on S by $\rho(u, v) = |1/u - 1/v|$, then it is easy to show ρ is a metric. We leave to you the task of showing that ρ induces the same topology as σ on S. However, (S, ρ) is a complete metric space because if (u_n) is a Cauchy sequence in (S, ρ) then $(1/u_n)$ is a Cauchy sequence in $[1, \infty)$ with the usual metric, so it has a limit point v. But then $u = 1/v$ is in $(0, 1]$ and must be the limit of (u_n).

The following facts may be established by adapting the arguments used in Example 9.1 (a).

Theorem 9.2:

(a) Suppose (x_n) is a Cauchy sequence in a metric space (S, ρ).

(1) (x_n) is bounded in the sense that there is y in S and $M > 0$ such that $\rho(x_n, y) < M$ for each n.

(2) If any subsequence of (x_n) converges, then (x_n) converges to the limit of the subsequence.

(b) If C is a compact subset of (S, ρ), then C is complete in the relative topology. ■

Example 9.3:

For (S, ρ) a metric space let $C(S)$ denote the space of bounded continuous scalar valued functions on S with the uniform norm

$$\|f-g\| = \sup\{|f(x)-g(x)|: x \in S\}.$$

To show that $C(S)$ is complete let (f_n) be a Cauchy sequence. Since for all x in S we have

$$|f_n(x)-f_m(x)| \leqslant \|f_n - f_m\|$$

it follows that $(f_n(x))$ is a Cauchy sequence of numbers. Define the function f on S by the equation

$$f(x) = \lim_n f_n(x).$$

To show f is continuous let x in S and $\varepsilon > 0$ be given. Let M be such that $\|f - f_M\| < \varepsilon/3$. Let δ be such that $|f_M(x) - f_M(y)| < \varepsilon/3$ when $\rho(x, y) < \delta$. Then if $\rho(x, y) < \delta$ we have

$$\begin{aligned} |f(x)-f(y)| &\leqslant |f(x)-f_M(x)| + |f_M(x)-f_M(y)| + |f_M(y)-f(y)| \\ &< \|f-f_M\| + \frac{\varepsilon}{3} + \|f_M - f\| < \varepsilon. \end{aligned}$$

This shows that f is continuous at x. To show that $\lim_n \|f-f_n\| = 0$ let $\varepsilon > 0$ be given. Since (f_n) is Cauchy there is M such that $\|f_n - f_m\| < \varepsilon$ when $m, n > M$. Thus, we have $|f_n(x) - f_m(x)| < \varepsilon$ when $n, m > M$ so that

$$|f(x) - f_m(x)| = \lim_n |f_n(x) - f_m(x)| \leqslant \varepsilon$$

when $m > M$, which implies that $\|f - f_m\| \leqslant \varepsilon$ when $m > M$.

If (S, ρ) and (T, σ) are two metric spaces, a function f from S into T is called an *isometry* if and only if for each x, y in S we have $\sigma(f(x), f(y)) = \rho(x, y)$. If there is an isometry from (S, ρ) onto (T, σ), then (S, ρ) and (T, σ) are called *isometric*. The reader may prove that every isometry is also a homeomorphism. The following theorem asserts that every metric space (S, ρ) can be embedded isometrically in a complete metric space $(\tilde{S}, \tilde{\rho})$, which is called the *completion* of S. We shall present the proof in outline form and leave the details to you.

Theorem 9.4: For every metric space (S, ρ) there is a complete metric space $(\tilde{S}, \tilde{\rho})$ and an isometry f from S into $\tilde{S}$ such that $f(S)$ is dense in $\tilde{S}$. Moreover, if (T, σ) is any complete metric space such that S is isometric to a dense subset of (T, σ), it then follows that (T, σ) is isometric to $(\tilde{S}, \tilde{\rho})$.

Proof: Let $\mathscr{U}$ denote the collection of all Cauchy sequences in S. (1) If (x_n) and (y_n) are in $\mathscr{U}$, then $\rho(x_n, y_n)$ is a convergent sequence of real numbers. Two sequences (x_n) and (y_n) are called equivalent if $\lim_n \rho(x_n, y_n) = 0$. For each (x_n) in $\mathscr{U}$ let $W(x_n)$ be the set of all (y_n) in $\mathscr{U}$ that are equivalent to (x_n). (2) Two sets $W(x_n)$ and $W(y_n)$ are either equal or disjoint. Let $\tilde{S}$ denote the set of all sets $W(x_n) = W$. (3) Let V and W be in $\mathscr{U}$. If (u_n) and (v_n) are in V while (x_n) and (y_n) are in W, then we have

$$\lim_n \rho(u_n, x_n) = \lim_n \rho(v_n, y_n).$$

Thus for W and V in $\tilde{S}$ we can define $\tilde{\rho}$ by

$$\tilde{\rho}(V, W) = \lim_n \rho(v_n, w_n) \quad \text{where } (v_n) \in V \quad \text{and} \quad (w_n) \in W.$$

(4) The function $\tilde{\rho}$ is a metric on $\tilde{S}$. (5) Suppose (V_n) is a Cauchy sequence in $(\tilde{S}, \tilde{\rho})$ and $(x_{nj})_j \in V_n$ for each n. Then we can find an increasing sequence $M(m)$ of integers such that $(x_{mM(m)})$ is a Cauchy sequence in S and $\lim_n \tilde{\rho}(W(x_{mM(m)}), V_n) = 0$. Thus, (V_n) converges to $W(x_{mM(m)})$. This proves that $\tilde{S}$ is complete.

For each x in S define $f(x)$ to be $W(x)$ where (x) denotes the constant sequence $(x, x, \ldots)$. (6) Then f is an isometry from S into $\tilde{S}$. (7) If $V = W(x_n)$ is any member of $\tilde{S}$, then the sequence $W(x_1), W(x_2), \ldots$ in $f(S)$ converges to V. Here for fixed i, $W(x_i)$ denotes the set of sequences determined by the constant sequence $(x_i, x_i, \ldots)$. Therefore, $f(S)$ is dense in $\tilde{S}$.

Suppose (T, σ) is a complete metric space and g is an isometry from (S, ρ) into T such that $g(S)$ is dense in T. (8) If both (x_n) and (y_n) are in the set V in $\tilde{S}$ then $\lim_n g(x_n) = \lim_n g(y_n)$ in T. Define G from $\tilde{S}$ into T by $G(V) = \lim_n g(x_n)$ where $(x_n) \in V$. (9) G is an isometry from $\tilde{S}$ into T. If y is any member of T, there is a sequence (x_n) in S such that $\lim_n g(x_n) = y$ since $g(S)$ is dense in T. (10) Then $(x_n) \in \mathscr{U}$ and $G(W(x_n)) = y$ so the range of G is all of T. ■

For example, the completion of the set Q of rationals endowed with the usual metric is the space $\mathbb{R}$.

The following theorem is an example of a fixed point theorem. It is called the *Contraction Mapping Theorem.*

Theorem 9.5: Suppose f is a function from a complete metric space (S, ρ) into itself such that there is $0 < t < 1$ for which

$$\rho(f(x), f(y)) \leqslant t\rho(x, y) \quad \text{for all } x, y \text{ in } S.$$

Then there is a unique point x_0 in S such that $f(x_0) = x_0$.

Proof: You may show that the contraction condition above implies that f must be continuous, and that the x_0 is unique if it exists.

Let x_1 be any point in S. If x_n has been defined let x_{n+1} be $f(x_n)$. Since we have

$$\rho(x_n, x_{n+1}) = \rho(f(x_{n-1}), f(x_n)) \leqslant t\rho(x_{n-1}, x_n)$$

for each n it follows that $\rho(x_n, x_{n+1}) \leqslant t^{n-1}\rho(x_1, x_2)$. If $m > n$ we have

$$\begin{aligned}\rho(x_n, x_m) &\leqslant \rho(x_n, x_{n+1}) + \rho(x_{n+1}, x_{n+2}) + \cdots + \rho(x_{m-1}, x_m)\\ &\leqslant \rho(x_1, x_2)\{t^{n-1} + t^n + \cdots + t^{m-2}\}\\ &= \rho(x_1, x_2)t^{n-1}\frac{(1-t^{m-n})}{(1-t)} \leqslant t^{n-1}\rho\frac{(x_1, x_2)}{(1-t)}.\end{aligned}$$

Since $t^{n-1} \to$ ("goes to") 0 as $n \to \infty$ it follows that (x_n) is a Cauchy sequence. Since S is complete (x_n) converges to some point x_0. Since $\lim_n x_n = x_0$ it follows that $f(x_0) = \lim_n f(x_n) = \lim_n x_{n+1} = x_0$. ■

Example 9.6:

(a) We can use the Contraction Mapping Theorem to prove that the equation, $x - \arctan(x) = 1$ has precisely two real solutions. Define $F(x) = 1 + \arctan(x)$. Then $F(0) = 1$ and F is continuous so there is $r > 0$ such that $F(x) > x$ for $|x| < r$. On the interval $[r, \infty)$ F is a contraction. To verify this suppose $r \leqslant x_1 \leqslant x_2$. Then by the Mean Value Theorem there is u between x_1 and x_2 such that $F(x_1) - F(x_2) = F'(u)(x_1 - x_2)$. Thus, we have

$$|F(x_1) - F(x_2)| = |F'(u)||x_1 - x_2| = \frac{|x_1 - x_2|}{(1+u^2)} < (1 + r^2)^{-1}|x_1 - x_2|.$$

By the Contraction Mapping Theorem there is a unique x_0 on $[r, \infty)$ such that $F(x_0) = x_0$. Similarly F is a contraction on $(-\infty, -r]$ so there is a unique negative number y_0 such that $F(y_0) = y_0$. The numbers x_0 and y_0 are the solutions of the given equation.

(b) Next we use the Contraction Mapping Theorem to show that there is a unique continuous function f on $[0, 1]$ that satisfies the nonlinear integral equation

$$f(t) - \left\{\int_0^t \frac{f(s)}{2}\,ds\right\}^2 = 1 \qquad 0 \leqslant t \leqslant 1.$$

Let B denote the set of all functions f in $C[0, 1]$ such that $1 \leqslant f(t) \leqslant 1 + t$ for all t in $[0, 1]$. Then B is a closed subset of $C[0, 1]$, thus it is a complete metric space with the uniform norm. Define T on B by the formula

$$(Tf)(t) = 1 + \left\{\int_0^t \frac{f(s)}{2}\,ds\right\}^2 \qquad 0 \leqslant t \leqslant 1.$$

In order to apply the Contraction Mapping Theorem it is first necessary to show that T maps B into B, and next to show that T is a contraction. It is clear that $(Tf)(t) \geqslant 1$. That $(Tf) \leqslant 1 + t$ follows because

$$\left\{\int_0^t \frac{f(s)}{2}\,ds\right\}^2 \leqslant \left\{\int_0^t \frac{(1+s)}{2}\,ds\right\}^2 = \frac{t}{2} + \frac{t^2}{4} < t.$$

That T is a contraction on B follows because if f and g are in B, then for all t in $[0, 1]$ we have

$$
\begin{aligned}
(Tf)(t)-(Tg)(t) &= \left\{\int_0^t \frac{f(s)}{2}\,ds\right\}^2 - \left\{\int_0^t \frac{g(s)}{2}\,ds\right\}^2 \\
&= \left\{\int_0^t \frac{f(s)+g(s)}{2}\,ds\right\}\left\{\int_0^t \frac{f(s)-g(s)}{2}\,ds\right\} \\
&\leqslant \left\{\int_0^1 (1+s)\,ds\right\}\frac{\|f-g\|}{2} = \frac{3}{4}\|f-g\|
\end{aligned}
$$

where $\|\ \|$ denotes the uniform norm on $C[0, 1]$. Therefore, we have

$$\|Tf-Tg\| \leqslant (\tfrac{3}{4})\|f-g\|$$

so that T is a contraction on B. Therefore, there is f in B such that $Tf = f$. This function f is a solution to the given equation. The uniqueness of f over all $C[0, 1]$ follows because if g is any other solution, then $g(t) \geqslant 1$ for all t, and we would have

$$\left\{\int_0^t f(s)+g(s)\,ds\right\}\left\{\int_0^t f(s)-g(s)\,ds\right\} = 0$$

from which it follows that $\int_0^t (f(s)-g(s))\,ds = 0$ for all t in $[0, 1]$; and since f and g are both continuous $f(s) = g(s)$ for all t in $[0, 1]$.

The next theorem, known as the *Baire Category Theorem*, may at first seem to be purely technical. However, its numerous applications show that it is one of the most fundamental results in analysis.

Theorem 9.7: Suppose (S, ρ) is a complete metric space and (C_n) is a sequence of closed subsets of S such that $\bigcup C_n = S$. Then at least one set C_n has a nonempty interior.

Proof: We may assume that all of the sets C_n are necessary; that is, C_m is not a subset of $\bigcup_{n=1}^{m-1} C_n$ for any m. Otherwise, we could throw out those C_m that are subsets. If for any m, $\bigcup_{n=1}^{m} C_n$ covers a nonempty open set O while $\bigcup_{n=1}^{m-1} C_n$ does not, it then follows that $O \cap (\bigcup_{n=1}^{m-1} C_n)^\sim$ is nonempty, open, and is contained in C_m.

Thus, if $\operatorname{int}(C_n)$ is empty for each n, it follows that for each m, $\operatorname{int}(\bigcup_{n=1}^{m} C_n)$ is empty as well. We shall now prove that under such circumstances it is impossible for $\bigcup_n C_n$ to be all of S. Let x_1 be any point in $C_1^\sim$. Let ε_1 between 0 and 1 be such that $B(x_1, \varepsilon_1) \subset C_1^\sim$. Let x_2 be any point in $B(x_1, \varepsilon_1/2)\backslash(C_1 \cup C_2)$ and let ε_2 be such that $0 < \varepsilon_2 < \frac{1}{2}$ and $B(x_2, \varepsilon_2) \subset B(x_1, \varepsilon_1/2)\backslash(C_1 \cup C_2)$. In general, if we have determined $x_1, x_2, \ldots, x_{m-1}$ in S and positive real numbers $\varepsilon_1, \varepsilon_2, \ldots, \varepsilon_{m-1}$, we let x_m be any point in $B(x_{m-1}, \varepsilon_{m-1}/2)\backslash(\bigcup_{n=1}^{m} C_n)$ and let ε_m be such that $0 < \varepsilon_m < 1/m$ and

$$B(x_m, \varepsilon_m) \subset B(x_{m-1}, \varepsilon_{m-1}/2) \setminus \left(\bigcup_{n=1}^{m-1} C_n\right).$$

Thus by induction we have defined a sequence (x_n) in S and a sequence (ε_n) of reals.

The sequence (x_n) is a Cauchy sequence because if p and $q > m$, x_p and x_q are both contained in $B(x_m, \varepsilon_m/2)$ so that $\rho(x_p, x_q) < 1/m$. Since (S, ρ) is complete there is a point x in S such that $\lim_n x_n = x$. We conclude the proof by showing that x cannot be in $\bigcup_n C_n$. For each m, the set $\{x_n: n > m\}$ is contained in the closed set $\{y: \rho(x_m, y) \leqslant \varepsilon_m/2\}$, which we denote by $CB(x_m, \varepsilon_m/2)$. Therefore, x must also be in $CB(x_m, \varepsilon_m/2)$. Since $CB(x_m, \varepsilon_m/2)$ is contained in $B(x_m, \varepsilon_m)$, which is contained in $(\bigcup_{n=1}^{m} C_n)^\sim$, it follows that x is not in $\bigcup C_m$. ■

A subset B of a topological space is *nowhere dense* if $\bar{B}$ has an empty interior. A set C is said to be of the *first category* if it is the countable union of nowhere dense sets, that is, $C = \bigcup_n B_n$ where $\operatorname{int}(\bar{B}_n) = \varnothing$ for all n. A set C is of the *second category* if it is not of the first category. The following statements are either direct consequences of their definitions or of the Baire Category Theorem.

Corollary 9.8: Suppose (S, ρ) is a complete metric space.

(a) If C is of the first category, then $C^\sim$ is of the second category.

(b) The countable union of subsets of the first category is of the first category.

(c) If C is of the first category and $B \subset C$ then B is of the first category. ■

The Baire Category Theorem can be used to prove the existence, in a nonconstructive way, of an object that has a property P and belongs to a set S: (1) Show that S forms a complete metric space. (2) Show that the objects without property P form a subset B of the first category. (3) Then the complement of B must be nonempty by the Baire Category Theorem so that an object with property P must exist. We illustrate this procedure in the following example.

Example 9.9:

We shall show that on the interval $[0, 1]$ there is a continuous function that is not differentiable at any point. Denote by $C[0, 1]$ the space of all continuous functions on $[0, 1]$ with the uniform norm. For each n let D_n be the set of all f in $C[0, 1]$ for which there is some x in $[0, 1 - 1/n]$ such that

$$|f(u) - f(x)| \leqslant n(u - x) \quad \text{for all } u \text{ with } x \leqslant u < 1.$$

To show that each D_n is closed in $C[0, 1]$ suppose (f_k) is a sequence in D_n that converges to f. For each k let x_k be the number in $[0, 1 - 1/n]$ that works in the definition of D_n. Since $[0, 1 - 1/n]$ is compact there is a subsequence $(x_{k(j)})$ that converges to x in $[0, 1 - 1/n]$. Since, as you can show, we then have

$$\lim_j f_{k(j)}(x_{k(j)}) = f(x),$$

it follows that

$$|f(u) - f(x)| \leqslant \lim_{k(j)} |f_{k(j)}(u) - f_{k(j)}(x_{k(j)})|$$

$$\leqslant \limsup n|u - x_{k(j)}| = n|u - x|.$$

The next task is to show that each D_n is nowhere dense in $C[0, 1]$. Let any function f in $C[0, 1]$ and $\varepsilon > 0$ be given. By Theorem 8.9 there is $\delta > 0$

such that $|f(u)-f(v)|<\varepsilon/2$ when $|u-v|<\delta$. Let $M \in \mathbb{N}$ be such that $1/2M < \min(\delta, \varepsilon/2n)$, and let $x_j = j/2M$ for $j=0, 1, \ldots, 2M$. Define g on $[0, 1]$ so that $g(x_j)=f(x_j)-\varepsilon/2$ if j is even and $f(x_j)+\varepsilon/2$ if j is odd and linear between x_j and x_{j+1} for each $j=0, \ldots, 2M$. To prove that g is in $B(f, \varepsilon)$ let x in $[0, 1]$ be given. There is j such that $x = ax_j + (1-a)x_{j+1}$ where $0 \leqslant a < 1$ so that $g(x) = ag(x_j) + (1-a)g(x_{j+1})$. Therefore, we have

$$\begin{aligned} |f(x)-g(x)| &= a|f(x)-g(x_j)| + (1-a)|f(x)-g(x_{j+1})| \\ &< a\varepsilon + (1-a)\varepsilon = \varepsilon. \end{aligned}$$

To prove that g is not in D_n we calculate the slope on each interval $[x_j, x_{j+1})$ to be

$$\frac{|f(x_{j+1})+\varepsilon/2-f(x_j)-\varepsilon/2|}{(x_{j+1}-x_j)} \geqslant \frac{(\varepsilon-|f(x_{j+1})-f(x_j)|)}{(\varepsilon/2M)} > \frac{(\varepsilon/2}{(\varepsilon/2n)} = n.$$

This proves that D_n cannot contain any neighborhood of any point in $C[0, 1]$ so that D_n is nowhere dense. You may now show that if f has a derivative at some point x in $[0, 1]$, that is, a derivative from the right, it must be in D_n for some n. However, since each D_n is nowhere dense there are functions in the complement of $\bigcup_n D_n$. These functions cannot have a derivative at any point in $[0, 1]$.

While the argument above does not produce a specific function having no derivative in $[0, 1]$ it does show that the class of such functions are of the second category. Thus, functions having no derivative in $[0, 1]$, in a sense, occur in greater abundance than functions that have a derivative at some point.

Exercises

A

1. Show that every set with its discrete topology is a complete metric space.
2. *(a)* Explain why Q, the set of rationals, is not complete with the usual metric on $\mathbb{R}$.
   (b) Show that $\mathbb{R}$ is the completion of Q.
3. Prove that a subset B of a complete metric space is complete if and only if B is closed.
4. Show that in Example 1 (b), the metric ρ induces the same topology on S as does σ.
5. Prove Theorem 9.2.
6. Suppose S is a metric space and (f_n) is a sequence of bounded functions on S that converges to the function f in the uniform norm. If each f_n is continuous at $s \in S$, then f is continuous at s. (Imitate the argument in Example 9.3.)
7. If (S, ρ) and (T, σ) are two metric spaces that are isometric show that S is complete if and only if T is.
8. *(a)* Show that a function f that satisfies the condition of the Contraction Mapping Theorem is continuous.
   (b) Show that the point x_0 in the Contraction Mapping Theorem is unique.

9. Use the Contraction Mapping Theorem to prove that if $0 < a < \frac{1}{3}$ and $0 < b < 1 - a$, then there is a unique x in $[0, 1]$ with $x = ax^3 + b$. [*Hint*: $|ax^3 + b - (ay^3 + b)| = a|x^3 - y^3| = a|x^2 + xy + y^2||x - y|$.]

10. Show that the following statement is equivalent to the Baire Category Theorem: If $\{O_n\}$ is a sequence of open subsets of a complete metric space S such that each O_n is dense in S, then $\bigcap_n O_n$ is nonempty.

11. *(a)* Prove Corollary 9.9.
(b) Is it true that if C is of the second category then $C^{\sim}$ is of the first category?

12. In Example 9.8 verify that $\lim_j f_{k(j)}(x_{k(j)}) = f(x)$.

13. In Example 9.8 verify that $\bigcup D_n$ contains all f in $C[0, 1]$ that have a derivative at some point n.

B

14. Show that the following metric spaces are complete
(a) $\mathbb{R}^n$ with the Euclidean norm;
(b) $\mathbb{C}^n$ with the Euclidean norm;
(c) $\mathbb{R}^n$ with an arbitrary p-norm;
(d) ω with the Fréchet norm;
(e) 1^1;
(f) 1^∞;
(g) 1^p, $1 < p < \infty$;
(h) bs;
(i) cs;
(j) $B(S)$ for any set S;
(k) c_0;
(l) bv.

15. Verify Statements (1)–(10) in the Proof of Theorem 9.4.

16. For each $n = 1, 2, \ldots$, let f_n denote the function on $[0, 1]$, which is 0 for $0 \leqslant t \leqslant \frac{1}{2} - 1/(2+n)$; 1 for $\frac{1}{2} + 1/(2+n) \leqslant t \leqslant 1$, and whose graph on $[\frac{1}{2} - 1/(2+n), \frac{1}{2} + 1/(2+n)]$ is the line that joins $(\frac{1}{2} - 1/(2+n), 0)$ to $(\frac{1}{2} + 1/(2+n), 1)$.
(a) Show that the sequence (f_n) is a Cauchy sequence in $C[0, 1]$ with the p-norm for each p with $1 \leqslant p < \infty$.
(b) Prove that the sequence (f_n) is not a Cauchy sequence in $C[0, 1]$ with the uniform norm.
(c) Show that if f is the function that is 0 for $0 \leqslant t \leqslant \frac{1}{2}$ and 1 for $\frac{1}{2} < t \leqslant 1$, then $\lim_n \|f - f_n\|_p = 0$.
(d) Show that there is no g in $C[0, 1]$ such that $\lim_n \|g - f_n\|_p = 0$. Therefore, $C[0, 1]$ is not complete with respect to a p-norm.

17. Show that a metric space (S, ρ) is compact if and only if it is (a) complete and (b) for each $\varepsilon > 0$ there is a finite subset $F(\varepsilon)$ of S such that for each x in S there is u in $F(\varepsilon)$ such that $\rho(u, x) < \varepsilon$. A space that satisfies condition (b) is said to be *totally bounded*. The set $F(\varepsilon)$ is called an *ε-net* in S. Show that the completion of a metric space is compact if and only if the space is totally bounded.

18. *(a)* Use the Mean Value Theorem to show that if x and y are in the interval $[0, a]$ where $a < \pi/2$ then $|\cos x - \cos y| < (\sin a)|x - y|$.
(b) Show that $|\cos x - \cos y| < |x - y|$ for all x in $[0, \pi/2]$.

(c) Use the Contraction Mapping Theorem to prove there is a unique x such that $\cos x = x$.

(d) Use a scientific calculator and approximate this x by first entering any number and pressing cos repeatedly.

(e) This procedure also works for sin but it is not as interesting since $\sin 0 = 0$ is the unique solution of $\sin x = x$. Find other functions on the calculator for which there is a unique solution for $f(x) = x$.

19. Find a Cauchy sequence $\{p_n\}$ of polynomials in $C[0, 1]$ whose limit is not a polynomial. This shows that the space $P[0, 1]$ of polynomials on $[0, 1]$ is not complete with the uniform metric.

20. Use the Contraction Mapping Theorem to prove that there is unique function y with continuous derivative y' such that $y(0) = 0$ and $y' = (y(t))^2$ for $0 \leqslant t \leqslant 1$. (Consider the integral equation $y(t) = \int_0^t (y(s))^2 \, ds$ on the space $C[0, r]$ for each $0 < r < 1$.)

21. Suppose f_n: $n = 1, 2, \ldots$ are continuous real valued functions defined on a complete metric space S such that for each x in S, $\sup_n |f_n(x)| < \infty$. Prove that there is an open subset U of S and $M > 0$ such that for all n and for all x in U, $|f_n(x)| < M$.

22. *(a)* Show that if O is an open subset of a topological space, then $\bar{O} \backslash O$ is closed and has an empty interior.

(b) Prove that if O is an open subset of a complete metric space, then O is of the second category.

C

23. The purpose of this exercise is to establish the following theorem: If f is a real valued function on a complete metric space S that is continuous on a dense subset D of S, then f is continuous on a subset of S of the second category. For each s in S define $U(s)$ by $U(s) = \inf\{\sup\{f(t): t \in B(s, \varepsilon)\}: \varepsilon > 0\}$ and define $L(s)$ by $L(s) = \sup\{\inf\{f(t): t \in B(s, \varepsilon)\}: \varepsilon > 0\}$. Let O_n consist of all s in S such that $U(s) - L(s) < 1/n$.

(a) Each O_n is open and dense in S so $O_n^{\sim}$ is nowhere dense in S.

(b) $\bigcup O_n^{\sim}$ is of the first category in S so $\bigcap O_n$ is of the second category in S.

(c) If s is in $\bigcap O_n$, then f is continuous at s.

(d) There is no real valued function on $\mathbb{R}$ that is continuous precisely at the rationals.

(e) The function defined by $f(s) = 0$ for s irrational and $f(s) = 1/q$ if $s = p/q$ with $q > 0$ and p having no common divisor with q, is continuous only at the irrational numbers.

10

Product Spaces

In Section 3 we defined the product $\prod_{\alpha \in \mathscr{A}} S_\alpha$ of a collection $\{S_\alpha: \alpha \in \mathscr{A}\}$ of sets to be the set of all functions f from $\mathscr{A}$ into $\bigcup_{\alpha \in \mathscr{A}} S_\alpha$ such that $f(\alpha) \in S_\alpha$ for all $\alpha \varepsilon \mathscr{A}$.

In order to define the term "function" in the first place we defined the product $S \times T$ of two sets S and T to be the set of all ordered pairs (s, t) with $s \in S$ and $t \in T$. Thus, from a formal point of view there are two interpretations for a finite product $S_1 \times S_2 \times \cdots \times S_n$.

(π1) All n-tuples $(s_1, s_2, \ldots, s_n)$ with $s_i \in S_i$ for each i.

(π2) All functions f from the set $\{1, 2, \ldots, n\}$ into $\bigcup_{i=1}^n S_i$ such that $f(i) \in S_i$ for each i.

A function such as described in (π2) is a set of ordered pairs $\{(1, s_1), (2, s_2), \ldots, (n, s_n)\}$ where $s_i \in S_i$ for each i. The correspondence

$$(s_1, s_2, \ldots, s_n) \quad \text{to} \quad \{(1, s_1), (2, s_2), \ldots, (n, s_n)\}$$

is a natural identification of the two concepts; that is, we shall regard the n-tuple as an abbreviated way of writing the function. When $\mathscr{A}$ is infinite we can identify f in $\prod_{\alpha \in \mathscr{A}} S_\alpha$ with the "generalized sequence" $(a_\alpha)_{\alpha \in \mathscr{A}}$ where $f(\alpha) = a_\alpha \in S_\alpha$ for each α. If each S_α is the same set S, then we write $S^{\mathscr{A}}$ instead of $\prod_{\alpha \in \mathscr{A}} S_\alpha$.

In this section we shall develop a theory of topologies on product spaces; first we shall discuss the product of two topological spaces and then we shall discuss the product of arbitrarily many spaces. If $(S_1, \mathscr{T}_1)$ and $(S_2, \mathscr{T}_2)$ are two topological spaces, then we define a subset U of $S_1 \times S_2$ to be *open in the product topology* if, for each ordered pair (s_1, s_2) in U, there are open sets $U_1 \in \mathscr{T}_1$ and $U_2 \in \mathscr{T}_2$ such that $s_1 \in U_1, s_2 \in U_2$, and $U_1 \times U_2 \subset U$. The functions P_1, P_2 defined from $S_1 \times S_2$ onto S_1 and S_2 by

$$P_1(s_1, s_2) = s_1 \quad \text{and} \quad P_2(s_1, s_2) = s_2$$

are called the *projections* from $S_1 \times S_2$ onto S_1 and S_2.

Example 10.1:

A subset O of $\mathbb{R} \times \mathbb{R}$ is open in the product topology if and only if for each point $x = (x_1, x_2)$ in O there are a_1, b_1, a_2, b_2 with $a_1 < x_1 < b_1$, $a_2 < x_2 < b_2$ such that the product of the open intervals, $(a_1, b_1) \times (a_2, b_2)$ is contained in O. See Figure 10.1.

Theorem 10.2: If $(S_1, \mathscr{T}_1)$ and $(S_2, \mathscr{T}_2)$ are topological spaces then

(a) the collection $\mathscr{T}_\times$ of sets that are open in the product topology is a topology on $S_1 \times S_2$;

(b) the projection P_i is continuous from $S_1 \times S_2$ onto S_i for $i = 1, 2$ and maps open sets onto open sets.

Proofs:

(a) The empty set $\varnothing$ is in $\mathscr{T}_\times$ by default since there is no pair (s_1, s_2) in $\varnothing$ to test. That $\mathscr{T}_\times$ contains $S_1 \times S_2$ follows from the fact that $S_1 \in \mathscr{T}_1$ and $S_2 \in \mathscr{T}_2$. To show $\mathscr{T}_\times$ is closed under unions suppose $\mathscr{G} \subset \mathscr{T}_\times$ and $(s, t) \in \bigcup \mathscr{G}$. There is some O in $\mathscr{G}$ with $(s, t) \in O$. Since O is open in the product topology there are $U_1 \in \mathscr{T}_1$ and $U_2 \in \mathscr{T}_2$ such that $(s, t) \in U_1 \times U_2 \subset O$. But $O \subset \bigcup \mathscr{G}$ so

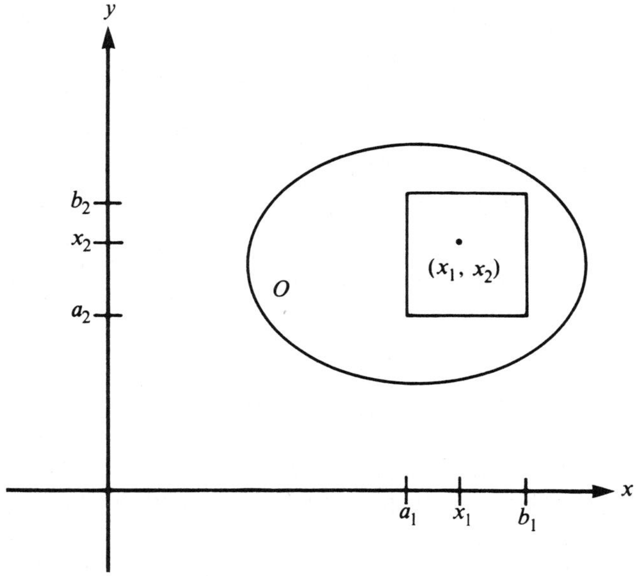

Figure 10.1

we also have

$$(s, t) \in U_1 \times U_2 \subset \bigcup \mathcal{G}.$$

To show that $\mathcal{T}_\times$ is closed under finite intersections suppose O_1 and O_2 are in $\mathcal{T}_\times$ and $(s, t) \in O_1 \cap O_2$. Let U_1, V_1 in $\mathcal{T}_1$ and U_2, V_2 in $\mathcal{T}_2$ be such that

$$(s, t) \in U_1 \times V_1 \subset O_1 \quad \text{and} \quad (s, t) \in U_2 \times V_2 \subset O_2.$$

Then $s \in U_1 \cap U_2$, and $t \in V_1 \cap V_2$ so that $(s, t) \in (U_1 \cap U_2) \times (V_1 \cap V_2)$. The reader can show that

$$(U_1 \cap U_2) \times (V_1 \cap V_2) = (U_1 \times V_1) \cap (U_2 \times V_2)$$

so that $(U_1 \cap U_2) \times (V_1 \cap V_2) \subset O_1 \cap O_2$. This proves that $O_1 \cap O_2$ is in $\mathcal{T}_\times$.

(b) If O is an open subset of S_1 then $P_1^{-1}(O) = O \times S_2$, which is clearly open in the product topology. Therefore, P_1 is continuous. A corresponding argument shows that P_2 is continuous. Suppose O is in $\mathcal{T}_\times$ and $s \in P_1(O)$. Then there is t in S_2 such that $(s, t) \in O$. Since O is open there is U in $\mathcal{T}_1$ and V in $\mathcal{T}_2$ such that $(s, t) \in U \times V \subset O$. This means that $s \in U \subset P_1(O)$ so that $P_1(O)$ is open. Similarly $P_2(O)$ is open. ■

Theorem 10.3: Suppose $(S_1, \mathcal{T}_1)$ and $(S_2, \mathcal{T}_2)$ are topological spaces. For each t in S_2 the function $Q_t s = (s, t)$ is continuous from S_1 into $S_1 \times S_2$ and a homeomorphism from S_1 onto its range $T_1 = \{(s, t): s \in S_1\}$ where T_1 bears the subspace topology.

Proof: We first observe the relation between Q_t and the projection P_1. For each (s, r) in $S_1 \times S_2$ we have $Q_t P_1(s, r) = (s, t)$ and for each s in S_1 we have

$P_1 Q_t(s) = s$. If O is open in $S_1 \times S_2$ then $Q_t^{-1}(O) = P_1(O)$ so that $Q_t^{-1}(O)$ is open. This proves that Q_t is continuous. If U is open in S_1, then $Q_t(U) = U \times (S_2 \cap Q_t(S_1))$ so that U maps open sets onto sets that are relatively open in the range of Q_t. Therefore, Q_t is a homeomorphism onto its range. ■

Theorem 10.4: Suppose (S, ρ) *and* (T, σ) are metric spaces.

(a) The product topology on $S \times T$ is determined by the metric

$$\tau((s_1, t_1), (s_2, t_2)) = \max\{\rho(s_1, s_2), \sigma(t_1, t_2)\}.$$

(b) If C is a compact subset of S and D is a compact subset of T, then $C \times D$ is a compact subset of $S \times T$.

Proof:

(a) You can see that τ satisfies (M1), (M2), and (M3) because both ρ and σ do. We first show the topology determined by τ is stronger than the product topology by showing that if O is open in the product topology then O is open in the topology induced by the metric τ. Suppose $(u, v) \in O$. Then there are open subsets U of S and V of T such that $(u, v) \subset U \times V \subset O$. Since both S and T are metric spaces there are ε_1 and $\varepsilon_2 > 0$ such that $B_\rho(u, \varepsilon_1) \subset U$ and $B_\sigma(v, \varepsilon_2) \subset V$ where $B_\rho(u, \varepsilon_1)$ denotes the open ball $\{s: \rho(u, s) < \varepsilon_1\}$ and $B_\sigma(v, \varepsilon_2)$ has an analogous meaning to $\{s: \sigma(v, s) < \varepsilon_2\}$. If $\varepsilon = \min(\varepsilon_1, \varepsilon_2)$ then we have

$$B_\tau((u, v), \varepsilon) = B_\rho(u, \varepsilon) \times B_\sigma(v, \varepsilon)$$

because $\tau((u, v), (s, t)) < \varepsilon$ if and only if both $\rho(u, s)$ and $\sigma(v, t)$ are less than ε. Therefore, we have $B_\tau((u, v), \varepsilon) \subset U \times V \subset O$. That the metric topology is weaker than the product topology follows from the equality of $B_\tau((u, v), \varepsilon)$ and $B_\rho(u, \varepsilon) \times B_\sigma(v, \varepsilon)$ for all $\varepsilon > 0$ since this shows that each ball is open in the product topology.

(b) It suffices to prove that $C \times D$ is sequentially compact. See Theorem 8.8 (d) on page 51. Suppose $((s_n, t_n))$ is a sequence in $C \times D$. Since C is compact there is a subsequence $(s_{k(n)})$ of (s_n) that converges, say to s, and since D is compact there is a subsequence $(t_{h(n)})$ of $(t_{k(n)})$ that converges, say to t. Then $(s_{h(n)}, t_{h(n)})$ converges to (s, t) in $C \times D$. For if M is such that

$$\rho(s, s_{h(n)}) < \varepsilon \quad \text{and} \quad \sigma(t, t_{h(n)}) < \varepsilon \quad \text{when } n > M,$$

it follows that

$$\tau((s, t), (s_{h(n)}, t_{h(n)})) < \varepsilon \quad \text{when } n > M. \ ■$$

It is also true that the product of compact topological spaces is compact, but this is more difficult to prove. See the Tychonoff Theorem below.

Example 10.5:

Important examples of product spaces are Euclidean n-spaces $\mathbb{R}^n$ and *unitary n-space* $\mathbb{C}^n$. It follows by application of Theorem 10.4 (a) that the product topology on both $\mathbb{R}^n$ and $\mathbb{C}^n$ is determined by the Minkowski ∞-norm

$$\|(x_1, x_2, \ldots, x_n)\|_\infty = \max\{|x_j|: j=1, \ldots, n\}.$$

Since for each x in $\mathbb{R}^n$ or $\mathbb{C}^n$ we have the relation

$$\|x\|_\infty \leqslant \|x\|_2 \leqslant \|x\|_1 \leqslant n\|x\|_\infty.$$

it follows from Theorem 8.7 that the three norms $\|\ \|_1$, $\|\ \|_2$, and $\|\ \|_\infty$ determine the same topology on $\mathbb{R}^n$ or $\mathbb{C}^n$.

A subset B of $\mathbb{R}^n$ is called *bounded* if there is $M>0$ such that $\max_i |x_i| < M$ for all x in B. Hence, B is bounded if and only if it is contained in the closed n-cube

$$C_M = [-M, M] \times [-M, M] \times \cdots \times [-M, M] \quad (n \text{ factors}).$$

By Theorem 10.4 (b) the set C_M is compact. Thus every closed and bounded subset of $\mathbb{R}^n$ is compact. That compact subsets are closed and bounded follows by an argument resembling that used for $\mathbb{R}$. Thus, we have derived the following result known as the *Heine-Borel Theorem for* $\mathbb{R}^n$.

Theorem 10.6: A subset C of $\mathbb{R}^n$ is compact if and only if it is closed and bounded. ∎

The rest of this section is devoted to the theory of product topologies for arbitrary products. This theory is considerably more difficult than the theory for finite products and may be omitted from the standard course.

If for each $\alpha \in \mathscr{A}$, S_α is a topological space, we can define the *product topology* $\mathscr{T}_\Pi$ on $\prod_\alpha S_\alpha$ by calling a subset O open if for each function f in O there is a set U of functions with $f \in U \subset O$ such that $U = \prod_\alpha U_\alpha$ where each U_α is a nbd of $f(\alpha)$, and for all but finitely many α, $U_\alpha = S_\alpha$. The set U is referred to as a *basic nbd*. For each α the function from $\prod_\alpha S_\alpha$ defined by $P_\alpha f = f(\alpha)$ is called the *projection* from $\prod_\alpha S_\alpha$ onto S_α. For $B \subset \prod_\alpha S_\alpha$ the set

$$P_\alpha(B) = \{f(\alpha): f \in B\}$$

is called the *α-cross section* of B. Suppose $g \in \prod_{\alpha \neq \beta} S_\alpha$. We can define the function Q_g from S_β into the product set $\prod_\alpha S_\alpha$ by

$$Q_g(s) = f_{g,s} \quad \text{where } f_{g,s}(\alpha) = g(\alpha) \text{ for } \alpha \neq \beta \quad \text{and} \quad f_{g,s}(\beta) = s.$$

Throughout the remainder of this section we shall write $\prod$ for the topological space $(\prod_\alpha S_\alpha, \mathscr{T}_\Pi)$ if there is no chance of ambiguity.

Theorem 10.7:

(a) The collection $\mathscr{T}_\Pi$ is a topology.

(b) For each α the projection P_α is continuous from $\prod$ onto S_α and maps open sets onto open sets.

(c) For each β and each g in $\prod_{\alpha \neq \beta} S_\alpha$ the function Q_g is a continuous function from S_β into $\prod$, which is a homeomorphism onto its range.

Proof:

(a) The empty set $\varnothing$ is in $\mathscr{T}_\Pi$ by default since there is no function in $\varnothing$ to test. It is clear that $\mathscr{T}_\Pi$ contains the entire product space $\prod$. Suppose $\mathscr{G} \subset \mathscr{T}_\Pi$

and $f \in \bigcup \mathscr{G}$. If O in $\mathscr{G}$ contains f, then there is a basic nbd U of f such that $U \subset O$. But then U is also contained in $\bigcup \mathscr{G}$ since O is. Therefore, $\mathscr{T}$ satisfies ($\mathscr{T}2$) on page 39. Next suppose the sets $O_1, \ldots, O_n$ are in $\mathscr{T}_\Pi$ and f is in $O = \bigcap_{i=1}^n O_i$. For each i let $U^{(i)} = \prod_\alpha U_\alpha^{(i)}$ be a basic nbd of f with $U^{(i)} \subset O_i$. Let ϕ_i be the set of all α such that $U_\alpha^{(i)}$ is not all of S_α and let $\phi = \bigcup_{i=1}^n \phi_i$. Since each ϕ_i is finite so is ϕ. Let $V = \prod_\alpha V_\alpha$ where, for each α in ϕ, we have $V_\alpha = \bigcap_{i=1}^n U_\alpha^{(i)}$ and $V_\alpha = S_\alpha$ for α not in ϕ. Then V is a basic nbd of f, which is contained in O. Therefore, $\mathscr{T}_\Pi$ satisfies ($\mathscr{T}3$).

The proof of (b) resembles the proof of Theorem 10.2 (b), and the proof of (c) resembles the proof of Theorem 10.3. ■

Our next result is that the countable product of metric spaces is a metric space; this will not necessarily be true for uncountable products.

Theorem 10.8: Suppose (S_i, ρ_i) is a metric space for $i = 1, 2, \ldots$. The product topology $\mathscr{T}_\Pi$ on $\prod_i S_i$ is determined by the metric

$$\sigma(x, y) = \sum_i \frac{2^{-i}\rho_i(x_i, y_i)}{1 + \rho_i(x_i, y_i)}$$

where $x = (x_i)$ and $y = (y_i)$ are in $\prod_i S_i$.

Proof: The first task is to verify that σ is a metric. That σ satisfies (M1) and (M2) on page 47 follows because each ρ_i does. To show σ satisfies the triangle inequality, suppose $x = (x_i)$, $y = (y_i)$, and $z = (z_i)$ are in $\prod_i S_i$. The real valued function $f(u) = u/(1 + u)$ is strictly increasing when $u > 0$ since $f'(u) > 0$. Let $a = \rho_i(x_i, y_i)$, $b = \rho_i(y_i, z_i)$, and $c = \rho_i(x_i, z_i)$. Then we have

$$\frac{a}{(1+a)} + \frac{b}{(1+b)} = \frac{(a+b+2ab)}{(1+a+b+ab)} \geqslant \frac{(a+b+ab)}{(1+a+b+ab)} \geqslant \frac{(a+b)}{(1+a+b)} \geqslant \frac{c}{(1+c)}.$$

The last inequality holds because $a + b \geqslant c$. This shows that (M3) holds for each summand so (M3) holds for the sum that defines σ.

Next we verify that the topology determined by the metric σ is stronger than the product topology. Let O be a nbd of f in the product topology. By definition of the product topology there is a basic nbd $U = \prod_i U_i$ of f contained in O. Since U is a basic nbd there is M such that $U_i = S_i$ for $i > M$ while U_i is a nbd of $f(i)$ for $i \leqslant M$. For $i \leqslant M$ let $\varepsilon_i > 0$ be such that $B(f(i), \varepsilon_i) \subset U_i$, and let

$$\delta = 2^{-(M+1)} \inf\{\varepsilon_i, 1\colon i = 1, 2, \ldots, M\}.$$

If $\sigma(f, g) < \delta$ then for each $i \leqslant M$, we have

$$\frac{\rho_i(f(i), g(i))}{2^i}\{1 + \rho_i(f(i), g(i))\} < \delta,$$

$$\rho_i(f(i), g(i)) < 2^i\delta(1 + \rho_i(f(i), g(i)),$$

$$(1 - 2^i\delta)\rho_i(f(i), g(i)) < 2^i\delta, \quad \text{and}$$

$$\rho_i(f(i), g(i)) < 2^{i+1}\delta \leqslant \varepsilon_i \quad \text{since } (1 - 2^i\delta) > \tfrac{1}{2}.$$

This means that $B(f, \delta) \subset U$ so that O is open in the metric topology.

We conclude by showing that every open set in the metric topology is open in the product topology. It suffices to show that for each open ball $B(f, \delta)$ there is a basic open set U such that $f \in U \subset B(f, \delta)$. Let M be such that $\sum_{n=M}^{\infty} 2^{-n} < \delta/2$. For $i < M$ let $U_i = B(f(i), \delta/2)$; for $i \geqslant M$ let $U_i = S_i$. Then $U = \prod_i U_i$ is a basic nbd of f. If $g \in U$ then we have

$$\sigma(f, g) = \sum_n \rho_i \frac{(f(i), g(i))}{\{2^i(1 + \rho_i(f(i), g(i))\}} \leqslant \sum_{n=1}^{M-1} \frac{\delta}{2^{i+1}} + \sum_{n=M}^{\infty} \frac{1}{2^i} < \delta.$$

This proves that $U \subset B(f, \delta)$. ■

Example 10.9:

The space ω of all sequences with the Fréchet metric as described in Example 8.10 is the countable product of scalar fields $\mathbb{R} \times \mathbb{R} \times \ldots$ or $\mathbb{C} \times \mathbb{C} \times \ldots$. Indeed, the metric used in Theorem 10.8 is an obvious generalization of the Fréchet metric.

The next theorem asserts that the product of connected spaces is connected.

Theorem 10.10: If for each $\alpha \in \mathscr{A}$, $(S_\alpha, \mathscr{T}_\alpha)$ is a connected topological space, then $\prod_\alpha S_\alpha$ is connected in the product topology.

Proof: Suppose U and V are nonempty open subsets of $\prod$ whose union is $\prod$. We shall prove that $U \cap V$ is nonempty so that (U, V) cannot be a separation of $\prod$. Suppose f is in U and g is in V. Since U and V are open there are basic open sets $O = \prod_\alpha O_\alpha$ and $Q = \prod_\alpha Q_\alpha$ such that $f \in O \subset U$ and $g \in Q \subset V$, where for all but finitely many subscripts $\alpha_1, \ldots, \alpha_n$, we have $O_\alpha = Q_\alpha = S_\alpha$. For $i = 1, 2, \ldots, n$, let A_i be the set of all h in $\prod$ such that

(a) $h(\alpha) = f(\alpha)$ for $\alpha \neq \alpha_1, \alpha_2, \ldots, \alpha_n$;
(b) $h(\alpha_i)$ is any member of S_{α_i};
(c) $h(\alpha_j) = f(\alpha_j)$ for all j with $1 \leqslant j < i$ if $i > 1$, and $h(\alpha_j) = g(\alpha_j)$ for all j with $i < j \leqslant n$ if $i < n$.

Each set A_i is homeomorphic to S_{a_i} by means of the projection function P_{α_i}. We leave the proof of this fact to you. Thus each A_i is connected. Since $A_i \cap A_{i+1}$ contains the function k that is equal to $f(\alpha)$ for $\alpha \neq \alpha_1, \alpha_2, \ldots, \alpha_n$; equal to $f(\alpha_j)$ for $j \leqslant i$; and equal to $g(\alpha_j)$ for $j > i$; it follows that $A_i \cap A_{i+1} \neq \varnothing$. Therefore, by Theorem 7.4 on page 40, the set $\bigcup_{i=1}^{n} A_i = A$ is connected. Since O and Q are open, and have a nonempty intersection with A and $A = (A \cap O) \cup (A \cap Q)$, it follows that the intersection of $A \cap O$ and $A \cap Q$ is nonempty. From this it follows that U and V have a nonempty intersection. ■

Example 10.11:

Suppose $\sum_n a_n$ is a series of positive numbers that converges to a. Suppose $b \in [0, a]$. Can we find c_n with $0 \leqslant c_n \leqslant 1$ for each n such that $\sum_n c_n a_n = b$? Yes, because $[0, 1] \times [0, 1] \ldots$ is a connected set and the function defined on the product by

$$T(x_1, x_2, \ldots) = \sum_n x_n a_n$$

is continuous. Therefore the image of this product under T is an interval containing 0 and a and hence contains all points between them.

The following result, known as Tychonoff's Theorem, has important applications both for finite and infinite products.

Theorem 10.12: If $(S_\alpha, \mathcal{T}_\alpha)$ is compact for each α, then $\prod_\alpha S_\alpha$ is compact in the product topology.

Proof: We shall prove that if $\mathcal{G}$ is a collection of closed subsets of $S = \prod_\alpha S_\alpha$, which has the finite intersection property (fip), then $\bigcap$ is nonempty. It then follows by Theorem 7.7 (b) on page 43 that S is compact.

The first step is to prove that there is a collection $\mathcal{A}$ of sets (not necessarily closed) that contains $\mathcal{G}$, has the fip, and is maximal in the sense that if $\mathcal{B}$ contains $\mathcal{G}$, has the fip, and contains $\mathcal{A}$, it follows that $\mathcal{B} = \mathcal{A}$. For this we use Postulate 3.6 on page 15. Let $\mathcal{X}$ denote the set of all collections $\mathcal{B}$ of S that have the fip and contain $\mathcal{G}$. If $\mathcal{Y}$ is a chain in $\mathcal{X}$ and $\mathcal{B} = \bigcup \mathcal{Y}$ then $\mathcal{B}$ is also in $\mathcal{X}$ because if $A_1, A_2, \ldots, A_n$ are in $\mathcal{B}$ there is, for each i, some $\mathcal{B}_i$ in $\mathcal{Y}$ such that $A_i \in \mathcal{B}_i$. Since $\mathcal{Y}$ is a chain there is $\mathcal{B}_k$ that contains $\mathcal{B}_i$ for each i. Since $\mathcal{B}_k$ is in $\mathcal{X}$, $\mathcal{B}_k$ has the fip so that $\bigcap_{i=1}^n A_i \neq \varnothing$. By Postulate 3.6, $\mathcal{X}$ contains a maximal member that we shall call $\mathcal{A}$.

Since $\mathcal{A}$ has the fip, it follows that for each α, the collection $\{A(\alpha)\colon A \in \mathcal{A}\}$ has the fip. Recall $A(\alpha)$ is the α-section of A. Thus, by Theorem 7.7 (b), there is a point x_α in $\bigcap \{\overline{A(\alpha)}\colon A \in \mathcal{A}\}$. Define f on the set of all α by $f(\alpha) = x_\alpha$. Then f is certainly in S, and it remains to show that f is in $\bigcap \mathcal{G}$.

If O is a nbd of x_α in S_α we show that $P_\alpha^{-1}(O)$ has a nonempty intersection with every A in $\mathcal{A}$. Since x_α is in $\overline{A(\alpha)}$ there is g in A such that $g(\alpha) \in O$. This means that $g \in A \cap P_\alpha^{-1}(O)$. Thus the collection $\mathcal{A} \cup \{A \cap P_\alpha^{-1}(O)\colon A \in \mathcal{A}\}$ has the finite intersection property. Since $\mathcal{A}$ is maximal it must include $\{A \cap P_\alpha^{-1}(O)\colon A \in \mathcal{A}\}$. Proceeding inductively we conclude that if $O_{\alpha(i)}$ is a nbd of $x_{\alpha(i)}$ in $S_{\alpha(i)}$ for $i = 1, 2, \ldots, n$ and $A \in \mathcal{A}$, it follows that $A \cap \{\bigcap_{i=1}^n P_{\alpha(i)}^{-1}(O_{\alpha(i)})\}$ is in $\mathcal{A}$.

If C is in $\mathcal{G}$ and U is any nbd of f there is a basic nbd V of f such that $f \in V \subset U$. This basic nbd has the form $\bigcap_{i=1}^n P_{\alpha(i)}^{-1}(O_{\alpha(i)})$ for a finite set $\alpha(1), \alpha(2), \ldots, \alpha(n)$ of indices. Therefore, $C \cap V$ and thus $C \cap U$ are nonempty, which implies f is in C. Consequently, we have $f \in \bigcap \mathcal{G}$. ■

Example 10.13:

$2^{\mathbb{N}} = \{0, 1\} \times \{0, 1\} \times \ldots$ is a compact metric space if $\{0, 1\}$ is given the metric topology for each factor.

Example 10.14:

Suppose $\sum_n a_n$ is a series of complex numbers such that $\sum_n n a_n$ converges absolutely. Is there a sequence (c_n) of real numbers such that $-n \leqslant c_n \leqslant n$ and

$|\sum_n c_n a_n| = \sup\{|\sum_n x_n a_n|\colon -n \leqslant x_n \leqslant n\}$ for each n? The difficulty here is that we require (c_n) to be real numbers. If complex numbers were allowed we could simply choose c_n such that $c_n a_n = n|a_n|$. The answer is still affirmative, however, because $[-1, 1] \times [-2, 2] \times [-3, 3] \times \dots$ is a compact metric space and the function $T(x_n) = |\sum_n x_n a_n|$ is a continuous function from this product space into the real line. Therefore, T attains its maximum somewhere on the space.

Exercises

A

1. Let $A = \{a, b, c\}$, $B = \{d, e\}$, and $C = \{f, g\}$. List the members of $A \times B \times C$ as 3-tuples and as functions from $\{1, 2, 3\}$ into $A \cup B \cup C$, and identify each 3-tuple with the corresponding function under the natural identification.
2. If both A and B have more than one point, show there is a proper subset C of $A \times B$ such that the projection $P_A(C)$ of C into A is equal to A and $P_B(C) = B$.
3. Let $A = \{a, b, c\}$ and $B = \{d, e, f\}$. Define the topology $\mathscr{T}_1$ on A by $\{A, \varnothing, \{a\}, \{a, b\}\}$ and the topology $\mathscr{T}_2$ on B by $\{B, \varnothing, \{e\}, \{d, e\}, \{e, f\}, \{d, e, f\}\}$.
   (a) Determine the open sets in the product topology on $A \times B$.
   (b) Determine $P_1(O)$ and $P_2(O)$ for each open subset of $A \times B$.
4. Let $\mathbb{N}$ denote the set of all positive integers. Show that a member x of $R^{\mathbb{N}}$ can be represented either as a sequence $x = (x_1, x_2, \dots)$ or as a real valued function on $\mathbb{N}$. Show that a sequence (x_n) in $R^{\mathbb{N}}$ (i.e., a sequence of sequences) converges to the sequence x if and only if $\lim_n x_n(i) = x(i)$ for $i \in \mathbb{N}$.
5. Prove that every product of Hausdorff spaces is Hausdorff.
6. Show that the function P_2 defined in Theorem 10.2 is continuous.
7. Show that if $C \times D$ is a compact subset of a product space $S \times T$, then C is compact in C, and D is compact in T. (Use the continuity of the projection functions.)
8. Show that the function τ defined in Theorem 10.4 satisfies conditions (M1) and (M2).
9. Show that a sequence $((x_n, y_n))$ in the product $S \times T$ of two topological spaces converges to (x, y) if and only if $x_n \to x$ in S and $y_n \to y$ in T.
10. *(a)* Show that the product of two complete metric spaces is complete with the product metric τ.
    (b) Use this fact to show that $\mathbb{R}^n$ is complete with the ∞-metric.
11. Describe the various metrics defined on $\mathbb{R}^n$ and $\mathbb{C}^n$ by means of norms and product of norms, and verify that all of these norms determine the same topology.
12. Show that if (S, ρ) and (T, σ) are metric spaces, then the following functions are metrics that determine the product topology on $S \times T$.

$$\tau_2((s_1, t_1), (s_2, t_2)) = \{\rho(s_1, s_2)^2 + \sigma(t_1, t_2)^2\}^{1/2};$$

$$\tau_1((s_1, t_1), (s_2, t_2)) = \rho(s_1, s_2) + \rho(t_1, t_2).$$

13. State and prove a similar result to Exercise 12 for normed spaces.

14. Show that the closed unit ball of l^1, that is, the set of all $x=(x_n)$ in l^1 such that $\sum_{n=1}^{\infty} |x_n| \leqslant 1$ is closed and bounded in l^1 but not compact. (Show the sequence (e_m) has no convergent subsequence.) Thus the Heine-Borel Theorem fails for l^1.

15. Prove that if $A \times B$ is a connected subset in the product space $S \times T$, then A is connected in S and B is connected in T.

B

16. *(a)* If x and y are in $\mathbb{R}^n$, the *line segment* $L(x, y)$ joining x and y consists of all points $x + t(y - x)$ for $0 \leqslant t \leqslant 1$. Show that $L(x, y)$ is connected by showing that the function f defined from $[0, 1]$ into $\mathbb{R}^n$ by $f(t) = x + t(y - x)$ is continuous.

(b) A subset C of $\mathbb{R}^n$ is called *convex* if whenever x and y are in C, then $L(x, y) \subset C$. Show that convex subsets of $\mathbb{R}^n$ are connected.

(c) Define line segments and convex sets in l^1 and l^p and show that they are connected.

17. Denote the set $\{0, 1\}$ by 2.

(a) If S is any set, show there is a one-to-one function from $\mathscr{P}(S)$, the collection of all subsets of S, onto 2^S given by the correspondence from $A \subset S$ to the characteristic function χ_A of A defined by $\chi_A(x) = 1$ if $x \in A$ and $\chi_A(x) = 0$ if x is not in A.

(b) If 2 is given the topology $\mathscr{T}_{\max}$, show that 2^A is a compact, Hausdorff space for every set A.

(c) If $A = \{a, b, c\}$, list all of the members of 2^A and the corresponding subsets of A.

18. Prove the product topology is the weakest topology on $\prod_\alpha S_\alpha$ for which each projection P_α is continuous from $\prod$ onto S_α. Show that a function g from a topological space R into $\prod$ is continuous if and only if $P_\alpha \circ g$ is continuous from R into S_α for all α.

19. Suppose R_α is a subspace of S_α for all α. Show that the mapping T from $\prod_\alpha R_\alpha$ into $\prod_\alpha S_\alpha$ given by $(Tf)(\alpha) = f(\alpha)$ is continuous and is a homeomorphism from $\prod_\alpha R_\alpha$ with the product topology onto $\prod_\alpha R_\alpha$ taken as a subspace of $\prod_\alpha S_\alpha$.

20. Prove that a countable product of separable topological spaces is separable. {This is also true for products with cardinality c.}

21. If A_α is a subset of S_α for each α, show that $\overline{\prod_\alpha A_\alpha}$ is equal to $\prod_\alpha \bar{A}_\alpha$.

22. Suppose $\{M_n\}$ is any sequence of positive numbers and $\{a_n\}$ is a sequence such that $\sum_n |a_n M_n| < \infty$. Prove that the set of all numbers of the form $\sum_n a_n t_n$ where $-M_n \leqslant t_n \leqslant M_n$ is precisely the interval

$$\left[-\sum_n |a_n M_n|, \sum_n |a_n M_n|\right].$$

23. Prove that the set C of all numbers of the form $\sum_n c_n/3^n$ where c_n is either 0 or 2 forms a compact subset of $[0, 1]$. C is called the *Cantor Set*. Prove that C is homeomorphic to 2^N where 2 is given discrete topology.

24. A topological space S is called sequentially compact if every sequence in S contains a convergent subsequence. Show that if S_n is sequentially compact for each n, then $\prod_n S_n$ is sequentially compact. {Let (f_k) be a sequence in $\prod$. For each i, let $(f_{k(i)})$ be a subsequence of $(f_{k(i-1)})$ that converges. Consider the diagonal sequence $(f_{k(k)})$.} Show, without using any form of the Axiom of Choice, that if S_n is a compact metric space for each n, then $\prod_n S_n$ is also a compact metric space.

25. Without using any form of the Axiom of Choice show that if S is compact so is $S \times [0, 1]$. {Let $\{O_\alpha\}$ be an open covering of $S \times [0, 1]$. Let b be the least upper bound of all $t \leqslant 2$ such that a finite subcollection of $\{O_\alpha\}$ covers $S \times [0, t]$. Show $b > 1$.}

26. Suppose U_α is a continuous function from R_α into S_α for all α.

(a) Show that the function T defined from $\prod_\alpha R_\alpha$ into $\prod_\alpha S_\alpha$ by $(Tf)(\alpha) = R_\alpha(f(\alpha))$ is continuous.

(b) Prove that if R_α is homeomorphic to S_α for each α, then $\prod_\alpha R_\alpha$ is homeomorphic to $\prod_\alpha S_\alpha$.

27. For any topological space S show $S \times S^{\mathbb{N}}$ is homeomorphic to $S^{\mathbb{N}}$.

28. Prove that if each of the metric spaces S_i in Theorem 10.8 is complete, then $\prod_{i=1}^{\infty} S_i$ is complete.

29. Prove Theorem 10.7 (b) and (c).

30. *(a)* Suppose $\prod = \prod_\alpha S_\alpha$ is a product of topological spaces. Prove that a sequence (f_n) in $\prod$ converges to f if and only if $(f_n(\alpha))$ converges to $f(\alpha)$ for each α.

(b) State and prove a special case of (a) for the space ω of all sequences.

C

31. Let $X = [0, 1]^{[0, 1]}$, that is, the collection of all functions from $[0, 1]$ into $[0, 1]$.

(a) Show X is a compact topological space with the product topology (easy).

(b) Let D consist of all f in X such that $f(x)$ is rational and $f(x) = 1$ for all except finitely many x. Show D is countable and dense in X.

(c) Prove that there is no metric on X that determines the product topology.

3

Measure and Integration

"Wir müssen wissen
Wir werden wissen." tombstone of David Hilbert

The problem of representing lengths, areas, and volumes by a number dates back to prehistoric times. The first notion was that a measure such as the area of a plane figure was a number that clearly existed and must be found. In the nineteenth and twentieth centuries, it was determined that it is also necessary to define area and other concepts of measure precisely, and to choose those objects that can be measured. Henri Lebesgue originated the modern theory of measure, and it was developed by Caratheodory, Hahn, Radon, Saks, and others to its present state.

Measure theory begins with the definition of an abstract space of objects that can be measured (a "measure space") much the same way as topology. The axiomatic definition of measure space is in Section 11 along with the main example, namely length (also called Lebesgue measure) on the real line $\mathbb{R}$. The concept of "measurable set" is crucial in this context. In Section 11 we verify that the real line allows a rich collection of measurable sets, and we conclude with an example of a nonmeasurable subset of $\mathbb{R}$.

Section 12 contains a theory of integration of functions. The theory proceeds from "simple" functions that take on only a finite number of values, to bounded functions with support of finite measure, and finally to functions that can be unbounded and have support of infinite measure. The important properties of linearity are verified for the various integrals. The section concludes with a discussion of the powerful convergence theory possible in the setting of Lebesgue integration.

In Section 13 we prove the Caratheodory Extension Theorem, which we use to develop a theory of product measures and Lebesgue-Stieltjes integration. The most efficient way to extend Lebesgue measure and integration from $\mathbb{R}$ to $\mathbb{R}^n$ is by means of the general theory of product measures and integrals. The Theorems of Fubini and Tonelli provide a way to reduce product integrals to iterated integrals of lower dimension.

Section 14 discusses "signed" measures that can assume negative as well as positive values. The main results are the theorems of Hahn and of Lebesgue, which show how signed measures can be decomposed into positive and negative parts. The Radon-Nikodym Theorem in Section 15 gives conditions under which a signed measure arises from integrating a function. There is also a short discussion of complex valued measures.

A discussion of differentiation and its relation to integration appears in Section 16. Using the Vitali Covering Theorem we show that an increasing function is differentiable almost everywhere. The main result is the Fundamental Theorem of Calculus for Lebesgue Integration, which asserts that a function f is the integral of its derivative if and only if f is absolutely continuous.

11

Measure Theory

The modern theory of measure and integration presented in this chapter was constructed in the late nineteenth century by Henri Lebesgue (1875–1941) and published in two monographs dated 1901 and 1902. Our approach to Lebesgue's theory follows the elegant treatment given by C. Caratheodory (1873–1950) in a textbook published in 1918.

A collection $\mathscr{A}$ of subsets of a set S is called an *algebra* (on S) if $\mathscr{A}$ satisfies the following conditions:

($\mathscr{A}$1) $\mathscr{A}$ contains S and $\varnothing$.

($\mathscr{A}$2) If $A \in \mathscr{A}$, then so is $A^{\sim}$, the complement of A in S.

($\mathscr{A}$3′) If $A_1, A_2, \ldots, A_n$ are in $\mathscr{A}$ so is the union $\bigcup_{i=1}^{n} A_i$.

Briefly stated, an algebra is a collection of subsets of a set S, which contains S and is closed under complement and finite union. The members of $\mathscr{A}$ are called *$\mathscr{A}$-measurable sets* or simply *measurable sets*. $\mathscr{A}$ is called a *σ-algebra*, "*sigma-algebra,*" if it is closed under countable unions, that is,

($\mathscr{A}$3) If $A_i \in \mathscr{A}$ for $i = 1, 2, \ldots$, then $\bigcup_{i=1}^{\infty} A_i \in \mathscr{A}$.

Using the DeMorgan laws we observe that ($\mathscr{A}$3′) is equivalent to the condition that $\mathscr{A}$ is closed under finite intersections and that ($\mathscr{A}$3) is equivalent to the condition that $\mathscr{A}$ is closed under countable intersections.

A function μ from an algebra $\mathscr{A}$ into the nonnegative extended real numbers is called a *finitely additive set function* (fasf) if μ satisfies the following two conditions:

(FA1) $\mu(\varnothing) = 0$.

(FA2) $\mu(A \cup B) = \mu(A) + \mu(B)$; $A, B \in \mathscr{A}$; and $A \cap B = \varnothing$.

A fasf on an algebra $\mathscr{A}$ is called a *measure* if it is *countably additive*, that is, if for every countable subcollection $\{A_1, A_2, \ldots\}$ of pairwise disjoint subsets of $\mathscr{A}$, we have

(CA) $\mu(\bigcup_{i=1}^{\infty} A_i) = \sum_{i=1}^{\infty} \mu(A_i)$.

provided that $\bigcup_{i=1}^{\infty} A_i$ is in $\mathscr{A}$. If $\mathscr{A}$ is a σ-algebra on a set S and μ is a measure on $\mathscr{A}$, the triple $(S, \mathscr{A}, \mu)$ is called a *measure space.* The members of $\mathscr{A}$ are called *$\mathscr{A}$-measurable* or *μ-measurable* sets. A measure space $(S, \mathscr{A}, \mu)$ is called *complete* if whenever $B \in \mathscr{A}$ and $\mu(B) = 0$ then $A \in \mathscr{A}$ for all $A \subset B$. We say that S is *σ-finite*, or μ is *σ-finite* if there is a countable subcollection $\{A_1, A_2, \ldots\}$ of $\mathscr{A}$ such that $\mu(A_n) < \infty$ for each n and $S = \bigcup_n A_n$.

A function μ^* from the collection of all subsets of a set S into the set $\mathbb{R}^*$ of extended reals is called an *outer measure* if μ^* satisfies the following three conditions.

(OM1) $\mu^*(\varnothing) = 0$.
(OM2) If $A \subset B$, then $\mu^*(A) \leqslant \mu^*(B)$.
(OM3) If $A = \bigcup_{i=1}^{\infty} A_i$, then $\mu^*(A) \leqslant \sum_{i=1}^{\infty} \mu^*(A_i)$.

Condition (OM2) is expressed by saying that μ^* is *monotone.* Condition (OM3) is expressed by saying that μ^* is *subadditive.*

Example 11.1:

This is the first step in the main example, which is Lebesgue measure on $\mathbb{R}$. For every open interval $I = (a, b)$, we denote by $l(I)$, the length of I, which is $b - a$. For $B \subset \mathbb{R}$ we define $m^*(B)$ by the formula

$$m^*(B) = \inf\left\{ \sum_{i=1}^{\infty} l(a_i, b_i) \colon B \subset \bigcup_i (a_i, b_i) \right\}.$$

We leave you the easy task of verifying that m^* satisfies (OM1) and (OM2). A small difficulty arises with (OM3) because m^* of the union might be infinite. We handle this in the following lemma.

Lemma 11.2: If $\{A_i\}$ is a sequence of subsets of $\mathbb{R}$ then

$$m^*\left(\bigcup_i A_i \right) \leqslant \sum_i m^*(A_i).$$

Proof: If for some i, $m^*(A_i) = \infty$, then the term on the right is ∞ so the lemma holds. Thus we are left with the case $m^*(A_i) < \infty$ for each i. In this case let $\varepsilon > 0$ be given, and for each i let $\{(a_{ij}, b_{ij})\}$ be a sequence of intervals such that

$$A_i \subset \bigcup_j (a_{ij}, b_{ij}) \quad \text{and} \quad \sum_j |b_{ij} - a_{ij}| < m^*(A_i) + \frac{\varepsilon}{2^i}.$$

Then we have $\bigcup_i A_i \subset \bigcup_i \bigcup_j (a_{ij}, b_{ij})$, and

$$\sum_i \sum_j |b_{ij} - a_{ij}| < \varepsilon + \sum_i m^*(A_i). \ \blacksquare$$

It is necessary that $m^*(I)$ have the correct value for all intervals I of the form (a, b), $(a, b]$, $[a, b)$, and $[a, b]$ namely, $b - a$ if I is finite and ∞ otherwise. This is verified in the following lemma.

Lemma 11.3: The outer measure of any finite interval (a, b), $(a, b]$, $[a, b)$, and $[a, b]$ is $b-a$. The outer measure of any infinite interval is equal to ∞.

Proof: We first show that the outer measure of the open interval (a, b) is equal to that of the closed interval $[a, b]$. Let $\varepsilon > 0$ be arbitrary. Suppose $(a, b) \subset \bigcup_i (c_i, d_i)$ where $\sum_i (d_i - c_i) < m^*((a, b)) + \varepsilon$. Then $[a, b]$ is contained in the union of $\bigcup_i (c_i, d_i)$ and $(a-\varepsilon, a+\varepsilon) \cup (b-\varepsilon, b+\varepsilon)$, which has length at most $m^*((a, b)) + 5\varepsilon$. Therefore, we conclude $m^*([a, b]) < m^*((a, b)) + 5\varepsilon$. Since ε is arbitrary it follows that $m^*(([a, b]) \leqslant m^*(a, b))$. The reverse inequality is valid because $[a, b]$ includes (a, b). Since $(a, b]$ and $[a, b)$ lie between (a, b) and $[a, b]$ they have the same outer measure.

Since by definition we know that the measure of (a, b) is at most $b-a$, we need to prove that $m^*([a, b]) \geqslant b-a$. We do this by showing that $m^*([a, b]) \geqslant b-a-\varepsilon$ for all $\varepsilon > 0$. Suppose $\{(a_i, b_i)\}$ is a sequence of open intervals that covers $[a, b]$ and $\sum_i (b_i - a_i) < m^*([a, b]) + \varepsilon$ where $\varepsilon > 0$ is arbitrary. By Theorem 4.5 on page 19, $[a, b]$ is compact so there is a finite subcover $\{(a_1, b_1), (a_2, b_2), \ldots, (a_n, b_n)\}$ of $[a, b]$. We rename these intervals as follows: Let (c_1, d_1) be any of the intervals (a_i, b_i) that contain a. If $d_1 > b$ we are finished since $m^*([a, b]) > d_1 - c_1 - \varepsilon > b - a - \varepsilon$. Otherwise, let (c_2, d_2) be any of the intervals that contain d_1. We proceed in this way until we obtain intervals (c_1, d_1), (c_2, d_2), $\ldots$, (c_k, d_k) where $d_k > b$ and $c_i < d_{i-1} < d_i$ for $i \leqslant k$. Then we shall have

$$m^*([a, b]) + \varepsilon > \sum_{i=1}^{k} (d_i - c_i) \geqslant d_1 - c_1 + \sum_{i=2}^{k} (d_i - d_{i-1}) = d_k - c_1 > b - a. \blacksquare$$

Let μ^* be an outer measure on a set S. A subset A of S is called *μ^*-measurable* if for every subset B of S we have

$$\text{(MEAS)} \quad \mu^*(B) = \mu^*(B \cap A) + \mu^*(B \cap A^\sim).$$

Since $\mu^*(B) \leqslant \mu^*(B \cap A) + \mu^*(B \cap A^\sim)$ for all B by (OM3), then in order to prove that a set A is μ^*-measurable we need only prove the reverse inequality

$$\mu^*(B) \geqslant \mu^*(B \cap A) + \mu^*(B \cap A^\sim)$$

for all subsets B of S.

Theorem 11.4: Let μ^* be an outer measure on a set S, let $\mathscr{A}$ be the collection of all subsets of S that are μ^*-measurable, and let μ denote the restriction of μ^* to $\mathscr{A}$, that is, $\mu(A) = \mu^*(A)$ for all A in $\mathscr{A}$. Then $(S, \mathscr{A}, \mu)$ is a complete measure space.

Proof: That the collection of $\mathscr{A}$ satisfies ($\mathscr{A}$1) follows quickly from property (OM1). That $\mathscr{A}$ satisfies ($\mathscr{A}$2) is a consequence of the symmetry of A and $A^\sim$ in the definition of μ^*-measurable.

Next we show that $\mathscr{A}$ satisfies ($\mathscr{A}3'$) and thus forms an algebra. Suppose A_1 and A_2 are in $\mathscr{A}$ and B is any subset of S. Since A_1 is μ^*-measurable we have

$$\begin{aligned} \mu^*((A_1 \cup A_2) \cap B) &= \mu^*(A_1 \cap \{(A_1 \cup A_2) \cap B)\} + \mu^*(A_1^\sim \cap \{(A_1 \cup A_2) \cap B\}) \\ &= \mu^*(A_1 \cap B) + \mu^*(A_1^\sim \cap A_2 \cap B). \end{aligned}$$

Therefore, we conclude

$$\mu^*((A_1 \cup A_2) \cap B) + \mu^*((A_1 \cup A_2)^\sim \cap B)$$
$$= \mu^*(A_1 \cap B) + \mu^*(A_2 \cap A_1^\sim \cap B) + \mu^*(A_2^\sim \cap A_1^\sim \cap B)$$
$$= \mu^*(A_1 \cap B) + \mu^*(A_1^\sim \cap B) = \mu^*(B).$$

Consequently, $\mathscr{A}$ is closed under finite unions. If A_1 and A_2 are disjoint sets in $\mathscr{A}$ and $B \subset S$, then we have

$$\mu^*(\{A_1 \cup A_2\} \cap B) = \mu^*(A_1 \cap \{A_1 \cup A_2\} \cap B) + \mu^*(A_1^\sim \cap \{A_1 \cup A_2\} \cap B)$$
$$= \mu^*(A_1 \cap B) + \mu^*(A_2 \cap B).$$

If $\{A_1, \ldots, A_n\}$ are pairwise disjoint, we can likewise show that

$$\mu^*\left(\left\{\bigcup_{j=1}^{n} A_j\right\} \cap B\right) = \sum_{j=1}^{n} \mu^*(A_j \cap B).$$

We now use the fact that $\mathscr{A}$ forms an algebra to help us show it also forms a σ-algebra. Suppose $\{A_n: n = 1, 2, \ldots\}$ is a sequence in $\mathscr{A}$ and $B \subset S$. For each n let $A_n' = A_n \backslash \{\bigcup_{j=1}^{n-1} A_j\}$ so that $\{A_n'\}$ is a sequence of disjoint sets and $\bigcup_{n=1}^{k} A_n' = \bigcup_{n=1}^{k} A_n$ for all k. Since each A_n is in $\mathscr{A}$ so is each A_n'. For each k, the set $\bigcup_{n=1}^{k} A_n$ is in $\mathscr{A}$ so we have

$$\mu^*(B) = \mu^*\left(\left\{\bigcup_{n=1}^{k} A_n\right\} \cap B\right) + \mu^*\left(\left\{\bigcup_{n=1}^{k} A_n\right\}^\sim \cap B\right)$$
$$= \mu^*\left(\left\{\bigcup_{n=1}^{k} A_n\right\} \cap B\right) + \mu^*\left(\left\{\bigcap_{n=1}^{k} A_n^\sim\right\} \cap B\right)$$
$$\geqslant \sum_{n=1}^{k} \mu^*(A_n' \cap B) + \mu^*\left(\left\{\bigcap_{n=1}^{\infty} A_n^\sim\right\} \cap B\right). \qquad \text{(OM2)}$$

Since the inequality above is true for all k we conclude

$$\mu^*(B) \geqslant \sum_{n=1}^{\infty} \mu^*(A_n' \cap B) + \mu^*\left(\left\{\bigcap_{n=1}^{\infty} A_n^\sim\right\} \cap B\right) \qquad \textbf{(11–1)}$$

By (OM3) we have

$$\mu^*\left(\left\{\bigcup_{n=1}^{\infty} A_n\right\} \cap B\right) = \mu^*\left(\left\{\bigcup_{n=1}^{\infty} A_n'\right\} \cap B\right) \leqslant \sum_{n=1}^{\infty} \mu^*(A_n' \cap B)$$

we conclude

$$\mu^*(B) \geqslant \mu^*\left(\left\{\bigcup_{n=1}^{\infty} A_n\right\} \cap B\right) + \mu^*\left(\left\{\bigcup_{n=1}^{\infty} A_n\right\}^\sim \cap B\right),$$

which proves that $\bigcup_n A_n \in \mathscr{A}$.

If $\{A_n: n = 1, 2, \ldots\}$ is a sequence of pairwise disjoint sets in $\mathscr{A}$, then $\mu^*(\bigcup_n A_n) \leqslant \sum_n \mu^*(A_n)$ by (OM3). By (11–1) with $B = \bigcup_n A_n$ and $A_n' = A_n$ for each n we obtain the reverse inequality. Therefore, μ^* restricted to $\mathscr{A}$ is a measure. This completes the proof that $(S, \mathscr{A}, \mu)$ is a measure space.

We conclude by showing that $(S, \mathscr{A}, \mu)$ is complete. Suppose $\mu^*(C)=0$, and $A \subset C$. If B is any subset of S we have $\mu^*(A \cap B)=0$ by (OM2). Again by (OM2) we have $\mu^*(B \cap A)+\mu^*(B \cap A^\sim)=\mu^*(B \cap A^\sim) \leqslant \mu^*(B)$. This implies that A is μ^*-measurable, and we have already noted why $\mu(A)=0$. This proves that $(S, \mathscr{A}, \mu)$ is a complete measure space. ■

The implication of Theorem 11.4 is that if you judiciously choose the outer measure μ^* you will obtain a useful measure space. On the other hand, there is no guarantee that $\mathscr{A}$ will contain any sets other than $\varnothing$ and S. The following theorem shows that the outer measure m^* defined in 11.1 has been well chosen.

Theorem 11.5: Every interval is measurable with respect to the measure m^* on $\mathbb{R}$. Therefore, the σ-algebra $\mathscr{M}$ of m^*-measurable sets contains all open and closed sets.

Proof: Let I denote the interval (a, ∞), and let B be an arbitrary subset of $\mathbb{R}$. We shall verify that $m^*(B) \geqslant m^*(B \cap I)+m^*(B \cap I^\sim)$. If $m^*(B)=\infty$ the conclusion is obvious. Otherwise, let $\varepsilon>0$ be arbitrary, and let $\{(a_n, b_n)\colon n=1, 2, \ldots\}$ be a sequence of intervals that covers B for which we have

$$\sum_n (b_n-a_n)<m^*(B)+\frac{\varepsilon}{2}.$$

For each n, let

$$J_n=(a_n, b_n) \cap(-\infty, a+2^{-(n+1)}\varepsilon), \quad \text{and} \quad \text{let } K_n=(a_n, b_n) \cap(a, \infty).$$

Then each J_n and K_n is either empty or an interval. The intervals $\{J_n\colon J_n \neq \varnothing\}$ cover $B \cap I^\sim$, and the intervals $\{K_n\colon K_n \neq \varnothing\}$ cover $B \cap I$. Therefore, we have

$$\begin{aligned} m^*(B \cap I)+m^*(B \cap I^\sim) &\leqslant \sum_n l(J_n)+l(K_n) \leqslant \frac{\varepsilon}{2}+\sum_n (b_n-a_n) \\ &\leqslant m^*(B)+\varepsilon. \end{aligned}$$

Since the inequality holds for arbitrary $\varepsilon>0$ it follows that

$$m^*(B \cap I)+m^*(B \cap I^\sim) \leqslant m^*(B).$$

A similar argument shows that intervals of the form $(-\infty, a)$ are also measurable. Since every interval of the form (a, b) is the intersection of such intervals it follows from Theorem 11.4 that every open interval is measurable. From Theorem 11.4 and Theorem 4.8 on page 21 it follows that every open set is measurable. Applying Theorem 11.4 again we conclude that closed sets are measurable as well. ■

If $\mathscr{M}$ denotes the collection of all m^*-measurable subsets of $\mathbb{R}$ and m denotes the restriction of m^* to $\mathscr{M}$, then $(R, \mathscr{M}, m)$ is a complete measure space. Each member of $\mathscr{M}$ is called a *Lebesgue measurable subset* of $\mathbb{R}$ and m is called (one-dimensional) *Lebesgue measure.* We shall sometimes write "measurable" instead of "Lebesgue measurable" if the context implies we are discussing subsets of $\mathbb{R}$. The smallest σ-algebra $\mathscr{B}$ of $\mathbb{R}$ that contains all open subsets of $\mathbb{R}$ is called the collection of *Borel sets.* From Theorem 11.5 we see that $\mathscr{M}$ contains $\mathscr{B}$. However, it is not true that every Lebesgue measurable set is a Borel set. In fact, many sets of measure 0 are not Borel sets. Thus $(\mathbb{R}, \mathscr{B}, m)$ is an example of a measure space that is not complete. In the future, *whenever we say measure space we shall mean complete measure space unless we specifically state otherwise.*

Example 11.6:

A nonmeasurable set For x and y in $[0, 1)$ let $x \oplus y = x + y$ if $x + y < 1$ and $x + y - 1$ otherwise. The operation $\oplus$ is called addition *modulo one*. For any subset A of $[0, 1)$ and t in $[0, 1)$ we define $A \oplus t$ to be the set of all $s \oplus t$ as s ranges over A. Thus we have

$$A \oplus t = (A_1 + t) \cup (A_2 + t - 1)$$

where

$$A_1 = A \cap [0, 1 - t) \quad \text{and} \quad A_2 = A \cap [1 - t, 1).$$

If A is measurable, then so are A_1 and A_2 and $m(A) = m(A_1) + m(A_2)$ since A_1 and A_2 are disjoint. Since $m(A_1 + t) = m(A_1)$, $m(A_2 + t - 1) = m(A_2)$ while $A_1 + t$ is contained in $[t, 1)$ and $A_2 + t - 1$ is contained in $[0, t)$, it follows that $m(A \oplus t) = m(A_1) + m(A_2) = m(A)$.

For each t in $[0, 1)$ let $Z(t)$ consist of all x in $[0, 1)$ such that $x - t$ is in the set Q of rational numbers. Thus $Z(t) = Z(s)$ if and only if $s \in Z(t)$. Using the Axiom of Choice we can determine a set W that contains one member of each $Z(t)$. If W were measurable, then for each rational number r the set $W \oplus r$ would be measurable and we would have $m(W \oplus r) = m(W)$. If w is in $Z(t) \cap W$ and $z \neq w$ is in $Z(t)$ then $z - w \in Q$. If $z - w > 0$ then let $r = z - w$; otherwise, let $r = 1 + z - w$. In both cases $0 < r < 1$ and $z = w \oplus r$. This proves that $\bigcup_{n=1}^{\infty} W \oplus r_n$ contains each $Z(t)$ where $\{r_1, r_2, \ldots\}$ is any enumeration of $Q \cap [0, 1)$. Since the sets $W \oplus r_n$ are disjoint, it follows that if W were measurable we would either have $m(W) > 0$ from which we must conclude

$$m([0, 1)) = \sum_n m(W \oplus r_n) = \infty$$

or else we would have $m(W) = 0$ from which we must conclude

$$m([0, 1)) = \sum_n m(W \oplus r_n) = 0.$$

Therefore, the set W must not be measurable.

Example 11.7:

Application to probability theory A measure space $(S, \mathscr{A}, \mu)$ such that $\mu(S) = 1$ is sometimes called a *probability space*. The members of $\mathscr{A}$ represent events. For each $A \in \mathscr{A}$ the number $\mu(A)$ represents the probability of the event A. If A and B are in $\mathscr{A}$, the event (A and B) is represented by $A \cap B$ and the event (A or B) by $A \cup B$. If B is in $\mathscr{A}$ and $\mu(B) > 0$, then we can define a measure space $(B, \mathscr{B}, \nu)$ where $\mathscr{B} = \{A \cap B\colon A \in \mathscr{A}\}$ and $\nu(C) = \mu(C)/\mu(A)$ for each C in $\mathscr{B}$. If $C = A \cap B$, then $\nu(C)$ is defined as the *probability of A given B*, and the members in $\mathscr{B}$ represent events given that the event B has occurred.

Exercises

A

1. Prove that a collection $\mathscr{A}$ of subsets of S is an algebra if and only if it satisfies ($\mathscr{A}$1), ($\mathscr{A}$2), and the following condition:

   ($\mathscr{A}i'$) if $A_1, A_2, \ldots, A_n$ are in $\mathscr{A}$, so is $\bigcap_{i=1}^{n} A_i$.

   Prove $\mathscr{A}$ is a σ-algebra if and only if it satisfies ($\mathscr{A}$1), ($\mathscr{A}$2), and the following condition:

   ($\mathscr{A}i$) if $A_1, A_2, \ldots$ are in $\mathscr{A}$, so is $\bigcap_i A_i$.

2. Let $\mathscr{A}$ consist of all subsets A of $\mathbb{N} = \{1, 2, \ldots\}$ that are finite (including $\varnothing$) or that have finite complement. Show that $\mathscr{A}$ is an algebra that is not a σ-algebra. Define μ on $\mathscr{A}$ by $\mu(A) = 0$ if A is finite, $\mu(A) = 1$ if A has finite complement. Show that A is a fasf on $\mathscr{A}$ but is not countably additive.
3. Suppose μ is a measure defined on the σ-algebra $\mathscr{B}$ of all subsets of $\mathbb{N}$. Show there are nonnegative real numbers $\{a_i: i = 1, 2, \ldots\}$ such that for each subset M of $\mathbb{N}$, $\mu(M) = \sum \{a_i: i \in M\}$.
4. Show that if A is a countable subset of $\mathbb{R}$, then there is a sequence $\{(a_n, b_n): n = 1, 2, \ldots\}$ whose union contains A and has total length less than ε for each $\varepsilon > 0$. Conclude that $m^*(A) = 0$.
5. Using Exercise 4, give a new proof that $\mathbb{R}$ is uncountable.
6. If A is the set of rational numbers between 0 and 1 and $\{(a_i, b_i): i = 1, 2, \ldots, n\}$ is a finite collection of open intervals that covers A, show that $\sum_{i=1}^{n} (b_i - a_i) \geqslant 1$.
7. Show that if $m^*(B) = 0$, then $m^*(A \cup B) = m^*(A)$ for all $A \subset \mathbb{R}$.
8. For $A \subset \mathbb{R}$ and $s \in \mathbb{R}$ let $A + s$ denote the set $\{a + s: a \in A\}$. Prove that $m^*(A + s) = m^*(A)$ for all A, s.
9. If $\{A_n: n = 1, 2, \ldots\}$ is a sequence of μ-measurable sets, show that
   (a) $\mu(\bigcup_{n=1}^{\infty} A_n) = \lim_k \mu(\bigcup_{n=1}^{k} A_n)$.
   (b) $\mu(\bigcap_{n=1}^{\infty} A_n) = \lim_k (\bigcap_{n=1}^{k} A_n)$ provided $\mu(A_n) < \infty$ for some n.
   (c) If $A_n = (n, \infty)$ for $n = 1, 2, \ldots$ show $m(\bigcap_{n=1}^{k} A_n) = \infty$ for all k but $\bigcap_{n=1}^{\infty} A_n = \varnothing$.
10. If A and B are μ-measurable, show $\mu(A \cap B) + \mu(A \cup B) = \mu(A) + \mu(B)$.
11. Let $S = \{1, 2, 3, 4, 5\}$. Suppose the event $\{2\}$ is twice as likely as the event $\{1\}$, $\{3\}$ twice as likely as $\{2\}$, $\{4\}$ twice as likely as $\{3\}$, and $\{5\}$ twice as likely as $\{4\}$. Determine a measure μ on the collection 2^S of all subsets of S so that $(S, 2^S, \mu)$ is a measure space.

B

12. (a) Prove that a subset A of $\mathbb{R}$ is measurable if and only if for all $\varepsilon > 0$ there is an open set O such that $A \subset O$ and $m^*(O \backslash A) < \varepsilon$.
    (b) A subset G of $\mathbb{R}$ is called a G_δ if it is equal to a countable intersection of open sets. Show that A is measurable if and only if there is a G_δ, G containing A with $m^*(G \backslash A) = 0$.

(c) Prove a subset A of $\mathbb{R}$ is measurable if and only if for all $\varepsilon > 0$ there is a closed set C such that $C \subset A$ and $m^*(A \backslash C) < \varepsilon$.

(d) A subset F of $\mathbb{R}$ is called an F_σ if it equals a countable union of closed sets. Prove A is measurable if and only if there is an F_σ, F with $F \subset A$ and $m^*(A \backslash F) = 0$.

13. Using Exercise 4 prove that for all $\varepsilon > 0$ there is an open subset O of $\mathbb{R}$ such that (1) $m(O) < \varepsilon$ and (2) O is dense in $\mathbb{R}$.

14. (*The Cantor Set*) For any closed interval $I = [a, b]$ and $0 < t < 1$ let $L_t(I)$ be the set $I \backslash (t_1, t_2)$ where

$$t_1 = \frac{\{(1+t)a + (1-t)b\}}{2} \quad \text{and} \quad t_2 = \frac{\{(1-t)a + (1+t)b\}}{2}.$$

Thus $L_t(I)$ is I less the middle open interval whose proportion to $b - a$ is t. For any union $A = \bigcup_\alpha I_\alpha$ of intervals let $L_t(A) = \bigcup_\alpha L_t(I_\alpha)$. Since $L_t(A)$ is then a union of disjoint closed intervals it is then possible to define $L_t^2(A) = L_t^2(L_t(A))$, and in general $L_t^n(A) = L_t(L_t^{(n-1)}(A))$. The set $C = \bigcap_{n=1}^{\infty} L_{1/3}^n([0, 1])$ is called the Cantor set. Establish the following facts about C.

(a) C is a compact subset of $[0, 1]$.

(b) A number x is in C if and only if it has a nonterminating ternary expansion $\sum_{n=1}^{\infty} \varepsilon_n/3^n$ where each ε_n is 0 or 2.

(c) The correspondence $\sum_{n=1}^{\infty} \varepsilon_n/3^n$ to $\sum_{n=1}^{\infty} (\varepsilon_n/2)/2^n$ is one-to-one from C onto $[0, 1]$, provided we use only nonterminating expansions. Thus, for example we use $\frac{2}{9} + \frac{2}{27} + \frac{2}{81} \ldots$ instead of $\frac{1}{3}$. Therefore, C is uncountable.

(d) For each n, $m(L_{1/3}^n([0, 1])) = (\frac{2}{3})^n$ so that $m(C) = 0$. This is an example of an uncountable set having 0 measure.

(e) Show that $\bigcap_{n=1}^{\infty} L_t^n(I)$ has measure 0 for all I and all $0 < t$.

(f) Let $A_0 = [0, 1]$, $A_{n+1} = L_{c_n}(A_n)$ where $c_n = (1 - t^n)/(1 - t^{n+1})$. The measure of $\bigcap_{n=1}^{\infty} A_n$ is $(1 - t)$.

15. *(a)* Show that for each $\varepsilon > 0$ and each $t \in [0, 1)$, $Z(t) \cap [0, \varepsilon) \neq \varnothing$ where $Z(t)$ is the set in Example 11.6.

(b) Show that for each $\varepsilon > 0$ the set W in Example 11.6 can be constructed in such a way that $m^*(W) < \varepsilon$.

(c) Can W be constructed in such a way that $m^*(W) = 0$ or 1?

16. Construct a subset D of $[0, 1]$ such that

(a) D is dense in $[0, 1]$;

(b) For any interval (a, b) contained in $[0, 1]$, $m((a, b) \cap D) > 0$;

(c) $m(D) < 1$;

(d) D is open.

17. Suppose $(S, \mathscr{A}, \mu)$ is a measure space that is not necessarily complete. Let $\mathscr{B}$ consist of all subsets of S of the form $A \cup B$ where $A \in \mathscr{A}$ and B is a subset of a set C in $\mathscr{A}$ for which $\mu(C) = 0$.

(a) Show that $\mathscr{B}$ is a σ-algebra.

(b) Show that the function $\bar{\mu}$ defined on $\mathscr{B}$ by $\bar{\mu}(A \cup B) = \mu(A)$ is unambiguously defined and is a measure on $\mathscr{B}$ that coincides with μ on $\mathscr{A}$.

(c) Prove that $(S, \mathscr{B}, \bar{\mu})$ is a complete measure space. The measure space $(S, \mathscr{B}, \bar{\mu})$ is called the *completion* of $(S, \mathscr{A}, \bar{\mu})$.

12

Integration

Let $(S, \mathscr{A}, \mu)$ be a measure space. A function f from S into $\mathbb{R}$ is called *measurable* if for each open subset O of $\mathbb{R}$, $f^{-1}(O) \in \mathscr{A}$. A function f from S into the extended real numbers $\mathbb{R}^*$ is called measurable if $f^{-1}(+\infty)$ and $f^{-1}(-\infty)$ are in $\mathscr{A}$ and if $f^{-1}(O) \in \mathscr{A}$ for each open subset O of $\mathbb{R}$. If the measure space $(S, \mathscr{A}, \mu)$ is not complete, this definition must be adjusted. See Exercise 3 on page 98.

Lemma 12.1: Suppose $(S, \mathscr{A}, \mu)$ is a measure space. A function f from S into $\mathbb{R}^*$ is measurable if and only if for each $a \in \mathbb{R}$ the set

$$f^{-1}((a, \infty]) = \{s\colon s \in S,\ f(s) > a\}$$

is in $\mathscr{A}$.

Proof: If f is measurable, then $f^{-1}((a, \infty))$ is in $\mathscr{A}$ because the interval (a, ∞) is open. The set $f^{-1}(\infty)$ is in $\mathscr{A}$ by definition. Hence the set $f^{-1}((a, \infty]) = f^{-1}((a, \infty)) \cup f^{-1}(\infty)$ is in $\mathscr{A}$.

Suppose $f^{-1}((a, \infty])$ is in $\mathscr{A}$ for each a. Then $f^{-1}(\infty)$ is in $\mathscr{A}$ because it is the intersection $\bigcap_n f^{-1}((n, \infty])$ of a countable collection of measurable sets. Moreover, $f^{-1}((a, \infty))$ is in $\mathscr{A}$ for each a because

$$f^{-1}((a, \infty)) = \{f^{-1}(\infty)\}^\sim \cap f^{-1}((a, \infty]).$$

Also $f^{-1}(-\infty)$ is in $\mathscr{A}$ because it is the complement of the measurable set $\bigcup_{n=1}^{\infty} f^{-1}((-n, \infty])$.

Since $f^{-1}((a, \infty))$ is in $\mathscr{A}$ for each a it follows that for all b the set $f^{-1}((-\infty, b]) \in \mathscr{A}$ because $f^{-1}((-\infty, b]) = f^{-1}((b, \infty)^\sim) = \{f^{-1}((b, \infty))\}^\sim$ and $\mathscr{A}$ is closed under taking complements. Thus if $a < b$ we have

$$f^{-1}((a, b]) = f^{-1}((-\infty, b] \cap (a, \infty)) = f^{-1}((-\infty, b]) \cap f^{-1}((a, \infty))$$

so that $f^{-1}((a, b])$ is in $\mathscr{A}$ since $\mathscr{A}$ is closed under intersection. We use the fact that $\mathscr{A}$ is closed under countable union to verify that whenever $a < b$

$$f^{-1}((a, b)) = f^{-1}\left(\bigcup_n \left(a, b - \frac{1}{n}\right]\right) = \bigcup_n f^{-1}\left(\left(a, b - \frac{1}{n}\right]\right)$$

is also in $\mathscr{A}$. By Theorem 4.8 on page 21, every open subset O of $\mathbb{R}$ is the union of a countable union of open intervals. Thus if $O = \bigcup_n (a_n, b_n)$ it follows that $f^{-1}(O) = \bigcup_n f^{-1}((a_n, b_n)) \in \mathscr{A}$. ■

If $A \subset S$, the *characteristic function* χ_A of A is defined by the formula

$$\chi_A(s) = 1 \text{ if } s \in A, \qquad \chi_A(s) = 0 \text{ if } s \in A^\sim.$$

If A is a measurable subset of S, that is, $A \in \mathscr{A}$, then χ_A is measurable because $\chi_A^{-1}(a, \infty]$ is $\varnothing$ if $a \geqslant 1$, A if $0 \leqslant a < 1$, and S if $a < 0$. If A is not measurable, then χ_A is not measurable because $\chi_A^{-1}(0, \infty] = A$.

Theorem 12.2: Suppose $(S, \mathscr{A}, \mu)$ is a measure space and f, g are measurable functions from S into $\mathbb{R}^*$. Then the following functions are also measurable:

(a) af for all a in $\mathbb{R}$,
(b) $f+g$,
(c) fg,
(d) $\sup(f, g)$,
(e) $\inf(f, g)$,
(f) $|f|$,
(g) If the range of f is contained in $\mathbb{R}$ and if ϕ is any continuous function from $\mathbb{R}$ into $\mathbb{R}$ then $\phi \circ f$ defined by $\phi \circ f(x) = \phi(f(x))$ is measurable.

Proof:

(g) If O is an open subset of $\mathbb{R}$ then $\phi^{-1}(O)$ is open because ϕ is continuous (Theorem 7.4). Therefore, $f^{-1}(\phi^{-1}(O))$ is in $\mathscr{A}$. But $(\phi \circ f)^{-1}(O) = f^{-1}(\phi^{-1}(O))$ so $\phi \circ f$ is measurable.

(a) This follows from (g) because the function $x \to ax$ is continuous from $\mathbb{R}$ into $\mathbb{R}$.

(b) Since $\mathscr{A}$ is closed under intersection and countable union it suffices to show that for all a in $\mathbb{R}$ we have

$$\{f+g\}^{-1}((a, \infty]) = \bigcup_n f^{-1}((a-r_n, \infty]) \cap g^{-1}((r_n, \infty]) \qquad \textbf{(12–1)}$$

where $\{r_n : n = 1, 2, \dots\}$ is the set of rationals. The equality (12–1) is true because if $x \in \{f+g\}^{-1}((a, \infty])$, then $f(x)+g(x) > a$ so that $f(x) - a > -g(x)$. If r_n is any rational number such that $-g(x) < r_n < f(x) - a$ (Theorem 2.6), then x is in the set

$$H = f^{-1}(a-(-r_n), \infty] \cap g^{-1}(-r_n, \infty].$$

This shows that $\{f+g\}^{-1}((a, \infty))$ is contained in H. The inverse inclusion is obvious.

(c) If f is measurable then f^2 is measurable by (g). If f and g are measurable then fg is measurable by the previous consideration plus (a) and (b) because

$$fg = \tfrac{1}{2}\{(f+g)^2 - f^2 - g^2\}.$$

(f) Since the function $x \to |x|$ is continuous from $\mathbb{R}$ into $\mathbb{R}$ this assertion follows from (g).

(d) This assertion follows from (b), (f), and the clever formula

$$\sup(f, g) = \tfrac{1}{2}\{(f+g) + |f-g|\}.$$

(e) This assertion follows from (a) and the formula $\inf(f, g) = -\sup(-f, -g)$ or by a formula analogous to that found in (d). ■

Conditions (d), (e), and (f) of Theorem 12.2 assert that the space of measurable functions forms a linear lattice, that is a linear space of functions that is closed under taking maxima and minima of finite subsets. The following theorem asserts that the space of measurable functions is also a σ-lattice, that is, closed under taking suprema and infima of countable subsets. For $\{f_n: n=1, 2, \ldots\}$ a sequence of functions from a set S into $\mathbb{R}^*$ we define the functions $\sup_n f_n$ by

$$\left(\sup_n f_n\right)(x) = \sup_n (f_n(x)) \quad \text{for all } x \text{ in } S$$

and $\lim\sup_n f_n$ by

$$\left(\lim\sup_n f_n\right)(x) = \lim\sup_n f_n(x) \quad \text{for all } x \text{ in } S.$$

The functions $\inf_n f_n$ and $\lim\inf_n f_n$ are analogously defined.

Theorem 12.3: Suppose $(S, \mathscr{A}, \mu)$ is a measure space and $\{f_1, f_2, \ldots\}$ is a sequence of measurable functions from S into $\mathbb{R}$. Then the functions $\sup_n f_n$, $\inf_n f_n$, $\lim\sup_n f_n$, and $\lim\inf_n f_n$ are measurable.

Proof: For each a in $\mathbb{R}$ we have $(\sup_n f_n)^{-1}(a, \infty] = \bigcup_n f_n^{-1}(a, \infty]$. Thus, $(\sup_n f_n)^{-1}(a, \infty]$ is in $\mathscr{A}$. A similar argument is valid for $\inf_n f_n$.

Since $\lim\sup_n f_n$ is equal to $\inf_k \sup_{n>k} f_n$ it follows that $\lim\sup_n f_n$ is measurable. A similar argument applies to $\lim\inf_n f_n$. ■

Suppose $(S, \mathscr{A}, \mu)$ is a measure space and that f and g are two functions defined from S into the same set T. Then f and g are said to be equal *almost everywhere* (a.e.) if the set $\{x: f(x) \neq g(x)\}$ has measure 0. More generally, any property that is valid except on a set of measure 0 is said to hold almost everywhere.

A function f from a measure space $(S, \mathscr{A}, \mu)$ into $\mathbb{R}$ if called a *simple function* if it is measurable, takes on only finitely many values, and if $\mu(\{x: f(x) \neq 0\}) < \infty$. Thus the function f that assumes the values $\{a_1, a_2, \ldots, a_n\}$ has the form

$$f = \sum_{j=1}^{n} a_j \chi_j. \tag{12–2}$$

Here χ_j denotes the characteristic function of the set $f^{-1}(a_j)$; that is, $\chi_j(x) = 1$ if $f(x) = a_j$ and $\chi_j(x) = 0$ otherwise. The representation given by (12–2) is called the *canonical representation* of the simple function f.

Example 12.4:

The function f defined on $\mathbb{R}$ by

$$f(x) = 0 \quad \text{if } x < 0 \text{ or } x > 2,$$

$$f(x) = 1 \quad \text{for } 0 \leqslant x \leqslant 1 \quad \text{and}$$

$$f(x) = 2 \quad \text{for } 1 < x \leqslant 2$$

is a simple function having the canonical representation

$$f = \chi_{[0,1]} + 2\chi_{(1,2]} + 0\chi_{(-\infty,0)\cup(2,\infty)}.$$

The function f has many representations that are not canonical, for example, $f = \chi_{[0,2]} + \chi_{(1,2]}$. Note that the function $\chi_{(-\infty,0)\cup(2,\infty)}$ in the canonical representation does not affect the value of f since its coefficient is 0. However, it helps us to account for all of the points in $\mathbb{R}$ since the union of the subscripted sets is $\mathbb{R}$.

Let f be a simple function on a measure space $(S, \mathscr{A}, \mu)$ such that f has the form (12–2). The *integral* of f written $\int_S f\, d\mu$ or $\int f$ if S and μ are implicitly understood is defined by the formula

$$\int_S f\, d\mu = \sum_j a_j \mu(f^{-1}(a_j)). \qquad \textbf{(12–3)}$$

For example, the integral of the function f in Example 12.4 with respect to Lebesgue measure m is

$$\int_R \chi_{[0,1]} + 2\chi_{(1,2]}\, dm = m([0, 1]) + 2m((1, 2]) = 3.$$

Theorem 12.5: Suppose f and g are simple functions on the measure space $(S, \mathscr{A}, \mu)$.

(a) For each a in $\mathbb{R}$, $\int_S af\, d\mu = a \int_S f\, d\mu$, and
(b) $\int_S (f+g)\, d\mu = \int_S f\, d\mu + \int_S g\, d\mu$.

Proof: We omit the proof of (a).

(b) Assume f has the canonical form $\sum_{j=1}^m a_j \chi_{fj}$ and g has the canonical form $\sum_{k=1}^n b_k \chi_{gk}$. Let $A_j = \{x: f(x) = a_j\}$ and $B_k = \{x: f(x) = b_k\}$. Then $f+g$ can assume at most mn values $a_j + b_k$; $j = 1, \ldots, m$; $k = 1, \ldots, n$. It is clear that $f+g$ assumes the value $a_j + b_k$ on the set $A_j \cap B_k$ if this intersection is nonempty. Therefore, $f+g$ has canonical form

$$f + g = \sum_{i=1}^{p} c_i \chi_i$$

where χ_i is the characteristic function of the set

$$\bigcup \{A_j \cap B_k: a_j + b_k = c_i\}$$

and p is some integer equal to at most mn. By definition, the integral of $f+g$ is given by the equation

$$\int_S (f+g)\, d\mu = \sum_{i=1}^{p} c_i \mu\left(\bigcup \{A_j \cap B_k: a_j + b_k = c_i\}\right)$$

$$= \sum_{i=1}^{p} c_i \sum \{\mu(A_j \cap B_k): a_j + b_k = c_i\}.$$

If we add zero terms whenever $A_j \cap B_k = \varnothing$, we can make the last sum encompass all values of $\mu(A_j \cap B_k)$ and we then have

$$\begin{aligned}\int_S (f+g)\, d\mu &= \sum_{j=1}^{m} \sum_{k=1}^{n} (a_j + b_k)\mu(A_j \cap B_k) \\ &= \sum_{j=1}^{m} a_j \sum_{k=1}^{n} \mu(A_j \cap B_k) + \sum_{k=1}^{n} b_k \sum_{j=1}^{m} \mu(A_j \cap B_k) \\ &= \sum_{j=1}^{m} a_j \mu(A_j) + \sum_{k=1}^{n} b_k \mu(B_k) \\ &= \int_S f\, d\mu + \int_S g\, d\mu. \blacksquare\end{aligned}$$

Corollary 12.6: Suppose f is a simple function on a measure space $(S, \mathscr{A}, \mu)$ and

$$f = \sum_{j=1}^{n} c_j \chi_{B_j}$$

is any representation for f such that $c_j = 0$ if $\mu(B_j) = \infty$. Then

$$\int_S f\, d\mu = \sum_{j=1}^{n} c_j \mu(B_j).$$

Proof: This follows by extending Theorem 12.5 (b) to arbitrary finite sums and by understanding that each function $c_j \chi_{B_j}$ is a simple function. ■

Corollary 12.7: Suppose f and g are simple functions on the measure space $(S, \mathscr{A}, \mu)$ such that $f(x) \geqslant g(x)$ for almost all x in S. Then we have

$$\int_S f\, d\mu \geqslant \int_S g\, d\mu.$$

Proof: The difference $f - g$ is a simple function. Suppose it has the canonical form $f - g = \sum_{j=1}^{n} c_j \chi_j$ where χ_j is the characteristic function of the set C_j. Since $f \geqslant g$ a.e. it follows that if $c_j < 0$ we must have $\mu(C_j) = 0$. Therefore, we conclude that

$$\int_S f\, d\mu - \int_S g\, d\mu = \int_S (f-g)\, d\mu = \sum_{j=1}^{n} c_j \mu(C_j) \geqslant 0. \blacksquare$$

If f is a real valued function on the measure space $(S, \mathscr{A}, \mu)$ then the *support*, Supp(f), of f consists of all $x \in S$ with $f(x) \neq 0$. By $L(f)$ we denote the set of all simple functions l on S such that $l(x) \leqslant f(x)$ a.e., and by $U(f)$ we denote the set of all simple functions u on S such that $f(x) \leqslant u(x)$ a.e. A function f is called a *BBS-function* (for bounded function of bounded support) if f is bounded, measurable, and $\mu(\text{Supp}(f)) < \infty$. The following theorem is a basic existence result for the Lebesgue integral of a BBS-function.

Theorem 12.8: Suppose f is a BBS-function on the measure space $(S, \mathscr{A}, \mu)$.

(a) For each $\varepsilon > 0$ there are simple functions u_ε in $U(f)$ and l_ε in $L(f)$ such that $u_\varepsilon(x) - l_\varepsilon(x) < \varepsilon$ for all x in S and the functions l_ε, u_ε and f have the same support.

(b) $\sup\{\int_S l\, d\mu: l \in L(f)\} = \inf\{\int_S u\, d\mu: u \in U(f)\}$.

Proof:

(a) Since f is bounded there is $M > 0$ such that $-M \leqslant f(x) \leqslant M$ for all x in S. Let n be a positive integer such that $2M/n < \varepsilon$. For each integer j with $-n < j \leqslant n$, let A_j be the set of all x in $\operatorname{Supp}(f)$ such that $(j-1)M/n \leqslant f(x) < jM/n$ and let χ_j be the characteristic function of A_j. The functions

$$l_\varepsilon = \sum_{j=-n+1}^{n} \{(j-1)M/n\}\chi_j$$

in $L(f)$ and the function

$$u_\varepsilon = \sum_{j=-n+1}^{n} \{jM/n\}\chi_j$$

in $U(f)$ satisfy (a).

(b) Let K denote the measure of $\operatorname{Supp}(f)$. For arbitrary $\delta > 0$, let l_ε in $L(f)$ and u_ε in $U(f)$ be the functions constructed in (a) with $\varepsilon = \delta/K$. Then we have

$$\inf\left\{\int_S u\, d\mu: u \in U(f)\right\} - \sup\left\{\int_S l\, d\mu: l \in L(f)\right\}$$
$$\leqslant \int_S u_\varepsilon\, d\mu - \int_S l_\varepsilon\, d\mu = \int_S u_\varepsilon - l_\varepsilon\, d\mu$$
$$\leqslant \int_S (\delta/K)\chi_{\operatorname{Supp}(f)}\, d\mu < \frac{\delta}{K} K = \delta.$$

Since $\delta > 0$ is arbitrary, conclusion (b) is valid. ■

If T is a subset in $\mathscr{A}$ for which $\mu(T) < \infty$ and f is a bounded function on S, then $\chi_T f$ where χ_T denotes the characteristic function of T is a bounded function on S having support of finite measure. We shall write $\int_T f\, d\mu$ for the integral $\int_S \chi_T f\, d\mu$.

Now we shall extend the definition of integral to functions that may be unbounded or have support of infinite measure. First we consider functions that are nonnegative. Suppose f is a measurable function on a measure space $(S, \mathscr{A}, \mu)$ such that $f(x) \geqslant 0$ for almost all x in S. Let $B(f)$ be the set of all BBS-functions g such that $g(x) \leqslant f(x)$ a.e. The function f is said to be *integrable* if

$$A = \sup\left\{\int_S g\, d\mu: g \in B(f)\right\} < \infty;$$

and the *integral* $\int_S f\, d\mu$ is defined to be the number A. If $A = \infty$ we write $\int_S f\, d\mu = \infty$ and say the integral of f is infinite (but we do not consider f to be integrable).

If f is an arbitrary measurable function define f^+ and f^- on S by

$$f^+ = \sup(f, 0) \quad \text{and} \quad f^- = -\inf(f, 0).$$

Then we have

$$f = f^+ - f^- \quad \text{and} \quad |f| = f^+ + f^-.$$

We define f to be *integrable* if and only if both f^+ and f^- are integrable and we define the *integral* as the difference of the integrals, namely

$$\int_S f\, d\mu = \int_S f^+\, d\mu - \int_S f^-\, d\mu.$$

The integral with respect to Lebesgue measure m on the real line will be written $\int_a^b f(t)\, dt$ as well as $\int_R \chi_{[a,b]} f\, dm$.

Theorem 12.9: Suppose f and g are integrable functions on the measure space $(S, \mathscr{A}, \mu)$.

(a) For each a in $\mathbb{R}$, af is integrable and $\int_S af\, d\mu = a \int_S f\, d\mu$ and

(b) $\int_S (f+g)\, d\mu = \int_S f\, d\mu + \int g\, d\mu$.

Proof: We leave the proof of (a) to the reader.

(b) This is true for simple functions by Theorem 12.5 (b). Next we verify the formula for BBS-functions f and g. Suppose u and v are simple functions with $u \leqslant f$ and $v \leqslant g$. Then $u + v$ is a simple function $\leqslant f + g$ so that

$$\int_S u\, d\mu + \int_S v\, d\mu = \int_S (u+v)\, d\mu \leqslant \int_S (f+g)\, d\mu.$$

If we keep v fixed and take the supremum of $\int_S u\, d\mu$ for $u \in L(f)$ we obtain

$$\int_S f\, d\mu + \int_S v\, d\mu \leqslant \int_S (f+g)\, d\mu.$$

If we take the supremum of $\int_S v\, d\mu$ for $v \in L(g)$ we obtain

$$\int_S f\, d\mu + \int_S g\, d\mu \leqslant \int_S (f+g)\, d\mu.$$

The reverse inequality is obtained by applying the equality above to obtain

$$\int_S (f+g)\, d\mu + \int_S -g\, d\mu \leqslant \int_S f\, d\mu.$$

The next step is to establish (b) for f and g nonnegative integrable functions. Suppose w is any BBS-function with $0 \leqslant w \leqslant f + g$. Let $u = \inf(w, f)$ and $v = w - u$. Then u and v are nonnegative BBS-functions such that $u \leqslant f$ and $v \leqslant g$. The latter inequality follows since $w \leqslant g + \inf(w, f)$. Therefore, we have

$$\int_S w\, d\mu = \int_S u\, d\mu + \int_S v\, d\mu \leqslant \int_S f\, d\mu + \int_S g\, d\mu.$$

Since this inequality holds for all BBS-functions $w \leqslant f+g$ we conclude that $f+g$ is integrable if f and g are integrable and

$$\int_S (f+g)\, d\mu \leqslant \int_S f\, d\mu + \int_S g\, d\mu.$$

The reverse inequality is proved just as in the case when f and g are BBS-functions.

Next we consider the case when f and g are integrable with f nonnegative and g nonpositive. If h^+ and h^- denote the positive and negative parts of $f+g$, then $f+g=h^+-h^-$ so $f+h^-=h^+-g$. Since both sides of the last equation are positive we apply the previous argument to obtain

$$\int_S f\, d\mu + \int_S h^-\, d\mu = \int_S h^+\, d\mu - \int_S g\, d\mu$$

from which the conclusion quickly follows.

For f and g arbitrary integrable functions we can establish (b) by manipulating the positive and negative parts of f and g and using the previous result. We leave this to you. ■

A sequence (f_n) of real valued functions defined on a set S is said to *converge pointwise* to the function f from S into $\mathbb{R}$ if

(PWC) $\lim_n f_n(s) = f(s)$

for all s in S. If (PWC) holds almost everywhere, then we say that (f_n) *converges almost everywhere* to f. The next pair of results examine the relation between the almost everywhere convergence of functions and convergence of the integral. The first (Theorem 12.10) is called the *Monotone Convergence Theorem* and the next (Theorem 12.12), *Lebesgue's Convergence Theorem.*

Theorem 12.10: Suppose (f_n) is a sequence of nonnegative measurable functions on a measure space $(S, \mathscr{A}, \mu)$ such that $f_n(x) \leqslant f_{n+1}(x)$ almost everywhere. If f is any function on S such that $f(x) = \lim_n f_n(x)$ wherever the sequence converges, then we have

$$\int_S f\, d\mu = \lim_n \int_S f_n\, d\mu.$$

Proof: If A_n denotes the set of all x in S such that $f_n(x) > f_{n+1}(x)$, then $\mu(A_n) = 0$ so that the countable union $A = \bigcup_n A_n$ also has measure 0. Therefore, $f_n(x)$ increases to f for almost all x in S so f is measurable by Theorem 12.3.

Since $(\int_S f_n\, d\mu)$ is an increasing sequence it converges either to a finite limit by Theorem 5.4 (page 25) or to ∞. Suppose u is a BBS-function with $0 \leqslant u \leqslant f$. Let $\varepsilon > 0$ be arbitrary. For each n let B_n consist of all x in $\mathrm{Supp}(u)$ such that $u(x) < f_n(x) + \varepsilon/2K$ where $K = \mu(\mathrm{Supp}(u))$. Since $u \leqslant \lim_n f_n$ a.e. there is N such that

$\mu(B_{N^\sim}) < \varepsilon/2M$ where $M = \sup\{|u(x)|: x \in S\}$. Thus we shall have

$$\begin{aligned}\int_S u\, d\mu &= \int_{B_N} u\, d\mu + \int_{S\setminus B_N} u\, d\mu \\ &< \int_{B_N} (f_N + \varepsilon/2K)\, d\mu + \int_{S\setminus B_N} M\, d\mu \\ &< (\varepsilon/2K)\mu(B_N) + M\mu(B_N^\sim) + \int_S f_N\, d\mu \\ &\leqslant \varepsilon + \sup_n \int_S f_n\, d\mu.\end{aligned}$$

Hence, for each $\varepsilon > 0$ we have

$$\begin{aligned}\int_S f\, d\mu &= \sup\left\{\int_S u\, d\mu: u \text{ is BBS and } u \leqslant f\right\} \\ &\leqslant \varepsilon + \sup_n \int_S f_n\, d\mu.\end{aligned}$$

Since $\varepsilon > 0$ is arbitrary it follows that $\int_S f\, d\mu \leqslant \sup_n \int_S f_n\, d\mu$. The reverse inequality holds since $\int_S f_n\, d\mu \leqslant \int_S f\, d\mu$ for all n. ■

The following result is known as *Fatou's Lemma*.

Corollary 12.11: If (f_n) is a sequence of nonnegative measurable functions on a measure space $(S, \mathscr{A}, \mu)$ such that $f_n(x)$ converges to $f(x)$ a.e. then

$$\int_S f\, d\mu \leqslant \liminf_n \int_S f_n\, d\mu.$$

Proof: If $g_k = \inf\{f_n: n \geqslant k\}$ then $g_k(x)$ increases to $f(x)$ for almost every x. Thus by Theorem 12.10 $\int_S f\, d\mu \leqslant \lim_n \int_S g_n\, d\mu$. But $g_n \leqslant f_m$ for every $m \geqslant n$ so that we have

$$\int_S g_n\, d\mu \leqslant \int_S f_m\, d\mu \text{ for } m \geqslant n,$$

which implies that

$$\int_S g_n\, d\mu \leqslant \inf\left\{\int_S f_m\, d\mu: m \geqslant n\right\}$$

for all n. From this we conclude that

$$\int_S f\, d\mu \leqslant \sup_n \int_S g_n\, d\mu \leqslant \liminf_n \int_S f_n\, d\mu. \blacksquare$$

Theorem 12.12: Suppose (f_n) is a sequence of real valued measurable functions defined on a measure space $(S, \mathscr{A}, \mu)$ that converges almost everywhere to a function f. If there is an integrable function g on S such that $|f_n(x)| \leqslant g(x)$ for almost all x in S then f is integrable and

$$\int_S f\,d\mu = \lim_n \int_S f_n\,d\mu.$$

Proof: The function f is measurable because it is the limit almost everywhere of measurable functions. It is integrable because $|f(x)| \leqslant g(x)$ for almost all x. By Fatou's Lemma we have

$$\int_S (g+f)\,d\mu \leqslant \liminf_n \int_S (g+f_n)\,d\mu,$$

which implies that

$$\int_S f\,d\mu \leqslant \liminf_n \int_S f_n\,d\mu.$$

On the other hand, we also have

$$\int_S (g-f)\,d\mu \leqslant \liminf_n \int_S (g+f_n)\,d\mu$$

$$= \int_S g\,d\mu - \limsup_n \int_S f_n\,d\mu,$$

which implies that

$$\int_S f\,d\mu \geqslant \limsup_n \int_S f_n\,d\mu. \blacksquare$$

The following result concerns the approximation of a nonnegative measurable function by simple functions. It will be used in the next section. The proof is left to you.

Theorem 12.13: Let f be a nonnegative measurable function on a measure space $(S, \mathscr{A}, \mu)$. There exists an increasing sequence (v_n) of simple functions that converges pointwise to f. If S is σ-finite then each v_n can be chosen to vanish outside of a set of finite measure.

The integral usually encountered in elementary calculus is called the *Riemann integral*. Whereas the Lebesgue integral of a function f is determined by dissecting the range of f, the Riemann integral is determined by dissecting its domain. A *step function* is a simple function u having the form

$$u = \sum_{j=1}^{n} a_j \chi_{I_j}.$$

$\{I_j : j = 1, 2, \ldots, n\}$ is a collection of disjoint intervals. The integral of a step function is the number $\sum_{j=1}^{n} a_j m(I_j)$. If f is a bounded real function defined on an interval $[a, b]$, the *lower Riemann integral* $L \int_a^b f$ is the least upper bound of all integrals $\sum_{j=1}^{n} a_j \mu(I_j)$ of step functions no greater than f. The *upper Riemann integral*

$U \int_a^b f$ is the greatest lower bound of all integrals of step functions that are greater than or equal to f. If $L \int_a^b f = U \int_a^b f$ then we say f is Riemann integrable on $[a, b]$ and write $\int_a^b f(t)\, dt$ for their common value. Since every step function is a simple function it follows that in the case of the Riemann integral we are taking greatest lower bounds and least upper bounds over smaller subsets. Therefore, if f is bounded and $\int_a^b f(t)\, dt$ exists in the sense of Riemann it also exists in the sense of Lebesgue and has the same value.

Exercises

A

1. Let f be a function from a measure space $(S, \mathscr{A}, \mu)$ into $\mathbb{R}$. Show that the following statements are equivalent
   (a) f is measurable.
   (b) $f^{-1}((a, b)) \in \mathscr{A}$ for all a, b.
   (c) $f^{-1}((a, b]) \in \mathscr{A}$ for all a, b.
   (d) $f^{-1}([a, b)) \in \mathscr{A}$ for all a, b.
   (e) $f^{-1}([a, b]) \in \mathscr{A}$ for all a, b.
   (f) $f^{-1}((-\infty, a)) \in \mathscr{A}$ for all a, b.
2. Suppose f and g are measurable and $h = f/g$ where $g \neq 0$. Show that h is measurable.
3. *(a)* Show that if f is measurable and $g = f$ a.e. then g is measurable.
   (b) Suppose $(S, \mathscr{A}, \mu)$ is an incomplete measure space and E is a nonmeasurable subset of a set A of measure 0. Show that $\chi_A - \chi_E$ is not measurable but almost everywhere equal to 0.
   (c) In view of (b) suggest a way to define measurable function on an incomplete measure space.
4. Find the canonical representation of the function f defined by $f(x) = [x]$ (greatest integer function) for $-1 \leqslant x \leqslant 3$, $f(x) = 0$ otherwise. Determine $\int_R f\, dm$.
5. Show that the space of simple functions on $(S, \mathscr{A}, \mu)$ forms a linear lattice and an algebra. (You have to prove that the sum, scalar multiple, product, and sup of simple functions is simple.) Show by example that the space of simple functions does not form a σ-lattice.
6. Show that the space of BBS functions forms a linear lattice and an algebra.
7. Show that the space of integrable functions forms a linear σ-lattice and an algebra.
8. If $|f|$ is measurable is f necessarily measurable? If f^2 is measurable is f measurable?
9. Show that if f is a nonnegative integrable function and $\int_S f\, d\mu = 0$, then $f = 0$ a.e.
10. For each $n = 1, 2, \ldots$ let $f_n(x) = 1$ for $n \leqslant x \leqslant n+1$ and $f_n(x) = 0$ otherwise. Show $\lim_n \int_{\mathbb{R}} f_n\, dm = 1$ for each n but $\int_R \lim_n f_n\, dm = 0$.
11. Prove that if f is integrable on a measure space $(S, \mathscr{A}, \mu)$ then so is $|f|$ and $|\int_S f\, d\mu| \leqslant \int_S |f|\, d\mu$.

12. Suppose f is a nonnegative function on $\mathbb{R}$ such that $\int_{-\infty}^{\infty} f(t)\,dt = 1$. Define μ on $\mathcal{M}$, the collection of measurable subsets of $\mathbb{R}$, by $\mu(A) = \int_A f(t)\,dt$ for $A \in \mathcal{M}$. Show that $(\mathbb{R}, \mathcal{M}, \mu)$ is a probability space.

B

13. Let $r_1, r_2, \ldots$ denote the sequence of rationals in $[0, 1]$. For $i = 1, 2, \ldots$ let $f_i(r_j) = 1$ if $j \leqslant i$ and $f_i(x) = 0$ otherwise. If $f = \lim_n f_n$ show that the Riemann integral of f does not exist but that f is a simple function with Lebesgue integral 0. Thus observe that Theorem 12.12 does not hold for Riemann integration.

14. Define $f(x) = (-1)^n/n$ for $n \leqslant x < n+1$, $n = 1, \ldots$ and $f(x) = 0$ for $x < 1$. Show that $\lim_{r \to \infty} \int_{-\infty}^{r} f(t)\,dt$ is finite, that is, $\int_{-\infty}^{\infty} f(t)\,dt$ exists as an improper Riemann integral, but f is not integrable in the sense of Lebesgue. Present a similar argument for $g(x) = (\sin x)/x$ for $0 < x < \infty$.

15. *(a)* Show that if f is a nonnegative measurable function then there is a sequence (u_n) of nonnegative simple functions that increases to f.

(b) Prove that $\int_S f\,d\mu = \sup\{\int_S u\,d\mu \colon u$ is simple and $u \leqslant f\}$.

(c) Show that if f is integrable on S, then for each $\varepsilon > 0$ there is a simple function p such that $\int_S |f - p|\,d\mu < \varepsilon$.

(d) Verify that there is a step function q such that $\int_S |f - q|\,d\mu < \varepsilon$.

(e) Show that if S is a subset of $\mathbb{R}$, there is a continuous function g on S such that $\int_S |f - g|\,dx < \varepsilon$.

16. For each n let $f_n(x) = n$ if $0 < x < 1/n$ and $f_n(x) = 0$ otherwise. Show that $\int_R f_n\,dm = 1$ for each n but $\int_R \lim_n f_n\,dm = 0$. Which hypothesis of Theorem 12.12 does this sequence not satisfy?

17. Show that if f is an integrable function on $\mathbb{R}$ then the function F defined by $F(x) = \int_{-\infty}^{x} f\,dt$ is continuous.

18. If f is integrable on $\mathbb{R}$ then $\lim_{n \to \infty} \int_{-\infty}^{\infty} f(x) \cos nx\;dx = 0$. (This result is called the *Riemann-Lebesgue Lemma.* First establish it for step functions and then approximate f by a step function.)

19. Define f on $\mathbb{R}$ by $f(x) = \sqrt{n}$ if $n \leqslant x \leqslant n + 1/n^2$ for $n = 1, 2, \ldots$ and $f(x) = 0$ otherwise. Show that f is integrable with respect to Lebesgue measure but f^2 is not.

20. Show that if f and g are integrable on $\mathbb{R}$ with respect by Lebesgue measure, then so is the convolution $f * g$ of f and g defined by

$$(f * g)(x) = \int_{-\infty}^{\infty} f(x - t) g(t)\,dt.$$

21. Show that if f is integrable on $\mathbb{R}$ with respect to Lebesgue measure then

$$\lim_{x \to 0} \int_{-\infty}^{\infty} |f(t + x) - f(t)|\,dt = 0.$$

C

22. *(a)* Prove that the collection $\mathcal{B}$ of Borel subsets of $\mathbb{R}$ has the same cardinality as $[0, 1]$.

(b) Prove that the collection of all subsets of $\mathbb{R}$ having measure 0 has cardinality greater than $[0, 1]$.
(c) Therefore, show that there are subsets of $\mathbb{R}$ having measure 0 that are not Borel sets.
(d) Use (c) to show there are non-Borel subsets of $\mathbb{R}$ having measure > 0.

13

The Extension Theorem

So far we have presented only one useful example of the abstract theory described in Sections 11 and 12, namely Lebesgue measure and integration on $\mathbb{R}$. We can construct more examples via the *Caratheodory Extension Theorem* (13.2), which establishes a condition where a measure on an algebra $\mathscr{A}_0$ can be extended to a σ-algebra $\mathscr{A}$ containing $\mathscr{A}_0$. The most important use we shall make of the Extension Theorem is to extend Lebesgue integration to $\mathbb{R}^n$. In Section 11 we allowed a fasf μ° on an algebra $\mathscr{A}_0$ to be called a measure if whenever $A_1, A_2, \ldots$ is a sequence of disjoint sets in $\mathscr{A}_0$ with $\bigcup_n A_n$ also in $\mathscr{A}_0$ we then have

(CA) $\mu^\circ\{\bigcup_n A_n\} = \sum_n \mu^\circ(A_n)$.

Condition (CA) is clearly necessary for the existence of an extension of μ° to a σ-algebra containing $\mathscr{A}_0$. The Extension Theorem asserts that this condition is also sufficient.

Lemma 13.1: Suppose μ_0 is a measure on an algebra $\mathscr{A}_0$ of subsets of a set S. For each subset B of S define $\mu^*(B)$ by the formula

$$\mu^*(B) = \inf\left\{ \sum_{n=1}^{\infty} \mu_0(A_n) : \{A_n\} \subset \mathscr{A}_0, \bigcup_n A_n \supset B \right\}.$$

(a) The extended real valued function μ^* is an outer measure.
(b) For all A in $\mathscr{A}_0$, A is μ^*-measurable and $\mu^*(a) = \mu_0(A)$.

Proof:

(a) That μ^* satisfies properties (OM1) and (OM2) on page 81 is immediately obvious.

In order to show that μ^* satisfies (OM3) suppose $B = \bigcup_n B_n$. There is an issue only if $\mu^*(B_n) < \infty$ for each n. If this happens, then for each $\varepsilon > 0$ and each n we can find a sequence $\{A_{nk}\}$ in $\mathscr{A}_0$ such that

$$B_n \subset \bigcup_k A_{nk} \quad \text{and} \quad \mu^*(B_n) + \frac{\varepsilon}{2^n} > \sum_k \mu_0(A_{nk}).$$

Then we have $B \subset \bigcup_{n,k} A_{nk}$ so that

$$\mu^*(B) \leqslant \sum_{n,k} \mu_0(A_{nk}) \leqslant \varepsilon + \sum_n \mu_0(B_n).$$

Since $\varepsilon > 0$ is arbitrary it follows that (OM3) is valid.

Suppose A is in $\mathscr{A}_0$ and $\{A_n\}$ is a sequence in $\mathscr{A}_0$ whose union contains A. Let $C_1 = A \cap A_1$ and for each $n > 1$ let $C_n = A \cap (A_n \setminus \bigcup_{j<n} A_j)$. Since $\mathscr{A}_0$ is an algebra, each C_n is in $\mathscr{A}_0$, and since μ_0 is a measure we have

$$\mu_0(A) = \sum_n \mu_0(C_n) \leqslant \sum_n \mu_0(A_n).$$

From this we conclude that $\mu_0(A) \leqslant \mu^*(A)$ for all A in $\mathscr{A}_0$. That $\mu_0(A) \geqslant \mu^*(A)$ is obvious.

Finally we show that every A in $\mathscr{A}_0$ is μ^*-measurable. Suppose B is an arbitrary subset of S. For each $\varepsilon > 0$ there is a sequence $\{A_n\}$ in $\mathscr{A}_0$ such that

$$B \subset \bigcup_n A_n \quad \text{and} \quad \sum_n \mu_0(A_n) \leqslant \mu^*(B) + \varepsilon.$$

Since $B \cap A \subset \bigcup_n A \cap A_n$ and $B \cap A^\sim \subset \bigcup_n A^\sim \cap A_n$ it follows that

$$\varepsilon + \mu^*(B) \geqslant \sum_n \mu(A_n) = \sum_n \{\mu(A \cap A_n) + \mu(A^\sim \cap A_n)\}$$

$$\geqslant \mu^*(B \cap A) + \mu^*(B \cap A^\sim). \blacksquare$$

If $\mathscr{A}_0$ is an algebra of subsets of a set S, then there is at least one σ-algebra containing $\mathscr{A}_0$, namely the collection of all subsets of S. The intersection of all σ-algebras containing $\mathscr{A}_0$ is a σ-algebra, thus the smallest σ-algebra containing $\mathscr{A}_0$. This smallest σ-algebra will be called the σ-algebra *generated* by the algebra $\mathscr{A}_0$.

Theorem 13.2: Every measure μ_0 on an algebra $\mathscr{A}_0$ has an extension to a measure μ on the σ-algebra $\mathscr{A}$ generated by $\mathscr{A}_0$. If μ_0 is σ-finite, then this extension is unique.

Proof: Let μ^* be the outer measure determined by μ_0 according to Lemma 13.1. Let $\mathscr{A}^*$ denote the collection of all sets that are μ^*-measurable. By Theorem 11.4 (page 82), $\mathscr{A}^*$ is a σ-algebra and the restriction of μ^* to $\mathscr{A}^*$ is a measure. By Lemma 13.1, $\mathscr{A}^*$ contains $\mathscr{A}_0$ so that $\mathscr{A}^*$ also contains $\mathscr{A}$. Therefore, if μ is the restriction of μ^* to $\mathscr{A}$ it follows that μ is a measure that is an extension of μ_0.

To verify that μ is unique when μ_0 is σ-finite suppose μ_1 is another such extension. Let A be a set in $\mathscr{A}$. Assume first that A is contained in a subset B of $\mathscr{A}_0$ with $\mu_0(B) < \infty$. Suppose $\{B_n\}$ is a sequence in $\mathscr{A}_0$ such that $\bigcup_n B_n$ contains A. Since

$$\mu_1(A) \leqslant \sum_n \mu_1(B_n) = \sum_n \mu(B_n)$$

it follows that $\mu_1(A) \leqslant \mu^*(A) = \mu(A)$. Similarly we have $\mu_1(B \setminus A) \leqslant \mu(B \setminus A)$. Since

$$\mu_1(A) + \mu_1(B \setminus A) = \mu_1(B) = \mu(B) = \mu(A) + \mu(B \setminus A)$$

it is out of the question that $\mu_1(A) < \mu(A)$. Therefore, $\mu_1(A) = \mu(A)$.

If A is not contained in a set in $\mathscr{A}_0$ having finite measure, it is at least the case that $A = \bigcup_n C_n \cap A$ where each C_n is in $\mathscr{A}_0$ and $\mu_0(C_n) < \infty$ for each n. This is because μ_0 is σ-finite. We can assume that the sets C_n are disjoint; then we apply the result of the previous paragraph to conclude

$$\mu_1(A) = \sum_n \mu_1(C_n \cap A) = \sum_n \mu(C_n \cap A) = \mu(A). \blacksquare$$

Our principle application of the Extension Theorem will be to construct the product of two measure spaces $(R, \mathscr{A}, \mu)$ and $(S, \mathscr{B}, \nu)$ under the assumption that both $(R, \mathscr{A}, \mu)$ and $(S, \mathscr{B}, \nu)$ are complete. If $A \in \mathscr{A}$ and $B \in \mathscr{B}$, the set $A \times B$ is called a *measurable rectangle* (MR for short). The collection of all finite unions of MR's is denoted by $\mathscr{R}_0$. For every MR $A \times B$ we define $\lambda(A \times B) = \mu(A)\nu(B)$.

Theorem 13.3:

(a) Every member C of $\mathscr{R}_0$ can be written as a finite union of disjoint MR's.

(b) If $\{R_j\}$ and $\{S_j\}$ are two sequences of disjoint MR's that have the same union then we have

$$\sum_{j=1}^{\infty} \lambda(R_j) = \sum_{j=1}^{\infty} \lambda(S_j). \tag{13–1}$$

Thus the function λ can be extended to all of $\mathscr{R}_0$ by setting

$$\lambda(C) = \sum_{j=1}^{n} \lambda(R_j);$$

$R_1, \ldots, R_n$ are disjoint MR's with $\bigcup_{j=1}^{n} R_j = C$.

(c) The collection $\mathscr{R}_0$ forms an algebra and λ as defined by (b) is a measure on $\mathscr{R}_0$.

Proof:

(a) For each set C in $\mathscr{R}_0$ let $i(C)$, the *index* of C, be the smallest integer i such that C can be written as the union of i MR's. If $i(C) = 1$, then the result is obvious. Assume the result is true for those D in $\mathscr{R}_0$ with $i(D) < n$, and suppose $i(C) = n$. Let $C = \bigcup_{j=1}^{n} R_j$. Since $\bigcup_{j=1}^{n-1} R_j$ has index less than n it follows that there are disjoint sets $\{S_1, \ldots, S_k\}$ such that $C = \{\bigcup_{j=1}^{k} S_j\} \cup R_n$. The result will now follow if we can establish that whenever $S_j \cap R_n \neq \varnothing$ then $R_n \backslash S_j \cap R_n$ is a finite union of disjoint MR's. If $S_j = A \times B$ and $R_n = C \times D$ then

$$R_n \backslash S_j \cap R_n = (C \backslash A \times D) \cup (C \cap A \times D \backslash B).$$

See Figure 13.1. The two sets on the right-hand side of the equation are disjoint so our contention is established.

(b) If $\bigcup_{j=1}^{\infty} R_j = \bigcup_{j=1}^{\infty} S_j$, then for each $i = 1, 2, \ldots$ we have $R_i = \bigcup_{j=1}^{\infty} R_i \cap S_j$. Since the intersection of two MR's is either empty or it is an MR it suffices to prove (b) for R_1. Suppose $R_1 = A \times B$ and $S_j = A_j \times B_j$ for $j = 1, 2, \ldots$. Let x be a fixed point in A. For each y in B, the point (x, y) is in exactly one MR, $A_j \times B_j$. This implies

$$B = \bigcup \{B_j\colon (x, y) \in A_j \times B_j,\ y \in B\}.$$

Since the union is disjoint it follows that

$$\nu(B)\chi_A(x) = \nu(B) = \sum_{j=1}^{\infty} \nu(B_j)\chi_{A_j}(x).$$

The partial sums of the series $\sum_{j=1}^{\infty} \nu(B_j)\chi_{A_j}(x)$ form an increasing sequence of nonnegative measurable functions on $(R, \mathscr{A}, \mu)$ so that by the Monotone

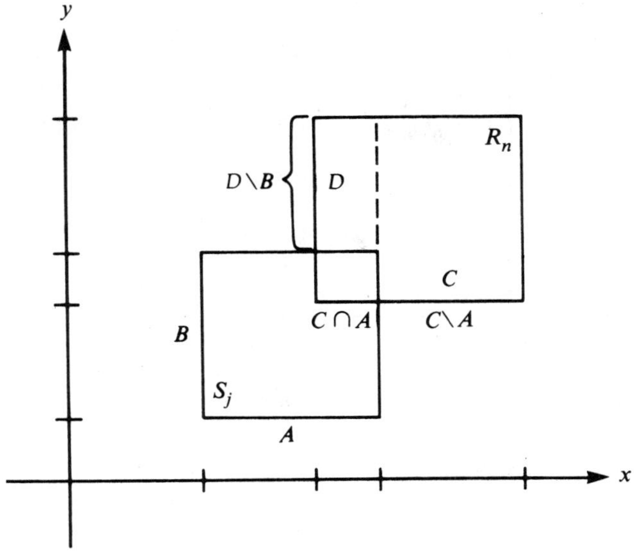

Figure 13.1

Convergence Theorem (12.10) we have

$$\mu(A)\nu(B) = \int \nu(B)\chi_A \, d\mu = \sum_{j=1}^{\infty} \int \nu(B_j)\chi_{A_j} \, d\mu = \sum_{j=1}^{\infty} \mu(A_j)\nu(B_j).$$

Formula (13–1) is more than sufficient to establish the second assertion of (b).

(c) It is obvious that $\mathscr{R}_0$ is closed under finite unions. If $S \in \mathscr{R}_0$ then $S = \bigcup_{j=1}^{n} R_j$ where each R_j is a MR. Thus, the complement of S is $\bigcap_{j=1}^{n} R_j^{\sim}$. If $R_j = A_j \times B_j$, then $R_j^{\sim} = S \times B_j^{\sim} \cup A_j^{\sim} \times B_j$ so $R_j^{\sim} \in \mathscr{R}_0$ for each j. Since the intersection of two MR's is also an MR it follows that $\mathscr{R}_0$ is closed under finite intersection. Therefore, $\mathscr{R}_0$ is closed under complement.

That λ is a measure on $\mathscr{R}_0$ now follows from (b). ■

Since λ is a measure on $\mathscr{R}_0$ it can be extended to the σ-algebra $\mathscr{R}$ determined by $\mathscr{R}_0$. The measure space $(S \times T, \mathscr{R}, \lambda)$ is called the *product* of $(S, \mathscr{A}, \mu)$ and $(T, \mathscr{B}, \nu)$; λ is called the *product* of the measures μ and ν, and we write $\lambda = \mu \times \nu$.

We shall now show how we can reduce measure and integration with respect to λ to measure, and integration with respect to μ. If A is a subset of $S \times T$ and $s \in S$ we define the s *cross-section* A_s by

$$A_s = \{t \colon (s, t) \in A\}.$$

See Figure 13.2. For $t \in T$ make a corresponding definition for the t cross-section A_t. It is not hard to show that

$$(A^{\sim})_s = (A_s)^{\sim} \quad \text{and} \quad \left(\bigcup_{\alpha} A_\alpha\right)_s = \bigcup (A_\alpha)_s$$

for every collection $\{A_\alpha\}$ of subsets of $S \times T$.

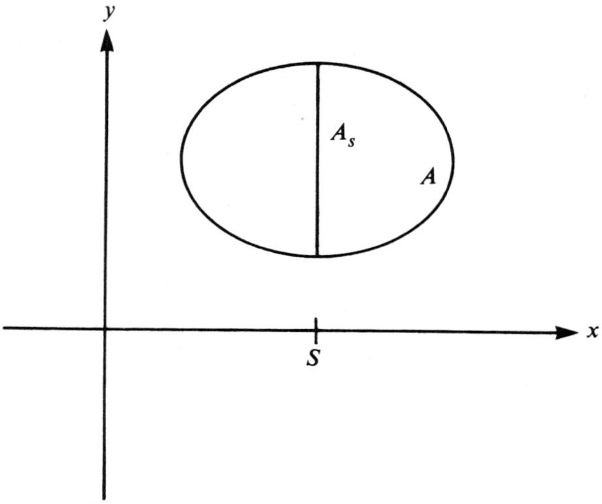

Figure 13.2

Let $\mathscr{R}_1$ denote the collection of all countable unions of MR's in $S \times T$ and let $\mathscr{R}_2$ denote the collection of all countable intersections of members of $\mathscr{R}_1$. Since $\mathscr{R}$ is a σ-algebra containing $\mathscr{R}_0$ it is clear that $\mathscr{R}$ contains $\mathscr{R}_1$ and $\mathscr{R}_2$. Since we originally defined $\lambda(A)$ for A in $\mathscr{R}$ to be the infimum of all numbers $\sum_{j=1}^{\infty} \mu(B_j)\nu(C_j)$ where $(B_j \times C_j)$ is a sequence of measurable rectangles with $\bigcup_j B_j \times C_j \supset A$, it follows that for each $\varepsilon > 0$ there is A_ε in $\mathscr{R}_1$ with $A_\varepsilon \supset A$ and $\lambda(A_\varepsilon) \leqslant \lambda(A) + \varepsilon$. If we let $A_0 = \bigcap_n A_{1/n}$, it follows that $A_0 \in \mathscr{R}_2$, A_0 contains A, and $\lambda(A_0) = \lambda(A)$. This fact indicates that we can draw conclusions about the measure of a set in $\mathscr{R}$ by approximating it by a set in $\mathscr{R}_2$.

The following theorem corresponds to the procedure for determining areas by cross section that you may have studied in elementary calculus.

Theorem 13.4: Let $(S, \mathscr{A}, \mu)$ and $(T, \mathscr{B}, \nu)$ be complete measure spaces. Suppose A is a measurable subset of $S \times T$ such that $\lambda(A)$ is finite. Then for almost all s in S the set A_s is a measurable subset of T. The function f defined for almost all s in S by

$$f(s) = \nu(A_s)$$

is measurable and

$$\int f \, d\mu = \lambda(A).$$

Proof: The proof proceeds according to the following plan. The conclusion is obvious by definition for every MR A and by the additivity of the integral for all A in $\mathscr{R}_0$. We next establish it for all A in $\mathscr{R}_1$, then for all A in $\mathscr{R}_2$, and finally for general measurable A.

$A \in \mathscr{R}_1$ Suppose $A = \bigcup A_n$ where $\{A_n\}$ is a disjoint collection of MR's. For each s in S and n_t in T we have

$$\chi_{A_s}(t) = \chi_A(s, t) = \sup_n \chi_{A_n}(s, t) = \sup_n \chi_{(A_n)_s}(t).$$

Since A_n is a MR it follows that $(A_n)_s$ is a measurable set for all n so that $\chi_{(A_n)_s}$ is a measurable function. By Theorem 12.3 we conclude that χ_{A_s} is also a measurable function. Therefore, A_s is a measurable set. If we define f_n on S by

$$f_n(s) = \nu((A_n)_s)$$

then each f_n is a nonnegative measurable function and

$$f = \lim_k \sum_{n=1}^{k} f_n.$$

Therefore, by the Monotone Convergence Theorem (12.10) we have

$$\int f\,d\mu = \sum_{n=1}^{\infty} \int f_n\,d\mu$$

$$= \sum_{n=1}^{\infty} \lambda(A_n) = \lambda(A).$$

Therefore, the conclusion holds for A in $\mathscr{R}_1$.

$A \in \mathscr{R}_2$ Suppose $A = \bigcap_n A_n$ where $A_n \in \mathscr{R}_1$ and $A_n \subset A_{n-1}$ for all n. It follows as in the preceding case that χ_{A_s} is the infimum of $\chi_{(A_n)_s}$ so that χ_{A_s} is a measurable function and thus A_s is a measurable set. Since $\lambda(A) < \infty$ we may assume $\lambda(A_1) < \infty$. If $f_n = \nu((A_n)_s)$, then each f_n is measurable by the preceding section so that $f = \lim_n f_n$ is measurable. Since f_1 is integrable the Lebesgue Dominated Convergence Theorem (12.12 on page 96) shows that

$$\int f\,d\mu = \lim_n \int f_n\,d\mu$$

$$= \lim_n \lambda(A_n)$$

$$= \lambda(A).$$

A measurable We have noted that for all measurable A there is B in $\mathscr{R}_2$ such that $B \supset A$ and $\lambda(A) = \lambda(B)$. If $\lambda(A) = 0$ then by the preceding paragraph we have $\int \nu(B_s)\,d\mu = 0$. This implies that $\nu(B_s) = 0$ for almost all s in S. Since $(T, \mathscr{B}, \nu)$ is a complete measure space and $A_s \subset B_s$ for all s in S it follows that $\nu(A_s) = 0$ for almost all s in S. This implies that conclusion holds when $\lambda(A) = 0$. If $\lambda(A) > 0$, then $\lambda(b \backslash A) = 0$ so that $\nu((B \backslash A)_s) = 0$ for almost all s. Since $A_s = B_s \backslash (B \backslash A)_s$ for all s it follows that A_s is measurable for almost all s. Moreover, we have

$$\int \nu(A_s)\,d\mu = \int \nu(B_s)\,d\mu = \lambda(B) = \lambda(A). \blacksquare$$

One implication of Theorem 13.4 is that if $f(x, y)$ is the characteristic function of a measurable subset A of $S \times T$ such that $\lambda(A) < \infty$, it follows that we can obtain $\lambda(A)$ from the formulas

$$\int f(s,t)\,d\lambda = \lambda(A) = \int_S \left\{ \int_T f(s,t)\,d\nu \right\} d\mu \qquad \textbf{(13–2)}$$

$$= \int_T \left\{ \int_S f(s,t)\,d\mu \right\} d\nu$$

because for fixed s in S we have

$$\nu(A_s) = \int_T f(s,t)\,d\nu,$$

and for t in T we have

$$\mu(A_t) = \int_S f(s,t)\,d\mu.$$

That formulas (13–2) hold more generally is the content of the following two theorems known respectively as *Fubini's Theorem* and *Tonelli's Theorem.* By additivity it is clear that formulas (13–2) are valid when f is a simple function, which vanishes outside of a set of finite measure. If f is a function defined on $S \times T$ then for each s in S we define f_s on T by $f_s(t) = f(s,t)$ and f_t on S by $f_t(s) = f(s,t)$.

Theorem 13.5: Suppose $(S, \mathscr{A}, \mu)$ and $(T, \mathscr{B}, \nu)$ are complete measure spaces and f is an integrable function on $S \times T$.

(a) For almost all s in S, f_s is an integrable function on T. The function F defined on S by

$$F(s) = \int_T f(s,t)\,d\nu(t)$$

is integrable and

$$\int_S F(s)\,d\mu(s) = \int_S \left\{ \int_T f(s,t)\,d\nu(t) \right\} d\mu(s) = \int_{S \times T} f\,d\lambda.$$

(b) For almost all t in T, f_t is an integrable function on S. The function G defined on T by

$$G(t) = \int_S f(s,t)\,d\mu(s)$$

is integrable and

$$\int_T G(t)\,d\nu(t) = \int_T \left\{ \int_S f(s,t)\,d\mu(s) \right\} d\nu(t) = \int_{S \times T} f\,d\lambda.$$

Proof: It suffices to prove (a) since a symmetric argument will then establish (b). If f is nonnegative, then by Theorem 12.13 on page 97 there is an increasing sequence $\{v_n\}$ of simple functions that converges to f. Since each v_n is integrable and simple it vanishes outside of a set of finite measure. Since f_s is the pointwise limit of the functions $\{(v_n)_s\}$ it follows that f_s is measurable, and by the Monotone Convergence Theorem (12.10) we have

$$F(s) = \int_T f(s, t)\, dv(t) = \lim_n \int_T v_n(s, t)\, dv(t).$$

Therefore, F is the limit of measurable functions and, therefore, measurable. Applying the Monotone Convergence Theorem again we calculate

$$\int_S F(s)\, d\mu = \lim_n \int_S \left\{ \int_T v_n(s, t)\, dv \right\} d\mu$$

$$= \lim_n \int_{S \times T} v_n(s, t)\, d\lambda = \int_{S \times T} f\, d\lambda. \blacksquare$$

Theorem 13.6: Suppose $(S, \mathscr{A}, \mu)$ and $(T, \mathscr{A}, v)$ are σ-finite measure spaces and f is a nonnegative measurable function on $S \times T$.

(a) For almost all s in S, f_s is a measurable function on T. The function F defined on S by

$$F(s) = \int_T f(s, t)\, dv(t)$$

is measurable and

$$\int_S F(s)\, d\mu(s) = \int_S \left\{ \int_T f(s, t)\, dv(t) \right\} d\mu(s) = \int_{S \times T} f\, d\lambda.$$

(b) For almost all t in T, f_t is a measurable function on S. The function G defined on T by

$$G(t) = \int_S f(s, t)\, d\mu(s)$$

is measurable and

$$\int_T G(t)\, dv(t) = \int_T \left\{ \int_S f(s, t)\, d\mu(s) \right\} dv(t) = \int_{S \times T} f\, d\lambda.$$

Proof: In this theorem we relax the requirement that f be integrable and require instead that it be positive and measurable. We also require that S and T be σ-finite. The conclusion is altered by allowing the possibility of infinite integrals. The details of the proof are the same except now we use the σ-finiteness of $S \times T$ rather than the integrability of f to assume that each v_n is integrable. $\blacksquare$

The following example shows that you should not draw hasty conclusions about a function's integrability on a product space.

Example 13.7:

Consider the product space of $(I, \mathcal{M}, m)$ where I is the interval $[0, 1]$ with Lebesgue measure m and $(N, \mathcal{A}, \mu)$ where N is the set of positive integers with counting measure μ. Define $F(x, n) = x$ for each x in I and each n in N. Thus $F(x, n) = f(x)$ where $f(x) = x$ is clearly integrable on I. However, F is not integrable on the product space because by Tonelli's Theorem we have

$$\int F(x, n)\, d\mu = \infty \quad \text{for each } x \neq 0 \text{ in } I.$$

We conclude this section with a discussion of the product of infinitely many measure spaces. This part of the section is optional because it is difficult, and subsequent material does not depend on it.

For each α in a set A let $(S_\alpha, \mathcal{A}, \lambda_\alpha)$ denote a measure space such that $\lambda_\alpha(S_\alpha) = 1$ for each α. A *cylindrical set* (CS) is a set of the form $\prod_\alpha A_\alpha$ where for all but finitely many α we have $A_\alpha = S_\alpha$. Define the projection function P_F from $\prod_\alpha S_\alpha$ onto $\prod_{\alpha \in F} S_\alpha$ by the formula

$$P_F f = g \quad \text{where } g(\alpha) = f(\alpha) \text{ for all } \alpha \text{ in } F.$$

As a special case we have $P_{\{\alpha\}} f$, which we write as $P_\alpha f$ and identify with the value $f(\alpha)$. In terms of these projections we characterize a cylindrical set C as one of the form

$$C = \prod\nolimits_{\alpha \in F} P_\alpha(C) \times \prod\nolimits_{\alpha \in A \setminus F} S_\alpha$$

where F is a finite subset of A. We define the nonnegative function λ on the collection of all cylindrical sets by the formula

$$\lambda\left(\prod_\alpha A_\alpha\right) = \prod_\alpha \lambda_\alpha(A_\alpha)$$

where the symbol $\prod$ on the right-hand side of the equation denotes product of numbers. Since $\lambda_\alpha(A_\alpha) = 1$ for all but finitely many α we do not have to consider the problem of convergence of infinite product. We denote by Γ_0 the collection of all finite unions of cylindrical sets. The following result is analogous to Theorem 13.3.

Theorem 13.8:

(a) Every member C of Γ_0 can be written as a finite union of disjoint CS's.

(b) If $\{R_j\}$ and $\{S_j\}$ are two sequences of disjoint CS's that have the same union, then we have

$$\sum_{j=1}^{\infty} \lambda(R_j) = \sum_{j=1}^{\infty} \lambda(S_j). \tag{13–3}$$

Thus the function λ can be extended to all of Γ_0 by setting

$$\lambda(C) = \sum_{j=1}^{n} \lambda(R_j);$$

$R_1, \ldots, R_n$ are disjoint CS's with union C.

(c) The collection Γ_0 forms an algebra, and λ as defined by (b) is a measure on Γ_0.

Proof:

(a) If we can show that whenever R and T are CS's then $R\backslash T\cap R$ is the union of disjoint CS's then we can imitate the reasoning in the proof of (a) of 13.3 to obtain the required result. We can write T as the intersection $\bigcap_{i=1}^{n} T(\alpha(i))$ where $T(\alpha(i))$ is the product $T(\alpha(i)) = T_{\alpha(i)} \times \prod_{\alpha\neq\alpha(i)} S_\alpha$. Therefore, we have

$$R\backslash T\cap R = R\cap T(\alpha(1)) \cup R\cap T(\alpha(1))^{\sim} \cap T(\alpha(2)) \cup \dots$$
$$\cup\, R\cap T(\alpha(1))^{\sim} \cap \dots T(\alpha(n-1))^{\sim} \cap T(\alpha(n))$$

and the right-hand side is the union of disjoint CS's.

(b) We shall prove that if $\{C_n\}$ is a sequence of pairwise disjoint cylindrical sets with union a cylindrical set C, then we have

$$\lambda(C) = \sum_{n=1}^{\infty} \lambda(C_n).$$

We leave you the task of using this fact to derive (b) as we did in the proof of Theorem 13.3 (b).

We first show how to derive the result for a finite union

$$C = \bigcup_{n=1}^{M} C_n.$$

In this case there is a finite subset F of A such that if α is not in F we have $P_\alpha C = S_\alpha$ and $P_\alpha C_n = S_\alpha$ for each $n = 1, 2, \dots$. Denote this set F by $\{1, 2, \dots, k\}$, and denote by λ_F the product measure $\lambda_1 \times \lambda_2 \times \cdots \times \lambda_k$. We shall now use the study of finite product measure spaces that we have just completed. Because of the way we defined λ we have

$$\lambda C = \lambda_F(P_F C) \quad \text{and} \quad \lambda C_n = \lambda_F(P_F C_n) \quad \text{for } n = 1, 2, \dots.$$

We show that $\{P_F C_n \colon n = 1, 2, \dots M\}$ is pairwise disjoint with union $P_F C$. If there were a function g in $P_F C_p \cap P_F C_q$ where $p \neq q$ then we could define f on A by $f(\alpha) = g(\alpha)$ for α in F and $f(\alpha)$ arbitrarily for α in $A\backslash F$. Then $f(i) \in P_i(C_p)$ for each i in F so $f \in C_p$. Similarly f is in C_q contradicting the assumption that these two sets are disjoint. Given in f in C there is p such that $f \in C_p$. This means that $P_F f \in P_F C_p$ so $P_F f$ is in $\bigcup_{n=1}^{M} P_F C_n$. Since

$$P_F C = \bigcup_{n=1}^{M} P_F C_n$$

and the sets in the union are pairwise disjoint we have by the theory of finite product measures

$$\lambda C = \lambda_F(P_F C) = \sum_{n=1}^{m} \lambda_F(P_F C_n) = \sum_{n=1}^{M} \lambda_n C_n.$$

In the case when there are infinitely many C_n the task is more complicated. In this case there is a countable subset F of indices in A such that $P_\alpha C = P_\alpha C_n = S_\alpha$ for all n if α is not in F. In order to simplify the notation we denote the members of F by

$$\{-k, -k+1, \dots, 0, 1, \dots\}$$

where

$$P_\alpha C = S_\alpha \quad \text{if } \alpha \text{ is not in } \{-k, -k+1, \ldots, 0\} \quad \text{and}$$

$$P_\alpha C_n = S_\alpha \quad \text{if } \alpha \text{ is not in } \{-k, -k+1, \ldots, 0, 1, \ldots, n\}.$$

We can force the positive indices to increase one at a time by adding empty sets to the union if necessary. We denote by Q_n the projection $P_{\{-k, -k+1, \ldots, 0, 1, \ldots, n\}}$, and by μ_n the product measure $\lambda_{-k} \times \lambda_{-k+1} \times \cdots \times \lambda_n$. For each M the set $\bigcup_{n=1}^{M} C_n$ has the form $D \times E$ where D is a subset of the product $S_{-k} \times S_{-k+1} \times \cdots \times S_M$, which is measurable with respect to the finite product measure μ_M and $E = \prod \{S_\alpha : \alpha \in A \setminus \{-k, -k+1, \ldots, 0, 1, \ldots, M\}\}$. Therefore $C \setminus D \times E$ has the form $B \times E$ where B is also measurable with respect to μ_M. We have

$$\lambda C = \mu_M Q_M C = \sum_{n=1}^{M} \mu_M Q_M C_n + \mu_M\left(Q_M\left(\bigcup_{n=M+1}^{\infty} C_n\right)\right)$$

$$= \sum_{n=1}^{M} \lambda C_n + \mu_M\left(Q_M\left(\bigcup_{n=M+1}^{\infty} C_n\right)\right)$$

If we denote $\bigcup_{n=M+1}^{\infty} C_n$ by R_M, we see from the above equation that it suffices to prove $\lim_M \mu_M(Q_M R_M) = 0$. Since $R_{M+1} \subset R_M$, $\mu_{M+1} Q_{M+1} R_{M+1} \leqslant \mu_M(Q_M R_M)$. Therefore, the sequence $\{\mu_M Q_M(R_M)\}$ is decreasing so that if its limit is not zero there is $\delta > 0$ such that $\mu_M(Q_M R_M) > \delta$ for each M. We shall show that this implies the false conclusion that $\bigcap_n R_n$ is non-empty. For each M we have

$$\mu_M(Q_M R_M) = \int_{S_{-k}} \int_{S_{-k+1}} \cdots \int_{S_M} \chi_{R_M}(x_{-k}, x_{-k+1}, \ldots, x_m)$$
$$\times\, d\lambda_{-k}(x_{-k})\, d\lambda_{-k+1}(x_{-k+1}) \ldots d\lambda_M(x_M).$$

By the Fubini Theorem the function $f_M(u)$ defined for each M by

$$f_M(u) = \int_{S_{-k+1}} \int_{S_{-k+2}} \cdots \int_{S_M} \chi_{R_M}(u, x_{-k+1}, \ldots, x_m)$$
$$\times\, d\lambda_{-k+1}(x_{-k+1})\, d\lambda_{-k+2}(x_{-k+2}) \ldots d\lambda_M(x_M)$$

exists for almost all u in S_{-k}, and we have

$$\mu_M(Q_M R_M) = \int_{S_{-k}} f_M(u)\, d\lambda_{-k}(u).$$

If it were true that $\lim_M f_M(u) = 0$ for each u in S_{-k}, we would have $\lim_M \mu_M(Q_M R_M) = 0$ by the Lebesgue Dominated Convergence Theorem that holds since $|f_M(u)| \leqslant 1$ for each M and each u. Since we are assuming that this is not the case there is c_{-k} in S_{-k} such that $\lim_M f_M(c_{-k}) \neq 0$. This means that the sequence of iterated integrals

$$\int_{S_{-k+1}} \int_{S_{-k+2}} \cdots \int_{S_M} \chi_{R_M}(c_{-k}, x_{-k+1}, \ldots, x_m)$$
$$d\lambda_{-k+1}(x_{-k+1})\, d\lambda_{-k+2}(x_{-k+2}) \ldots d\lambda_M(x_M)$$

does not converge to 0 as $M \to \infty$. If we define g_M by the formula

$$g_M(u) = \int_{S_{-k+2}} \int_{S_{-k+3}} \cdots \int_{S_M} \chi_{R_M}(c_{-k}, u, \ldots, x_m)$$
$$\times\, d\lambda_{-k+2}(x_{-k+2})\, d\lambda_{-k+3}(x_{-k+3}) \ldots d\lambda_M(x_M)$$

we argue as above that there is c_{-k+1} such that $\lim_M g_M(c_{-k+1})$ does not converge to 0. Proceeding by induction we can construct a sequence c_{-k}, $c_{-k+1}, \ldots, c_0, c_1, c_2, \ldots$ such that for each q the sequence of integrals

$$\int_{S_{q+1}} \int_{S_{q+2}} \cdots \int_{S_M} \chi_{R_M}(c_{-k}, c_{-k+1}, \ldots, c_q, x_{q+1}, \ldots, x_M)$$
$$d\lambda_{q+1}(x_{q+1})\, d\lambda_{q+2}(x_{q+2}) \ldots d\lambda_M(x_M),$$

which is defined for $M > q$, does not converge to 0. Since this sequence is decreasing we must have that each of these integrals is greater than 0. Consequently, if $M > q$ there are $x_{q+1}, \ldots, x_M$ in $S_{q+1} \times \cdots \times S_M$ such that $(c_{-k}, c_{-k+1}, \ldots, c_q, x_{q+1}, \ldots, x_M)$ is in $Q_M(R_M)$. Therefore, we conclude that

$$(c_{-k}, c_{-k+1}, \ldots, c_q) \in Q_q(R_M) \subset Q_q(R_q) \quad \text{for all } q.$$

If we now define f in S by $f(j) = c_j$ for $j = -k, -k+1, \ldots$ and $f(\alpha)$ arbitrarily for α not in F, then we have $Q_q f \in Q_q(R_q)$ for each q so that $f \in R_q$ for each q, which implies that $\bigcap R_n$ is nonempty.

(c) You can adapt the proof of Theorem 13.3 (c) to this result. ■

The extension of the measure λ on Γ_0 is called the *infinite product measure* $\lambda = \times_{\alpha \in A} \lambda_\alpha$.

Exercises

A

1. Let S denote a half-open interval $[a, b)$ and let $\mathscr{A}_0$ denote the collection of all finite unions $A = \bigcup_{j=1}^k J_j$ where each J_j is a half-open interval of the form $[a_j, b_j)$ contained in S.
   (a) $\mathscr{A}_0$ forms an algebra of sets. Define m on $\mathscr{A}_0$ by $m(A) = \sum_{j=1}^k (b_j - a_j)$ if A has the form above and the intervals are disjoint.
   (b) Show that m is well defined.
   (c) Show that m is a measure on $\mathscr{A}_0$.
   (d) Show that the σ-algebra generated by $\mathscr{A}_0$ is the collection $\mathscr{B}_{[a,b)}$ of Borel subsets of $[a, b)$.
   (e) Prove that the unique extension of m to $\mathscr{B}$ is Lebesgue measure on $\mathscr{B}_{[a,b)}$.
2. Prove that every open subset U of $\mathbb{R}^n$ is the countable union of products of open rectangles ($U = \bigcup_k R_k$ where $R_k = \prod_{j=1}^n (a_j, b_j)$). This implies that every open subset and hence every Borel subset of $\mathbb{R}^n$ is measurable.
3. Let S denote the set of all rationals r with $0 \leqslant r < 1$ and let $\mathscr{A}_0$ denote the collection of all finite unions $\bigcup_{n=1}^k J_n$ where J_n has the form $J_n = \{r : r \in Q, a \leqslant r < b\}$ for $0 \leqslant a < b \leqslant 1$. That is, each J_n is the set of all rational numbers in a half-open interval.

(a) Show that $\mathscr{A}_0$ forms an algebra of sets.
(b) Define μ on $\mathscr{A}_0$ by the formula $\mu(A)=\infty$ if $A \neq \varnothing$, and $\mu(\varnothing)=0$. Then μ is a consistent *fa* measure on $\mathscr{A}_0$.
(c) Prove that the smallest σ-algebra $\mathscr{A}$ containing $\mathscr{A}_0$ is the collection of all subsets of S.
(d) Show that the extension of μ to $\mathscr{A}$ is not unique.

4. Let μ be the counting measure on the collection $\mathscr{A}$ of all subsets of the set $\mathbb{N}=\{1, 2, \ldots\}$, that is, $\mu(A)=$ the number of members of A if A is finite and ∞ otherwise.
(a) Describe the meaning of integration in $(\mathbb{N}, \mathscr{A}, \mu) \times (\mathbb{N}, \mathscr{A}, \mu)$ in terms of double series, $\sum_{i,j} a_{i,j}$.
(b) Interpret the theorems of Fubini and Tonelli in this setting.
(c) Suppose $a_{i,i}=-1$, $a_{i,i+1}=1$, and $a_{i,j}=0$ otherwise. Show that

$$\sum_i \sum_j a_{i,j} = \sum_j \sum_i a_{i,j}, \quad \text{but } \sum_{i,j} a_{i,j} \text{ is not integrable.}$$

(d) Suppose $a_{i,i}=-1$ for each i, $a_{i-1,i}=1$ for $i>1$ and $a_{i,j}=0$ otherwise. Show that $\sum_i \sum_j a_{i,j} \neq \sum_j \sum_i a_{i,j}$. What hypotheses of Fubini's and Tonelli's theorems have been violated?

5. *(a)* Show that if R is in $\mathscr{A} \times \mathscr{B}$, then $R_s \in \mathscr{B}$ for all s in S.
(b) Show that if the function f on $S \times T$ is measurable with respect to $\mathscr{A} \times \mathscr{B}$, then f_s is measurable with respect to $\mathscr{B}$ for all s in S.

6. Suppose f and g are BBS-functions on $(S, \mathscr{A}, \mu)$ and $(T, \mathscr{B}, v)$ respectively. Then show that $F(x, y)=f(x)g(y)$ is integrable on $S \times T$ and

$$\int_{S \times T} F \, d\lambda = \left\{\int f \, d\mu\right\}\left\{\int g \, dv\right\}.$$

7. Show that if $(S, \mathscr{A}, \mu)$ and $(T, \mathscr{B}, v)$ are two σ-finite measure spaces, then $\lambda \equiv \mu \times v$ is the only measure on $\mathscr{A} \times \mathscr{B}$ such that $\lambda(A \times B) \equiv \mu(A)v(B)$ for every MR $A \times B$.

8. Show that if f is a nonnegative function from $\mathbb{R}$ into $\mathbb{R}$ then $\int_a^b f(t)\, dm$ is equal to $m_2(B)$ where B is the set of all (s, t) such that $0 \leqslant t \leqslant f(s)$ and $a \leqslant s \leqslant b$ and m_2 is the product measure $m \times m$.

B

9. A collection $\mathscr{A}_s$ of subsets of a set S is called a *semialgebra* if (1) whenever A, B are in $\mathscr{A}_s$ so is $A \cap B$; (2) If $A \in \mathscr{A}_s$ then the complement $A^{\sim}$ of A is a finite union of members of $\mathscr{A}_s$.
(a) Show that if $\mathscr{A}_s$ is a semialgebra then the collection $\mathscr{A}_0$ consisting of $\varnothing$ and all finite unions of sets in $\mathscr{A}_s$ forms an algebra. $\mathscr{A}_0$ is called the algebra generated by $\mathscr{A}_s$.
(b) Show that the collection of all half-open intervals $[a, b)$ forms a semialgebra of subsets of $[-\infty, \infty)$.
(c) Suppose $\mathscr{A}_s$ is a semialgebra of subsets of a set S and μ is a nonnegative function on $\mathscr{A}_s$ such that $\mu(\varnothing)=0$ if $\varnothing \in \mathscr{A}_s$. Show that there is a unique

extension of μ to a measure on the algebra $\mathscr{A}_0$ generated by $\mathscr{A}_s$ if and only if the following two conditions hold. (1) If $A_1, A_2, \ldots, A_n$ are pairwise disjoint subsets in $\mathscr{A}_s$ and $\bigcup_{j=1}^{n} A_j$ is in $\mathscr{A}_s$ then $\mu\{\bigcup_{j=1}^{n} A_j\} = \sum_{j=1}^{n} \mu(A_j)$. (2) If $A_1, A_2, \ldots$ is a sequence of disjoint subsets of $\mathscr{A}_s$ and $\bigcup_j A_j \in \mathscr{A}_s$, then $\mu\{\bigcup_j A_j\} \leqslant \sum_j \mu(A_j)$.

(d) Show that if $\mathscr{A}_s$ is the collection of half-open intervals and $\mu[a, b) = b - a$, then μ satisfies (1) and (2) of (c).

10. Let μ be a measure on an algebra $\mathscr{A}$, and μ^* the outer measure derived from μ. Let $\mathscr{A}_\sigma$ denote the collection of all countable unions of sets in $\mathscr{A}$ and let $\mathscr{A}_\delta$ denote the collection of all countable intersections of sets in $\mathscr{A}$. The following five statements are equivalent for a set A:

(a) Show that A is measurable.

(b) Show that for each $\varepsilon > 0$ there is B in $\mathscr{A}_\sigma$ such that $B \supset A$ and $\mu^*(B \setminus A) < \varepsilon$.

(c) Show that for each $\varepsilon > 0$ there is C in $\mathscr{A}_\delta$ such that $C \subset A$ and $\mu^*(A \setminus C) < \varepsilon$.

(d) Show that there is B in $\mathscr{A}_{\sigma\delta} = (\mathscr{A}_\sigma)_\delta$ such that $B \supset A$ and $\mu^*(B \setminus A) = 0$.

(e) Prove that there is C in $\mathscr{A}_{\delta\sigma}$ such that $C \subset A$ and $\mu^*(A \setminus C) = 0$;

11. Let $S = [0, 1]$; let m denote Lebesgue measure; let v denote counting measure and let $\mathscr{B}$ denote the class of Borel subsets of S.

(a) Show that the diagonal $D = \{(s, t) \in S \times S: s = t\}$ is a measurable subset of $(S, \mathscr{B}, m)$ $(S, \mathscr{B}, v)$. Let f denote the characteristic function of D.

(b) Show that $\int_S f(s, t)\, dv(t) = 1$ for all s.

(c) Show that for all t, $\int_S f(s, t)\, dm(t) = 0$.

(d) Prove that $\int_{S \times T} f(s, t)\, dm \times v = \infty$.

12. Suppose f and g are integrable functions on complete measure spaces $(S, \mathscr{A}, \mu)$ and $(T, \mathscr{B}, v)$ respectively.

(a) Show $F(x, y) = f(x)g(y)$ is integrable on the product space. (This is easy for f and g BBS functions. If f and g are positive, define $f_n(x)$ to be $f(x)$ if $1/n < f(x) < n$ and 0 otherwise. Define $g_n(x)$ in a similar way. Show f_n and g_n are BBS functions. Show $f_n g_n$ converges pointwise to F, and apply the Monotone Convergence Theorem. For arbitrary f and g, divide them into positive and negative parts.)

(b) Evaluate $\int_{S \times T} F\, d\lambda$ where $\lambda = \mu \times v$.

C

13. Fill in the details in the proof of Theorem 13.8.

14. Denote the set $\{0, 1\}$ by 2. Let λ be the measure on 2 for which $\lambda\{0\} = \lambda\{1\} = \frac{1}{2}$. Let μ be the infinite product measure on $2^N = 2 \times 2 \times \ldots$ where $\mu = \lambda \times \lambda \times \ldots$.

(a) What is the measure of the subset A of 2^N of all sequences (d_n) where $d_1 = 1$, $d_3 = 0$, and $d_{11} = 1$?

(b) Let $d_1, d_2, \ldots, d_n$ be a finite sequence of 0's and 1's. Prove that the set B of all sequences (b_i) in 2^N that contains no sequence of n terms equal to this given sequence has measure 0.

(c) Prove that a monkey typing steadily on a typewriter long enough will, with probability one, produce any finite number of the complete works of Shakespeare.

14

Signed Measures

The following result is a consequence of Lebesgue's Convergence Theorem (12.12).

Theorem 14.1: Suppose f is an integrable function on a measure space $(S, \mathscr{A}, \mu)$. If $\{A_n\}$ is a disjoint sequence of measurable subsets of S and $A = \bigcup_n A_n$, then we have

$$\int_A f\, d\mu = \sum_n \int_{A_n} d\mu$$

with the infinite series converging absolutely.

The implication of Theorem 14.1 is that the set function μ_f defined by $\mu_f(A) = \int_A f\, d\mu$ satisfies condition (CA) of Section 11. It is also true that $\mu_f(\varnothing) = 0$ so that if f is nonnegative, almost everywhere it follows that μ_f is a measure. Otherwise, μ_f is an example of a *signed measure* of *s*-measure, which is defined to be a set function v taking values in $\mathbb{R}^*$ that satisfies the following conditions

(SM1) $v(\varnothing) = 0$.

(SM2) If $\{A_n\}$ is a sequence of disjoint measurable sets, then the series $\sum_n v(A_n)$ converges absolutely to $v\{\bigcup_n A_n\}$ if $v\{\bigcup_n A_n\}$ is finite and diverges to $v\{\bigcup_n A_n\}$ otherwise.

(SM3) The range of v does not contain both $+\infty$ and $-\infty$.

The reason for requiring condition (SM3) is that the sum of ∞ and $-\infty$ is not defined.

If μ_1 and μ_2 are measures on a σ-algebra $\mathscr{A}$ and μ_1 is finite, that is, μ_1 does not assume the value ∞, then both $\mu_1 - \mu_2$ and $\mu_2 - \mu_1$ are signed measures. In this section we shall find that every signed measure can be decomposed in this way. Let v be a signed measure on a σ-algebra $\mathscr{A}$. A set A in $\mathscr{A}$ is called *v-positive* (*positive* if v is understood) if $v(b) \geqslant 0$ for every subset B of A. The set A is called *v-negative* if $v(B) \leqslant 0$ for every subset B of A. If $v(B) = 0$ for every subset B of A then A is called a *v-null* set. The following theorem known as the *Hahn Decomposition Theorem* shows how a signed measure induces a decomposition of a space into a positive and negative set.

Theorem 14.2: Suppose $\mathscr{A}$ is a σ-algebra of subsets of a set S and v is an *s*-measure on $\mathscr{A}$. Then there are positive set S^+ and negative set S^- such that $S = S^+ \cup S^-$ and $S^+ \cap S^- = \varnothing$.

Proof: Assume that v does not assume the value $+\infty$. If the set S itself is a negative set, then we are finished. Otherwise, we can find a set A in $\mathscr{A}$ such that $v(A) > 0$. Our first task is to verify the following statement.

(POS) If $v(A) > 0$, then A contains a positive set C with $v(C) > 0$.

If A itself is a positive set, then we choose A. If A is not a positive set, then let n_1 be the smallest positive integer such that there is a subset B_1 of A with

$\nu(B_1) < -1/n_1$. If $A \backslash B_1$ is not a positive set, let n_2 be the smallest positive integer such that there is a subset B_2 of $F \backslash B_1$ with $\nu(B_2) < -1/n_2$. If $C_k = A \backslash \{B_1 \cup B_2 \cup \ldots \cup B_k\}$ is not a positive set, let n_{k+1} be the smallest positive integer such that there is a subset B_{k+1} of C_k such that $\nu(B_{k+1}) \leqslant -1/n_{k+1}$. In this way we either obtain a positive set after a finite number of steps or else we obtain an infinite sequence $\{B_k\}$ such that $\nu(B_k) < -1/n_k$ for all k. If $C = A \backslash \bigcup_{k=1}^{\infty} B_k$ then $A = C \cup \{\bigcup_k B_k\}$. Since this is a union of disjoint sets we have

$$\nu(A) = \nu(C) + \sum_k \nu(B_k)$$

with the series on the right converging absolutely. Each term $\nu(B_k)$ is negative but $\nu(A)$ is positive so that $\nu(C)$ must be positive. The assumption that $\nu(B_k) < -1/n_k$ implies that $(1/n_k)$ converges to 0. Since C is contained in C_k for each k it follows that if D is a subset of C we must have $\nu(D) > -1/(n_k - 1)$ for each k. This implies $\nu(D) \geqslant 0$, which proves that C is positive. This completes the proof of (POS).

Next let M denote the supremum of all $\nu(A)$ where A is a positive subset of S. Let $\{A_n\}$ be a sequence of positive sets such that $\nu(A_n)$ increases to M. If $A = \bigcup_n A_n$ then A is a positive set because if $B \subset A$, we can write B as a disjoint union $\bigcup_n B_n$ where

$$B_n = B \cap A_n \backslash \{A_1 \cup \ldots \cup A_{n-1}\}.$$

For each n, $B_n \subset A_n$ so B_n is positive. Thus $\nu(B) = \sum \nu(B_n) \geqslant 0$. That A is a positive set means that $\nu(A) \leqslant M$, but for each n it is also true that

$$\nu(A) = \nu(A_n) + \nu(A \backslash A_n) \geqslant \nu(A_n).$$

Therefore, we conclude $\nu(A) = M$ and $M < \infty$.

We conclude the proof of the theorem by showing that the complement $A^{\sim}$ of A is a negative set. If not we could find a subset B of $A^{\sim}$ with $\nu(B) > 0$. Thus by (POS) B contains a positive set B_1 with $\nu(B_1) > 0$. Then $A \cup B_1$ is positive and $\nu(A \cup B_1) = \nu(A) + \nu(B_1) > M$. But this contradicts the definition of M. Therefore, $A^{\sim}$ must be a negative set. ■

Two measures μ_1 and μ_2 defined on a σ-algebra $\mathscr{A}$ of subsets of a set S are said to be *mutually singular* and we write $\mu_1 \perp \mu_2$ if there is a subset T of S such that $\mu_1(T^{\sim}) = \mu_2(T) = 0$. This can be expressed intuitively by saying that μ_1 and μ_2 "live in different worlds." The following result is called the *Jordan Decomposition Theorem.*

Corollary 14.3: Suppose ν is an s-measure on a σ-algebra $\mathscr{A}$ of subsets of S. Then there is a unique pair (ν^+, ν^-) of mutually singular measures, one of which is finite, such that $\nu = \nu^+ - \nu^-$.

Proof: Let S^+ and S^- be disjoint positive and negative subsets of S determined by Theorem 14.2. Define ν^+ and ν^- by

$$\nu^+(A) = \nu(A \cap S^+) \quad \text{and} \quad \nu^-(A) = -\nu(A \cap S^-) \quad \text{for } A \text{ in } \mathscr{A}.$$

The countable additivity of ν^+ and ν^- follows from that property for ν. The other properties are obvious. If ν cannot take on the value $+\infty$ then ν^+ is finite. Otherwise, ν^- is finite.

Suppose $\nu = \mu_1 - \mu_2$ is a decomposition of ν into mutually singular measures. We must show that for all A in $\mathscr{A}$ we have

$$\mu_1(A) = \nu(A \cap S^+) \quad \text{and} \quad \mu_2(A) = -\nu(A \cap S^-).$$

Let T in $\mathscr{A}$ be such that $\mu_1(T^\sim) = \mu_2(T) = 0$. Then $T^\sim$ is a ν-negative set so that $S^+ \cap T^\sim$ is a ν-null set. Consequently, for all A in $\mathscr{A}$ we have

$$\begin{aligned}\nu(A \cap S^+) &= \nu(A \cap S^+ \cap T) + \nu(A \cap S^+ \cap T^\sim) \\ &= \mu_1(A \cap S^+ \cap T) - \mu_2(A \cap S^+ \cap T) = \mu_1(A \cap T) \\ &= \mu_1(A).\end{aligned}$$

That $\mu_2(A) = -\nu(A \cap S^-)$ has a similar derivation. ■

If the s-measure ν has the decomposition $\nu = \nu^+ - \nu^-$ as described above we denote by $|\nu|$ the measure $\nu^+ + \nu^-$. The measure $|\nu|$ is called the *variational measure* induced by ν; and the number

$$\|\nu\| = |\nu|(S) = \nu^+(S) + \nu^-(S)$$

is called the *variation of* ν. A measure μ_1 is said to be *absolutely continuous* with respect to a measure μ on a σ-algebra $\mathscr{A}$, and we write $\mu_1 \ll \mu$ if $\mu_1(A) = 0$ for each A in $\mathscr{A}$ for which $\mu(A) = 0$. A s-measure ν is *absolutely continuous* with respect to a measure μ if both ν^+ and ν^- are absolutely continuous with respect to μ.

For the rest of this section we denote by $\mathscr{A}_0$ the collection of finite unions $A = \bigcup_{j=1}^n J_j$ where J_j is either a finite half-open interval of the form $(a, b]$ or an infinite interval having the form $(-\infty, a]$ or $(a, \infty]$. It is obvious that $\mathscr{A}_0$ is closed under finite unions. The intersection of two intervals $(a, b]$ and $(c, d]$ is either empty or else the interval $(\sup(a, c), \inf(b, d)]$. Since the intersection of two finite unions is the union of intersections it follows that $\mathscr{A}_0$ is closed under intersections. That $\mathscr{A}_0$ is also closed under complements if we use $\mathbb{R} \cup \{\infty\}$ as our space follows from the fact that

$$(a, b]^\sim = (-\infty, a] \cup (b, \infty],$$

and the fact that $\mathscr{A}_0$ is closed under intersections. Therefore, $\mathscr{A}_0$ forms an algebra of sets.

A real valued function f defined on $\mathbb{R}$ is said to be *a distribution* if

(D1) f is increasing.

(D2) $\lim_{x \to -\infty} f(x) = 0$.

(D3) f is *continuous from the right*.

An arbitrary function f is continuous from the right if for all $x \in \mathbb{R}$ and $\varepsilon > 0$, there is $\delta > 0$ such that $|f(x) - f(y)| < \varepsilon$ when $x < y < x + \delta$. In the special case when f is increasing it suffices to verify that $f(x + \delta) < f(x) + \varepsilon$.

Theorem 14.4:

(a) If f is a distribution on $\mathbb{R}$, then a function μ_f is defined on $\mathscr{A}_0$ by the formula

$$\mu_f(A)=\sum_{j=1}^{n}\{f(b_j)-f(a_j)\}$$

where A is the disjoint union of $\{(a_j, b_j]: j=1, 2, \ldots, n\}$.

(b) The function μ_f defined by (a) is a measure on $\mathscr{A}_0$.

(c) There is a unique measure μ defined on the σ-algebra $\mathscr{B}$ of Borel subsets of $\mathbb{R}$ such that

$$\mu(a, b]=f(b)-f(a) \quad \text{for all } a<b.$$

Proof:

(a) Here it is necessary to show that if A has two representations $A=\bigcup_{j=1}^{n}(a_j, b_j]$ and $A=\bigcup_{k=1}^{m}(c_j, d_j]$, we have

$$\sum_{j=1}^{n}(f(b_j)-f(a_j))=\sum_{k=1}^{m}(f(d_k)-f(c_k)). \tag{14–1}$$

If A has the form $(a, b]$ then (14–1) is true because if $A=\bigcup_{j=1}^{n}(a_j, b_j]$ it must result that $a_{j(1)}=a$ for some $j(1)$, and that for each $p<n$ we have $a_{j(p+1)}=b_{j(p)}$ while $b_{j(n)}=b$. Thus we have,

$$\sum_{j=1}^{n}(f(b_j)-f(a_j))=\sum_{p=1}^{n}(f(b_{j(p)})-f(a_{j(p)}))$$
$$=f(b)-f(a),$$

while $\sum_{k=1}^{m}(f(d_k)-f(c_k))$ also must equal $f(b)-f(a)$. To conclude the proof let $\{(s_r, t_r]: r=1, 2, \ldots, K\}$ denote the collection of all nonempty intersections of intervals $(a_j, b_j]$ and $(c_k, d_k]$. Then by the preceding argument we have

$$\sum_{r=1}^{K}(f(t_r)-f(s_r))=\sum_{j=1}^{n}\sum\{(f(t_r)-f(s_r)): (s_r, t_r]\subset(a_j, b_j]\}$$
$$=\sum_{j=1}^{n}(f(b_j)-f(a_j)).$$

Similarly we have

$$\sum_{r=1}^{K}(f(t_r)-f(s_r))=\sum_{k=1}^{m}(f(d_k)-f(c_k)).$$

Therefore, formula (14–1) is valid.

(b) Suppose $\{(a_i, b_i]: i=1, 2, \ldots\}$ is a sequence of disjoint intervals whose union is the interval $(a, b]$. For each n we have

$$\sum_{i=1}^{n}\{f(b_i)-f(a_i)\}\leqslant f(b)-f(a).$$

To see this, renumber the intervals so that $b_{i-1}\leqslant a_i$ for $i=2, \ldots, n$; then we have

$$\sum_{i=1}^{n} \{f(b_i)-f(a_i)\} = f(b_n) + \sum_{i=2}^{n} \{f(b_{i-1})-f(a_i)\} - f(a_1)$$
$$\leqslant f(b_n) - f(a_1) \leqslant f(b) - f(a)$$

because f is increasing. This implies that

$$\sum_{i=1}^{\infty} \{f(b_i)-f(b_{i-1})\} \leqslant f(b)-f(a).$$

In order to prove the reverse inequality we use the continuity of f from the right. Let $\varepsilon > 0$ be given. For each i, let δ_i be such that $f(b_i+\delta_i) \leqslant f(b_i) + \varepsilon/2^i$, and let δ be such that $f(a+\delta) \leqslant f(a) + \varepsilon$. Since $\{(a_i, b_i]: i = 1, 2, \ldots\}$ covers $(a, b]$ it follows that the collection of open intervals $\{(a_i, b_i+\delta_i): i = 1, 2, \ldots\}$ covers $[a+\delta, b]$. Thus there is a finite subcover $\{(a_i, b_i+\delta_i): i = 1, 2, \ldots, n\}$. We can renumber this subcover so that $a_1 < a+\delta$, $b_n+\delta_n > b$, and $a_i < a_{i+1} < b_i+\delta_i$ for $i = 1, 2, \ldots, n$; that is, we can chain the intervals together in order. When we do this we see that

$$\begin{aligned} f(b)-f(a)-\varepsilon &< f(b)-f(a+\delta) \\ &\leqslant f(b_n+\delta_n) + \sum_{i=1}^{n-1} \{f(b_i+\delta_i)-f(a_{i+1})\} - f(a_1) \\ &\leqslant \sum_{i=1}^{\infty} \{f(b_i+\delta_i)-f(a_i)\} \\ &\leqslant \varepsilon + \sum_{i=1}^{\infty} \{f(b_i)-f(a_i)\}. \end{aligned}$$

We conclude that for each $\varepsilon > 0$ we have

$$f(b)-f(a) \leqslant 2\varepsilon + \sum_{i=1}^{\infty} \{f(b_i)-f(a_i)\}.$$

This argument can be modified for infinite intervals.

If $\{(a_j, b_j]: j = 1, 2, \ldots\}$ is a countable collection of intervals that covers an arbitrary set $A = \bigcup_{i=1}^{K} J_i$ in $\mathscr{A}_0$, we can first assume the intervals $J_1, J_2, \ldots, J_K$ are disjoint and then apply the preceding argument to the collection $\{J_i \cap (a_j, b_j]: j = 1, 2, \ldots\}$ for each i in order to ascertain that

$$\mu_0\left\{\bigcup_{i=1}^{K} J_i\right\} = \sum_{j=1}^{\infty} \{f(b_j)-f(a_j)\}.$$

This shows that μ_0 is a consistent measure on the algebra $\mathscr{A}_0$ so (c) follows by Theorem 13.2. ■

Suppose f is a distribution function and μ is the unique measure such that $\mu(a, b] = f(b)-f(a)$ for all a, b. If g is a function on $\mathbb{R}$, which is integrable with respect to μ, the integral $\int g\, d\mu$ is also written $\int g\, df$ and is called the *Lebesgue-Stieltjes integral* of g with respect to f. Since the smallest σ-algebra containing $\mathscr{A}_0$ is the collection of Borel sets it follows that every continuous function is measurable with respect to μ. If f is any increasing function, then the function $\hat{f}$ defined by

$$\hat{f}(x) = \lim_{u \downarrow x} f(u) \tag{14–2}$$

is the unique increasing function that is continuous on the right and coincides with f at all x where f is continuous on the right. We can extend the definition of Lebesgue-Stieltjes integral by the formula

$$\int g\, df = \int g\, d\hat{f}$$

where f is an arbitrary increasing function.

Let f be a real valued function defined on an interval $[a, b]$. If $x < y$ are points in $[a, b]$ then the *variation* $V(f, x, y)$ of f between x and y is defined to be the supremum of all sums

$$\sum_{j=1}^{n} |f(u_i) - f(u_{i-1})|$$

as $\{u_0, \ldots, u_n\}$ ranges over all finite sets of reals with $x \leqslant u_0 \leqslant u_1 \leqslant \ldots \leqslant u_n \leqslant y$. It follows quickly from the definition of variation that if $x \leqslant v < w \leqslant y$ then $V(f, v, w) \leqslant V(f, x, y)$. If $V(f, a, b) < \infty$ then f is said to be of *bounded variation* on $[a, b]$.

Theorem 14.5: A real valued function f on an interval $[a, b]$ is of bounded variation on $[a, b]$ if and only if f is the difference of two increasing functions.

Proof: Every increasing function f on an interval $[a, b]$ is of bounded variation because if $a \leqslant u_0 \leqslant \ldots \leqslant u_n \leqslant b$ we have

$$\begin{aligned}\sum_{j=1}^{n} |f(u_j) - f(u_{j-1})| &= \sum_{j=1}^{n} f(u_j) - f(u_{j-1}) \\ &= f(u_n) - f(u_0) \leqslant f(b) - f(a).\end{aligned}$$

Hence, $V(f, a, b) \leqslant f(b) - f(a)$. If f_1 and f_2 are of bounded variation on $[a, b]$ so is $f_1 - f_2$ and $V(f_1 - f_2, a, b) \leqslant V(f_1, a, b) + V(f_2, a, b)$. For if $a \leqslant u_0 \leqslant \ldots \leqslant u_n \leqslant b$ we have

$$\begin{aligned}&\sum_{j=1}^{n} |f_1(u_j) - f_2(u_j) - \{f_1(u_{j-1}) - f_2(u_{j-1})\}| \\ &\qquad \leqslant \sum_{j=1}^{n} |f_1(u_j) - f_1(u_{j-1})| + \sum_{j=1}^{n} |f_2(u_j) - f_2(u_{j-1})| \\ &\qquad \leqslant V(f_1, a, b) + V(f_2, a, b).\end{aligned}$$

These two observations prove that the difference of two increasing functions is a function of bounded variation.

If f is of bounded variation on $[a, b]$, define F on $[a, b]$ by $F(x) = V(f, a, x)$. Then $F + f$ and $F - f$ are both increasing functions. In the case of $F + f$, we have for $x < y$

$$\begin{aligned} F(y)+f(y)-\{F(x)+f(x)\} &= V(f, a, y)-V(f, a, x)+f(y)-f(x) \\ &= V(f, x, y)+\{f(y)-f(x)\} \\ &\geqslant |f(y)-f(x)|+\{f(y)-f(x)\} \\ &\geqslant 0. \end{aligned}$$

Here we have used the relation

$$V(f, a, x)+V(f, x, y)=V(f, a, y),$$

which you may prove. Therefore we have, $f=f_1-f_2$ where $f_1=(F+f)/2$ and $f_2=(F-f)/2$. ■

Suppose f is a function of bounded variation on the interval $[a, b]$ and suppose $f=f_1-f_2$ where f_1 and f_2 are increasing. We can assume that f_1 and f_2 are nonnegative by adding $-\inf\{f_1(a), f_2(a)\}$ to both of them if necessary. Then we can extend f_1 and f_2 to all of R by letting $f_i(u)=0$ for $u<a$ and $f_i(u)=f_i(b)$for $u>b$. We can then define the distribution functions $\hat{f}_1$ and $\hat{f}_2$ by formula (14–2) and measures μ_1 and μ_2 by Theorem 14.3. The signed measure $\mu_1-\mu_2$ is called the *measure determined by the distribution function* f, and the integral $\int_a^b g\,d\mu$ is called the *Lebesgue-Stieltjes integral* of g with respect to f; it is usually written $\int_a^b g\,df$. The existence portion of the following theorem is valid because the μ-measurable sets include the Borel sets. The proof of the linearity results is left to you.

Theorem 14.6: If f is of bounded variation and g is continuous on an interval $[a, b]$ then the Lebesgue-Stieltjes integral $\int_a^b g\,df$ exists. For real numbers c, d, continuous functions g_1, g_2, and functions f_1, f_2 of bounded variation we have the identities

$$\int_a^b \{cg_1+dg_2\}\,df_1 = c\int_a^b g_1\,df_1 + d\int_a^b g_2\,df_1$$

and

$$\int_a^b g_1\,d\{cf_1+df_2\} = c\int_a^b g_1\,df_1 + d\int_a^b g_1\,df_2.$$

A function ν from a σ-algebra $\mathscr{A}$ on a set S into the field $\mathbb{C}$ of complex numbers is called a *complex measure* if it satisfies (SM1) and (SM2). In this case we explicitly require that the range of ν be restricted to noninfinite values. We leave the proof of the following result to you.

Theorem 14.7: Suppose $\mathscr{A}$ is a σ-algebra of subsets of a set S and ν is a complex measure on $\mathscr{A}$. Then there are measures λ_1, λ_2, λ_3, λ_4 on $\mathscr{A}$ such that

(CM) $\nu(A)=\lambda_1(A)-\lambda_2(A)+i(\lambda_3(A)-\lambda_4(A))$

for all A in $\mathscr{A}$.

It is clear that every complex function on $\mathscr{A}$ having form (CM) is a complex measure.

Exercises

A

1. For a real number t, t^+ denotes $\max\{t, 0\}$ and t^- denotes $-\min\{t, 0\}$.
   (a) Show that $t = t^+ - t^-$ and $|t| = t^+ + t^-$. If f is a function from a set S into the real numbers, the function f^+ is defined by $f^+(s) = (f(s))^+$ and f^-, by $f^-(s) = (f(s))^-$ for all s in S.
   (b) Show that $f = f^+ - f^-$ and $|f| = f^+ + f^-$.
   (c) Show that f is a continuous function from $\mathbb{R}$ into $\mathbb{R}$ if and only if f^+ and f^- are continuous from $\mathbb{R}$ into $\mathbb{R}$.
   (d) Show that if f is a function from a measure space $(S, \mathscr{A}, \mu)$ into $\mathbb{R}$ then f is measurable if and only if f^+ and f^- are measurable.
   (e) If $v(A) = \int_A f\, d\mu$, determine the measures v^+ and v^- in terms of f^+ and f^-.
2. Show that the function $\rho(v_1, v_2) = \|v_1 - v_2\|$ is a metric on the space of all bounded s-measures on a σ-algebra $\mathscr{A}$.
3. Let $v(A)$ be defined by $v(A) = \int_{A \cap C} x^3 - x\, dx$ for all Lebesgue measurable subsets A of $\mathbb{R}$, where $C = (-\infty, 1000]$. Determine the Jordan decomposition of v. Determine the Hahn decompositions for v.
4. For two s-measures v_1 and v_2 on a σ-algebra $\mathscr{A}$ of subsets of a set S we say that $v_1 \leqslant v_2$ if $v_1(A) \leqslant v_2(A)$ for all A in $\mathscr{A}$.
   (a) Show that with this ordering the s-measure v defined by
   $$v = \tfrac{1}{2}\{v_1 + v_2\} + \tfrac{1}{2}|v_1 - v_2|$$
   corresponds to the supremum of v_1 and v_2 because $v(A)$ is no smaller than $v_1(A)$ and $v_2(A)$ for all A in $\mathscr{A}$ and is the smallest measure having this property.
   (b) Give the formula for the infimum of two measures.
   (c) Prove that the set of all finite s-measures on $\mathscr{A}$ forms a linear space and a lattice with the definition for sup given in (a). Determine the Jordan decomposition of v.
5. (a) Show that if $S^+ \cup S^-$ and $T^+ \cup T^-$ are Hahn decompositions for the s-measure v, then $S^+ \cap T^-$ and $S^- \cap T^+$ are v-null sets.
   (b) Use (a) as a hint to find two different Hahn decompositions of the measure v in Exercise 3.
6. (a) Show that for any real number a and s-measure v, $|av| = |a|\,|v|$ and $\|av\| = |a|\,\|v\|$.
   (b) Show that for v_1 and v_2, two s-measures, we have $|v_1 + v_2| \leqslant |v_1| + |v_2|$ and $\|v_1 + v_2\| \leqslant \|v_1 + v_2\|$.
7. The greatest integer function $[x]$ of a real number x is defined to be the greatest integer m that is $\leqslant x$.
   (a) Determine the measure μ such that $\mu(a, b] = f(b) - f(a)$.
   (b) Calculate $\int f\, d[x]$ for f, a μ-measurable function.

8. *(a)* Show that if f has a bounded derivative on $[a, b]$, then f is of bounded variation on $[a, b]$.

(b) Test the following functions for bounded variation on $[-1, 1]$: (i) $f_1(x) = x^2 \sin(1/x)$ for $x \neq 0$ and $f_1(0) = 0$; (ii) $f_2(x) = x^2 \sin(1/x^2)$ for $x \neq 0$ and $f_2(0) = 0$.

B

9. Suppose f is an increasing function and $\hat{f}$ is determined by (14–1). Prove that $\hat{f}$ is the unique increasing function that is continuous on the right and coincides with f wherever f is continuous on the right.

10. *(a)* Show that if ν is a finite s-measure on a σ-algebra $\mathscr{A}$, then for each A in $\mathscr{A}$ we have $\nu^+(A) = \sup\{\nu(B): B \in \mathscr{A}, B \subset A\}$.

(b) State and prove a corresponding formula for ν^-.

(c) Show that $|n|(A) = \sup\{\sum_{i=1}^{k} |\nu(A_i)|: \{A_i\}$ is a partition of $A\}$.

11. Suppose integration with respect to a signed measure ν is defined by $\int f\,d\nu = \int f\,d\nu^+ - \int f\,d\nu^-$ where $\nu^+ - \nu^-$ is the Jordan decomposition of ν.

(a) Show that there is a measurable function g with $|g| = 1$ such that $\int f\,d\nu = \int fg\,d|\nu|$.

(b) Show that if $|f(t)| \leqslant M$ for all t then we have $|\int_A f\,d\nu| \leqslant M|\nu|(A)$ for all A in $\mathscr{A}$.

12. Let $(S, \mathscr{A}, \mu)$ be a measure space, and let f be a μ-measurable function on S.

(a) Show that if $\psi(A) = \int_A f\,d\mu$ for A in $\mathscr{A}$ is an s-measure then ψ is absolutely continuous with respect to μ.

(b) What condition on f is necessary for ψ to be an s-measure?

13. *(a)* Prove that an s-measure ν is absolutely continuous with respect to a measure μ if and only if for each $\varepsilon > 0$ there is $\delta > 0$ such that $\nu(A) < \varepsilon$ whenever $\mu(A) < \delta$.

(b) Show by example that $\nu \ll \mu$ does not imply there is $M > 0$ such that $|\nu(A)| < M\mu(A)$ for each μ-measurable set A.

14. *(a)* Show that if f is an increasing function on $[a, b]$, then for each x in $[a, b]$ the limits $f(x-) = \lim_{u \uparrow x} f(u)$ and $f(x+) = \lim_{u \downarrow x} f(u)$ both exist.

(b) Show that the set of all x such that $f(x+) - f(x-) > \varepsilon$ is finite for all $\varepsilon > 0$.

(c) Show that the set of all x at which f is discontinuous is countable.

(d) Establish (a)–(c) for f, a function of bounded variation on $[a, b]$.

15. *(a)* Prove that every set function of the form (CM) is a complex measure.

(b) Prove Theorem 14.7.

15

The Radon-Nikodym Theorem

If f is an integrable function on a measure space $(S, \mathscr{A}, \mu)$, it is a straightforward exercise to show that the s-measure ν defined by $\nu(A) = \int_A f\,d\mu$ is absolutely continuous with respect to μ. It is reasonable to ask whether every s-measure that is

μ-absolutely continuous arises in this way. When $(S, \mathscr{A}, \mu)$ is a σ-finite measure space then the answer is positive. This result is known as the *Radon-Nikodym Theorem.*

Theorem 15.1: Suppose $(S, \mathscr{A}, \mu)$ is a σ-finite measure space and ν is a finite measure defined on $\mathscr{A}$, which is μ-absolutely continuous. Then there is a nonnegative measurable function f such that for each A in $\mathscr{A}$ we have

$$\nu(A) = \int_A f \, d\mu. \tag{15–1}$$

If g is any other measurable function for which (15–1) is valid, then $g = f$ μ almost everywhere.

Proof: We first establish the theorem under the hypothesis that $(S, \mathscr{A}, \mu)$ is a finite measure space. Let $L(\nu)$ denote the set of all nonnegative measurable functions h on S such that for each $A \in \mathscr{A}$ we have

$$\int_A h \, d\mu \leqslant \nu(A).$$

Since ν is nonnegative the set $L(\nu)$ at least contains the zero function. If h_1 and h_2 are in $L(\nu)$ so is $\sup\{h_1, h_2\}$. For if A in $\mathscr{A}$ is arbitrary, let B denote the set of all x in A such that $h_1(x) \geqslant h_2(x)$ and let $C = A \backslash B$. Then we have

$$\begin{aligned}\int_A \sup\{h_1, h_2\} \, d\mu &= \int_B h_1 \, d\mu + \int_C h_2 \, d\mu \\ &\leqslant \nu(B) + \nu(C) = \nu(A).\end{aligned}$$

Let m denote the quantity $\sup\{\int_S h \, d\mu : h \in L(\nu)\}$ and let $\{h_n\}$ be a sequence in $L(\nu)$ such that

$$\lim_n \int_S h_n \, d\mu = m.$$

Since ν is finite we have $m \leqslant \nu(S) < \infty$.

If for each n we set $f_n = \sup\{h_1, h_2, \ldots, h_n\}$, then each f_n is in $L(\nu)$ and $\{f_n\}$ is an increasing sequence of functions that have an integral over S increasing to m. Let f be the pointwise limit of $\{f_n\}$. By the Monotone Convergence Theorem (12.10)

$$\int_A f \, d\mu = \lim_n \int_A f_n \, d\mu$$

for every A in $\mathscr{A}$ so that $f \in L(\nu)$ and $\int_S f \, d\mu = m$.

Define a measure ψ on $\mathscr{A}$ by

$$\psi(A) = \nu(A) - \int_A f \, d\mu.$$

That $\psi(A)$ is nonnegative follows because f is in $L(\nu)$. Suppose there were B in $\mathscr{A}$

with $\psi(B) > 0$. Then there is $\varepsilon > 0$ such that $\psi(B) > \varepsilon\mu(B)$, that is,

$$(\psi - \varepsilon\mu)(B) > 0.$$

Thus the set B contains a positive set C for the s-measure $\psi - \varepsilon\mu$ with $(\psi - \varepsilon\mu)(C) > 0$. Since $\psi - \varepsilon\mu$ is absolutely continuous with respect to μ we conclude that $\mu(C) > 0$.

If χ is the characteristic function of C then $f + \varepsilon\chi$ is in $L(v)$ because for each A in $\mathscr{A}$ we have

$$\int_A (f + \varepsilon\chi)\, d\mu = \int_{A \setminus C} f\, d\mu + \int_{A \cap C} (f + \varepsilon\chi)\, d\mu$$
$$\leqslant v(A \setminus C) + v(A \cap) = v(A).$$

But this contradicts the definition of m because

$$\int_S (f + \varepsilon\chi)\, d\mu = m + \varepsilon\mu(C) > m.$$

This means that ψ is identically 0 on $\mathscr{A}$ so that formula (15–1) is valid in the case when μ is finite.

Suppose g were another function that satisfied (15–1). Let A_n be the set of all x in S with $f(x) - g(x) > 1/n$. Then we have

$$\frac{\mu(A_n)}{n} \leqslant \int_{A_n} (x) - g(x)\, d\mu = v(A_n) - v(A_n) = 0$$

from which it follows that $\mu(A_n) = 0$. The set A of all x such that $f(x) > g(x)$ is equal to $\bigcup A_n$ so that $\mu(A) = 0$. Likewise the set of all x such that $f(x) < g(x)$ has measure 0. Therefore, f is equal to g almost everywhere.

Now we establish the theorem when μ is σ-finite. Let $\{S_n\}$ be an increasing sequence of sets in $\mathscr{A}$ with $\mu(S_n) < \infty$ for each n and $S = \bigcup Sn$. By the previous argument there is for each n a measurable function g_n such that $v(A) = \int_A g_n\, d\mu$ for each measurable subset A of S_n. By the uniqueness argument $g_n = g_{n=1}$ μ-almost everywhere on S_{n-1}. Let us set $g_n(x) = 0$ for x not in S_n and let $f = \sup_n \{g_n\}$. Then f is measurable and satisfies (15–1). The uniqueness of f is established just as for v finite. ∎

We obtain the following result by applying the Radon-Nikodym Theorem to v^+ and v^-.

Corollary 15.2: Suppose $(S, \mathscr{A}, \mu)$ is a σ-finite measure space and v is a finite s-measure defined on $\mathscr{A}$, which is μ-absolutely continuous. Then there is a measurable function f such that for each A in $\mathscr{A}$ we have

$$v(A) = \int_A f\, d\mu. \qquad \textbf{(15–2)}$$

If g is any other measurable function for which (15–2) is valid, then $g = f$ μ-almost everywhere.

The function f in formula (15–2) is called the *Radon-Nikodym derivative* of ν with respect to μ and written $[d\nu/d\mu]$. It can be shown that the Radon-Nikodym derivative of a measure has similar properties to the usual derivative of a function. The following example illustrates the calculation of a simple Radon-Nikodym derivative.

Example 15.3:

Define the measure μ on Borel subsets of $\mathbb{R}$ by the formula

$$\mu(S) = \int_S x\,dm + 1 \quad \text{if 0 is in } S,$$

$$\mu(S) = \int_S x\,dm \quad \text{if 0 is not in } S.$$

Here m denotes Lebesgue measure. The measure μ is not absolutely continuous with respect to m because $\mu\{0\} = 1$, but μ is absolutely continuous with respect to m_1 where $m_1(S) = m(S) + \mu_0(S)$, where $\mu_0(S) = 1$ if S contains 0 and $\mu_0(S) = 1$ otherwise. The Radon-Nikodym representation of μ is not $\mu(S) = \int_S x\,dm_1$ because $\int_{\{0\}} x\,dm_1 = 0$. Rather it is $\mu(S) = \int_S f(x)\,dm_1$ where $f(x) = x$ for $x \neq 0$, and $f(0) = 1$.

The following result is known as the *Lebesgue Decomposition Theorem.*

Theorem 15.4: Suppose $(S, \mathscr{A}, \mu)$ is a σ-finite measure space and ν is a σ-finite measure defined on $\mathscr{A}$. Then there is a unique pair of measures (ν_0, ν_1) such that $\nu_0 \perp \mu$, $\nu_1 \ll \mu$ and $\nu = \nu_0 + \nu_1$.

Proof: First assume μ and ν are finite measures. Let $\lambda = \mu + \nu$. Since both μ and ν are absolutely continuous with respect to λ, there are measurable functions f and g such that for all A in $\mathscr{A}$ we have

$$\mu(A) = \int_A f\,d\lambda \quad \text{and} \quad \nu(A) = \int_A g\,d\lambda.$$

Let $C = \{x\colon f(x) > 0\}$ and let $D = C^{\sim} = \{x\colon f(x) = 0\}$. If ν_0 is defined by $\nu_0(A) = \nu(A \cap D)$, then $\nu_0 \perp \mu$ because $\nu_0(C) = \mu(D) = 0$. Define ν_1 by $\nu_1(A) = \nu(A \cap C) = \nu(A) - \nu_0(A)$.

To show that $\nu_1 \ll \mu$ suppose $\mu(B) = \int_B f\,d\lambda = 0$. Then f is 0 λ-a.e. on B, which implies that $\lambda(B \cap C) = 0$. However, $\lambda(B \cap C) \geq \nu(B \cap C)$ so we have $\nu(B \cap C) = \nu_1(B) = 0$.

Now suppose that μ and ν are σ-finite measures. Then we can partition S into a countable union of disjoint measurable sets $\{S_i\}$ such that $(\mu + \nu)(S_i) < \infty$ for all i. For each i there is a pair (ν_{i0}, ν_{i1}) defined on the measurable subsets of S_i such that $\nu_{i0} \perp \mu$ with regard to subsets of S_i while $\nu_{i1} \ll \mu$. Define ν_0 and ν_1 on $\mathscr{A}$ by

$$\nu_0(A) = \sum_i \nu_{i0}(A \cap S_i),$$

$$\nu_1(A) = \sum_i \nu_{i1}(A \cap S_i).$$

We verify that $\nu_0 \perp \mu$, and leave to you the task of verifying that $\nu_1 \ll \mu$. Since each $\nu_{i0} \perp \mu$ with regard to S_i it follows that there is a subset C_i of S_i such that $\nu_{i0}(C_i) = \mu(S_i \backslash C_i) = 0$. If $C = \bigcup C_i$, then $C^\sim = \bigcap (S_i \backslash C_i)$ and so $\nu_0(C) = \mu(C^\sim) = 0$. ■

Exercises

A

1. Suppose $\{A_n: n = 1, 2, \ldots\}$ is a countable partition of the interval $[0, 1]$ such that each A_n is measurable, and $\{a_n\}$ is a sequence of real numbers such that $\sum_n a_n$ converges absolutely. Define a function ν on the σ-algebra $\mathscr{M}_I$ of all Lebesgue measurable subsets of $[0, 1]$ by $\nu(B) = \sum_n a_n m(B \cap A_n)$.
   (a) Verify that ν is a measure on $\mathscr{M}_I$.
   (b) Find the function on $[0, 1]$ that determines ν in the sense of Theorem 15.2.
2. *(a)* Show that if ν is an s-measure such that $\nu \perp \mu$ and $\nu \ll \mu$, then $\nu = 0$.
   (b) Show that if ν_1 and $\nu_2 \ll \mu$ then so is $c_1 \nu_1 + c_2 \nu_2$.
   (c) If ν_1 and $\nu_2 \perp \mu$, then so is $c_1 \nu_1 + c_2 \nu_2$.
3. Let I denote the interval $[0, 1]$, and let $\mathscr{M}_I$ denote the collection of Lebesgue-measurable subsets of I. Let μ be the counting measure on $\mathscr{M}_I$ and let m be Lebesgue measure.
   (a) Prove that m is absolutely continuous with respect to μ.
   (b) Prove that there is no μ-measurable function f on I such that $m(A) = \int_A f \, d\mu$ for all A in $\mathscr{M}_I$.
   (c) Why does the conclusion of the Radon-Nikodym fail in this case?
4. Prove Corollary 15.2.
5. Show that if f is a measurable function on a measure space $(S, \mathscr{A}, \mu)$, then the function on $\mathscr{A}$ defined by $\nu(A) = \int_A f \, d\mu$ is an absolutely μ-continuous measure.
6. Suppose μ is defined on the Borel subsets of $\mathbb{R}$ by

$$\mu(S) = \int_S x \, dm + c \quad \text{if } p \in S; \qquad \mu(S) = \int_S x \, dm \quad \text{otherwise.}$$

   Determine the Radon-Nikodym decomposition of μ in terms of m and μ_p if $\mu_p(A) = 1$ when $p \in A$ and $\mu(A) = 0$ otherwise.

B

7. *(a)* Show that if ν_1 and $\nu_2 \ll \mu$, then $[d(\nu_1 + \nu_2)/d\mu] = [d\nu_1/d\mu] + [d\nu_2/d\mu]$.
   (b) Show that if $\nu \ll \mu$ and f is a nonnegative ν-measurable function, then

$$\int f \, d\nu = \int f[d\nu/d\mu] \, d\mu.$$

   (c) Prove that if $\nu \ll \lambda$ and $\lambda \ll \mu$, then $[d\nu/d\lambda] = [d\nu/d\mu][d\mu/d\lambda]$.
8. *(a)* Suppose $(S, \mathscr{A}, \mu)$ is a measure space. Show that a finite measure ν on $\mathscr{A}$ is μ-absolutely continuous if and only if for each $\varepsilon > 0$ there is $\delta > 0$ such that $\nu(B) < \varepsilon$ whenever $B \in \mathscr{A}$ and $\mu(B) < \delta$.
   (b) If $\nu(A) = \int_A |t| \, dt$ show ν is absolutely continuous with respect to Lebesgue measure but does not satisfy the condition in (a).

9. Suppose that μ and ν are σ-finite measures on a σ-algebra $\mathscr{A}$ of subsets of a set S such that $\mu \ll \nu$ and $\nu \ll \mu$. Show that $[d\nu/d\mu] \neq 0$ and $[d\mu/d\nu] = [d\nu/d\mu]^{-1}$.
10. Let $\mathscr{A}$ denote a σ-algebra of subsets of a set S.
    (a) Show that the collection M of all finite s-measures on $\mathscr{A}$ forms a linear space.
    (b) Suppose μ is a measure on $\mathscr{A}$. Let M_1 denote the set of all s-measures on $\mathscr{A}$ that are absolutely continuous with respect to μ, and let M_2 denote the set of all s-measures on $\mathscr{A}$ that are singular with respect to μ. Then show that M_1 and M_2 are subspaces of M and $M = M_1 \oplus M_2$.
    (c) Show that M is a metric space with distance $\rho(\mu, \nu) = \|\mu - \nu\|$.
    (d) Show that M is complete with this metric.

16

Differentiation

If f is a real valued function defined on an interval containing a number x, we define the four *derivates* of f at x in the following way

$$D^+ f(x) = \lim_{h\downarrow 0} \sup \frac{f(x+h) - f(x)}{h}$$

$$D^- f(x) = \lim_{h\uparrow 0} \sup \frac{f(x+h) - f(x)}{h}$$

$$D_+ f(x) = \lim_{h\downarrow 0} \inf \frac{f(x+h) - f(x)}{h}$$

$$D_- f(x) = \lim_{h\uparrow 0} \inf \frac{f(x+h) - f(x)}{h}.$$

It is not hard to see that the four derivates are equal if and only if the limit

$$\lim_{h\to 0} \frac{f(x+h) - f(x)}{h}$$

exists. In this case we say that f is *differentiable* at x and has as its *derivative* f' the value of this limit.

The following result is called the *Vitali Covering Theorem.* It is useful for deriving facts about the existence of derivatives and has other applications as well.

Theorem 16.1: Suppose A is a subset of $\mathbb{R}$ with $m^*(A) < \infty$ and $\mathscr{U}$ is a collection of intervals that satisfies the following condition:

(V) For each x in A and each $\varepsilon > 0$ there is $I \in \mathscr{U}$ such that $x \in I$ and $m(I) < \varepsilon$.

Then for each $\varepsilon > 0$ there is a finite disjoint subcollection $\{I_1, I_2, \ldots, I_n\}$ of $\mathscr{U}$ such that

$$m^*\left[A \setminus \bigcup_{j=1}^{n} I_j\right] < \varepsilon. \tag{16–1}$$

Proof: We may assume that each interval in $\mathscr{U}$ is closed since removing the endpoints of the intervals does not change the value of the outer measure in (16–1).

If O is an open set of finite measure that contains A, then the collection of all I in $\mathscr{U}$, which are contained in O, also satisfies (V) so it suffices to prove the theorem under the assumption that $\bigcup \mathscr{U} \subset O$. We determine a sequence $\{I_n\}$ of disjoint intervals in $\mathscr{U}$ as follows: Let I_1 be any interval in $\mathscr{U}$. If $I_1, \ldots, I_m$ have been chosen, let k_m be the supremum of the lengths of the intervals in $\mathscr{U}$ that have an empty intersection with each of the intervals $I_j, j \leqslant m$. Since each member of $\mathscr{U}$ is in O it follows that $k_m < \infty$. Unless the set $\bigcup_{j=1}^m I_j$ contains A, and thus satisfies (16–1), there must be I_{m+1} in $\mathscr{U}$ disjoint from $\bigcup_{j=1}^m I_j$ and having length greater than $k_m/2$.

Since each of the disjoint intervals I_j is contained in the set O it follows that $\sum m(I_j) \leqslant m(O) < \infty$. Thus there is an integer M such that $\sum_{j>M} m(I_j) < \varepsilon/5$. If

$$D = A \setminus \bigcup_{j=1}^{M} I_j$$

it remains only to verify that $m^*(D) < \varepsilon$. For each x in D there is an interval I in $\mathscr{U}$ such that $x \in I$ and $I \cap I_j = \varnothing$ for $j = 1, 2, \ldots, M$ because the set $\bigcup_{j=1}^M I_j$ is closed. If $I \cap I_j = \varnothing$ for $j \leqslant m$ then $m(I) \leqslant k_m < 2m(I_{m+1})$. Since $\lim_m m(I_m) = 0$ the interval I has nonempty intersection with at least one interval I_m. If n is the smallest subscript of such an interval, then $n > M$ and $m(I) \leqslant k_{n-1} \leqslant 2m(I_n)$. Since x is in I, and I has a nonempty intersection with I_n, it follows that the distance from x to the midpoint of I_n is no more than $m(I) + m(I_n)/2 \leqslant \frac{5}{2}m(I_n)$. This implies that x belongs to the interval J_n, which has the same midpoint as I_n and 5 times the length. Therefore, we have

$$D \subset \bigcup_{j=M+1}^{\infty} J_j$$

from which it follows that

$$m^*(D) \leqslant \sum_{j=M+1}^{\infty} m(J_j) = 5 \sum_{j=M+1}^{\infty} m(I_j) < \varepsilon. \blacksquare$$

A covering $\mathscr{U}$ of a set A that satisfies condition (V) is called a *covering of A in the sense of Vitali* or a *Vitali covering of A*. The following surprising theorem of Lebesgue is our main application of the Vitali Covering Theorem.

Theorem 16.2: An increasing real valued function f on an interval $[a, b]$ is differentiable almost everywhere. The derivative f' is measurable and we have

$$\int_a^b f'(x)\, dx \leqslant f(b) - f(a).$$

Proof: The existence of f' is proved by showing that the set of all x for which two derivates are unequal has measure 0. Let

$$A = \{x\colon x \in [a, b],\ D^+ f(x) > D_- f(x)\}, \quad \text{and}$$

$$B = \{x\colon x \in [a, b],\ D^- f(x) > D_+ f(x)\}.$$

Since $D^+f(x) \geqslant D_+f(x)$ and $D^-f(x) \geqslant D_-f(x)$ for all x, it follows that for $y \in (A \cup B)^\sim$ we have

$$D_-f(y) \geqslant D^+f(y) \geqslant D_+f(y) \geqslant D^-f(y) \geqslant D_-f(y)$$

so that equality holds throughout. Therefore, it suffices to prove that A and B have measure 0. We carry out the proof for A, and leave to you the proof for B.

For each pair of rational numbers s, t, define $A_{s,t}$ to be the set of all x such that $D^+f(x) > s > t > \underline{D}f(x)$. Then

$$A = \bigcup_{s,t} A_{s,t}$$

so it suffices to prove that $m^*(A_{s,t}) = 0$ for all $s > t$. Denote $m^*(A_{s,t})$ by a; and let O be an open set such that $A_{s,t} \subset O$ and $m(O) < a + \varepsilon$ where $\varepsilon > 0$ is chosen arbitrarily. By the way we defined $D_-f(x)$, we know that for each x in $A_{s,t}$ there is an arbitrarily small interval $[x-h, x]$ contained in O such that

$$f(x) - f(x-h) < th.$$

By Theorem 16.1 there is a finite disjoint collection $\{I_1, \ldots, I_M\}$ of these intervals whose interiors cover a subset D of $A_{s,t}$ having outer measure greater than $a - \varepsilon$. If $I_j = [x_j - h_j, x_j]$ for $j = 1, 2, \ldots, M$, we have

$$\begin{aligned}\sum_{j=1}^{M} \{f(x_j) - f(x_j - h_j)\} &< t \sum_{j=1}^{M} h_j \\ &< tm(O) \\ &< t(a+\varepsilon).\end{aligned}$$

By the way $D^+f(x)$ is defined, we know that for each y in the set D defined above there is an arbitrarily small interval $(y, y+k)$ contained in some I_j such that

$$f(y+k) - f(y) > sk.$$

By the Vitali Covering Theorem there is a finite collection $\{J_1, J_2, \ldots, J_K\}$ of such intervals with $m^*(\bigcup_{i=1}^{K} J_i) > a - 2\varepsilon$. If J_i denotes the interval $(y_i, y_i + k_i)$ we have

$$\sum_{i=1}^{K} \{f(y_i + k_i) - f(y_i)\} > s \sum_{i=1}^{k} k_i > s(a - 2\varepsilon).$$

We have chosen each J_i to be contained in some interval I_j so if we sum over those i for which $J_i \subset I_j$ we find that

$$\sum \{f(y_i + k_i) - f(y_i)\} \leqslant f(x_j) - f(x_j - h_j)$$

because f is an increasing function. Now if we sum over all I_j we see

$$\begin{aligned}s(a - 2\varepsilon) &< \sum_{i=1}^{K} \{f(y_i + k_i) - f(y_i)\} \\ &\leqslant \sum_{j=1}^{M} \{f(x_j) - f(x_j - h_j)\} \\ &\leqslant t(a+\varepsilon)\end{aligned}$$

That $s(a-2\varepsilon) \leqslant t(a+\varepsilon)$ for each $\varepsilon > 0$ implies that $sa \leqslant ta$, but $s > t$ so that we must conclude that $a = 0$.

We have now shown that the limit

$$\lim_{h \to 0} \frac{f(x+h)-f(x)}{h}$$

exists for almost all x in $[a, b]$. Define $g(x)$ to be the value of this limit where it exists and 0 elsewhere. Set $f(x) = f(b)$ for $x > b$ and define $g_n(x)$ by

$$g_n(x) = n\{f(x+1/n) - f(x)\}.$$

Then each g_n is nonnegative because f is increasing and (g_n) converges to f' almost everywhere. Hence by Fatou's Lemma (Corollary 12.11 on page 96) we conclude

$$\begin{aligned}
\int_a^b g\, dx &\leqslant \liminf \int_a^b g_n\, dx = \liminf n \int_a^b \{f(x+1/n) - f(x)\}\, dx \\
&= \liminf n \left\{ \int_{a+1/n}^{b+1/n} f(x)\, dx - \int_a^b f(x)\, dx \right\} \\
&= \liminf n \left\{ \int_b^{b+1/n} f(x)\, dx - \int_a^{a+1/n} f(x)\, dx \right\} \\
&\leqslant \liminf n\{f(b)/n - f(a)/n\} = f(b) - f(a). \blacksquare
\end{aligned}$$

It is important to note that equality need not hold in Theorem 16.2. As a trivial example, let $f(x) = 0$ for $x < 0$ and $f(x) = 1$ for $x \geqslant 0$. Then $f'(x) = 0$ for $x \neq 0$ but $f(1) - f(-1) = 1$, which is strictly greater than $\int_{-1}^{1} f'(x)\, dx$.

A function F defined on an interval $[a, b]$ is called an *integral* if there is an integrable function f on $[a, b]$ and a constant A such that

(int) $F(x) = A + \int_a^x f(t)\, dt$

for all x in $[a, b]$. If F is an integral, then the s-measure μ determined by F in the sense of Section 14 must be absolutely continuous with respect to Lebesgue measure m, and the Radon-Nikodym derivative of μ must be almost everywhere equal to f. The function F must be of bounded variation on $[a, b]$ because if $a \leqslant s_0 < s_1 < \ldots < s_n \leqslant b$, we shall have

$$\begin{aligned}
\sum_{i=1}^n |F(s_i) - F(s_{i-1})| &= \sum_{i=1}^n \left| \int_{s_{i-1}}^{s_i} f(t)\, dt \right| \\
&\leqslant \sum_{i=1}^n \int_{s_{i-1}}^{s_i} |f(t)|\, dt = \int_{s_0}^{s_n} |f(t)|\, dt \\
&\leqslant \int_a^b |f(t)|\, dt.
\end{aligned}$$

Therefore, by the preceding theorem $F'(x)$ exists for almost all x in $[a, b]$. A natural question is: does $F'(x)$ coincide almost everywhere with f? The following result known as the *Fundamental Theorem of Calculus for Lebesgue Integration* answers this question positively and also provides a condition for a function F to be an integral. A function F is said to be *absolutely continuous* on an interval $[a, b]$ if it satisfies the following condition

(AC) For each $\varepsilon > 0$ there is $\delta > 0$ such that $\sum_{j=1}^{n} |F(y_j) - F(x_j)| < \varepsilon$ whenever $\{(x_j, y_j): j = 1, 2, \ldots, n\}$ is a finite set of pairwise disjoint subintervals of $[a, b]$ with total length less than δ.

Theorem 16.3: A function F on an interval $[a, b]$ is an integral if and only if it is absolutely continuous. If F is an integral given by (int) then $F' = f$ almost everywhere.

Proof: If F is a function on $[a, b]$ that satisfies (AC) then F is of bounded variation on $[a, b]$. For suppose that δ corresponds to $\varepsilon = 1$ in (AC). Let K be the smallest integer that is greater than $(b-a)/\delta$. If $a \leqslant s_0 < s_1 < \ldots < s_n \leqslant b$ is given, let $\{t_0, t_1, \ldots t_k\}$ consist of the set $\{s_j\} \cup \{a + j(b-a)/K: j = 1, 2, \ldots, K\}$. For $i = 0, 1, \ldots K$, let $u_i = a + i(b-a)/K$. Then we shall have

$$\sum_{j=1}^{n} |F(s_j) - F(s_{j-1})| \leqslant \sum_{j=1}^{k} |F(t_j) - F(t_{j-1})|$$
$$= \sum_{i=1}^{K} \sum \{|F(t_j) - F(t_{j-1})|: u_{i-1} \leqslant t_{j-1} < u_i\} < K.$$

Hence the variation of F is no greater than K. Moreover, the function $V(F, a, x)$ also satisfies (AC). To see this, suppose $\{(x_i, y_i): i = 1, 2, \ldots, n\}$ is a disjoint collection of intervals with length less than δ where δ corresponds to ε in (AC). If for each i, $\{(s_{ij}, t_{ij}): j = 1, 2, \ldots, m(i)\}$ is a partition of (x_i, y_i), the total length of $\{(s_{ij}, t_{ij}): i = 1, 2, \ldots, n; j = 1, 2, \ldots, m(i)\}$ is less than δ so that

$$\sum_{i=1}^{n} \sum_{j=1}^{m(i)} |F(t_{ij}) - F(s_{ij})| < \varepsilon$$

from which we conclude that

$$\sum_{i=1}^{n} |V(F, a, y_i) - V(F, a, x_i)| \leqslant \varepsilon.$$

Since $F(x) = V(F, a, x) - \{V(F, a, x) - F(x)\}$ while both $V(F, a, x)$ and $\{V(F, a, x) - F(x)\}$ are increasing and satisfy (AC) it suffices to consider the situation where F satisfies (AC) and is increasing.

Since F has bounded variation there is a unique measure μ such that $\mu(x, y] = F(y) - F(x)$ for all x, y in $[a, b]$. We shall prove that the condition (AC) implies that μ is absolutely continuous with respect to Lebesgue measure. Suppose A is a subset of $[a, b]$ with $m(A) = 0$. Let $\varepsilon > 0$ be given, and let δ correspond to ε in (AC). There is a finite collection $\{(a_i, b_i): i = 1, 2, \ldots, n\}$ of disjoint intervals that covers A and has total length less than δ. This implies that

$$|\mu(A)| \leqslant \sum_{i=1}^{n} |\mu(a_i, b_i)| \leqslant \sum_{i=1}^{n} |F(b_i) - F(a_i)| < \varepsilon.$$

Since $\varepsilon > 0$ was arbitrary it follows that $\mu(A) = 0$.

Since μ is absolutely continuous with respect to Lebesgue measure the Radon-Nikodym Theorem (15.1) implies there is a measurable function f such that

$$F(x) - F(a) = \mu(a, x] = \int_a^x f(t)\, dt$$

for all x in $[a, b]$. Therefore, F is an integral.

Now assuming F satisfies (int) we shall prove that $F' = f$ almost everywhere. We do this first under the assumption that f is bounded, say $|f(x)| \leqslant M$ for all x in $[a, b]$. Since F has bounded variation, its derivative $F'(x)$ exists for almost all x in $[a, b]$. For each $n = 1, 2, \ldots$ set

$$\begin{aligned} f_n(x) &= \{F(x + 1/n) - F(x)\}/(1/n) \\ &= n \int_x^{x+1/n} f(t)\, dt. \end{aligned}$$

Then $|f_n(x)| \leqslant M$ for all x in $[a, b]$ and $f_n(x)$ converges to $F'(x)$ almost everywhere. By the Bounded Convergence Theorem (12.12 on page 96) we have

$$\begin{aligned} \int_a^u F'(x)\, dx &= \lim_n \int_a^u f_n(x)\, dx \\ &= \lim_n n \int_a^u \{F(x + 1/n) - F(x)\}\, dx \\ &= \lim_n \left\{ n \int_u^{u+1/n} F(x)\, dx - n \int_a^{a+1/n} F(x)\, dx \right\}. \end{aligned}$$

Since F is continuous we have

$$\begin{aligned} |F(u) - n \int_u^{u+1/n} F(x)\, dx| &= \left| n \int_u^{u+1/n} \{F(u) - F(x)\}\, dx \right| \\ &< n \int_u^{u+1/n} |F(u) - F(x)|\, dx \to 0. \end{aligned}$$

Therefore, $\lim_n n \int_u^{u+1/n} F(x)\, dx = F(u)$. Similarly we have

$$\lim_n n \int_a^{a+1/n} F(x)\, dx = F(a)$$

so that

$$\int_a^u F'(x)\, dx = F(u) - F(a) = \int_a^u f(x)\, dx$$

for all u in $[a, b]$. By the uniqueness portion of the Radon-Nikodym Theorem (15.1) we conclude that $F'(x) = f(x)$ a.e.

Next we suppose f is an arbitrary nonnegative measurable function on $[a, b]$. Define $f_n(x) = \inf\{f(x), n\}$ for $n = 1, 2, \ldots$ and define $G_n(x)$ to be $\int_a^x f(t) - f_n(t)\, dt$. Then G_n is an increasing function of x so it has a nonnegative derivative almost everywhere. By the previous paragraph we have

$$d/dx \int_a^x f_n(t)\,dt = f_n(x) \text{ a.e.}$$

so that

$$F'(x) = \frac{d}{dx} G_n + \frac{d}{dx}\int_a^x f_n$$

$$\geqslant f_n(x) \text{ a.e.}$$

Since n is arbitrary it follows that $F'(x) \geqslant f(x)$ a.e. so that

$$\int_a^u F'(x)\,dx \geqslant \int_a^u f(x)\,dx = F(u) - F(a)$$

for all u in $[a, b]$. By Theorem 16.2 we know the reverse inequality also holds so that

$$\int_a^u F'(x)\,dx = F(u) - F(a)$$

for almost all u in $[a, b]$. Therefore, by the uniqueness portion of the Radon-Nikodym Theorem we conclude that $F'(x) = f(x)$ a.e.

If f is an arbitrary integrable function we apply the previous paragraph to f^+ and f^- to show that $F'(x) = f(x)$ a.e. ■

Exercises

A

1. Let $f(x) = -1$ if x is rational and $\leqslant 0$, 1 if x is irrational and $\leqslant 0$, -2 if x is rational and > 0 and 2 if x is irrational and > 0. Determine all of the derivates of f at 0.
2. Let $\{r_1, r_2, \ldots\}$ be the sequence of rational numbers in the interval (0, 1]. For each i, let f_i be the characteristic function of the interval $(r_i, \infty]$; that is $f_i(x) = 0$ if $x \leqslant r_i$ and $f_i(x) = 1$ if $x > r_i$. Define $f(x) = \sum_{i=1}^{\infty} 2^{-i} f_i$.
   (a) Show that the series converges uniformly to f so that f is Riemann integrable on [0, 1].
   (b) Show that f increases from 0 to 1 on the interval [0, 1].
   (c) Show that the derivative of f is 0 for almost all x in [0, 1].
   (d) What are the derivates of f at r_i?
3. Verify that the set B in Theorem 16.2 has measure 0.
4. *(a)* Prove that an absolutely continuous function is continuous.
   (b) Give an example of a continuous function that is not absolutely continuous.
5. Define F on $[-1, 1]$ by $F(x) = 1 + x$ if $-1 \leqslant x \leqslant 0$ and $F(x) = -x$ if $0 < x \leqslant 1$.
   (a) Show that for every measurable subset A of $[-1, 1]$ the measure of $F(A)$ is equal to the measure of A.

(b) Show that an F is not absolutely continuous.

(c) Show that if f is absolutely continuous and A has measure 0 then $f(A)$ also has measure 0.

6. Show that a function F of bounded variation on $[a, b]$ is absolutely continuous on $[a, b]$ if and only if the measure μ induced by F is absolutely continuous with respect to Lebesgue measure.

7. *(a)* Show that if f is absolutely continuous on $[a, b]$, then $V(f, a, b) = \int_a^b |f'(t)|\, dt$.

(b) Show that the function f defined by $f(x) = x \sin(1/x)$ for $x \neq 0$ and $f(0) = 0$ is continuous on $[0, 1]$ and absolutely continuous on $[\varepsilon, 1]$ for all $\varepsilon > 0$, but is not of bounded variation and hence not absolutely continuous on $[0, 1]$.

8. A function f on an interval $[a, b]$ is said to *satisfy the Lipschitz condition* if there is $M > 0$ such that $|f(s) - f(t)| \leqslant M|s - t|$ for all s, t in $[a, b]$.

(a) Show that if f satisfies the Lipschitz condition on $[a, b]$, then f is absolutely continuous on $[a, b]$.

(b) Show that an absolutely continuous function on $[a, b]$ satisfies the Lipschitz condition if and only if $|f'|$ is bounded.

B

9. Suppose F is of bounded variation on an interval $[a, b]$. Let μ be the measure derived from F and let μ_1 be defined by the formula $\mu_1(A) = \int_A F'(t)\, dt$ for A a measurable subset of $[a, b]$. Show that $\mu - \mu_1$ is singular with respect to Lebesgue measure m so that $\mu = \mu_1 + (\mu - \mu_1)$ is the decomposition of μ with respect to Lebesgue measure given by Theorem 15.3.

10. *(a)* Show that every increasing function is the sum of an absolutely continuous function f_1 and a function f_2 with $f_2' = 0$ a.e. A function with 0 derivative a.e. is called *singular*.

(b) Show there is a strictly increasing singular function on $[0, 1]$. (See Exercise 2.)

11. Let g be an increasing absolutely continuous function on an interval $[a, b]$ with $g(a) = c$, $g(b) = d$. Show that if f is a nonnegative function on $[c, d]$ then $(f \circ g)g'$ is measurable on $[a, b]$ and

$$\int_c^d f(y)\, dy = \int_a^b f(g(x))g'(x)\, dx. \tag{*}$$

Formula (*) is known as the *change of variables* formula. [*Hint*: Prove (*) first for the characteristic function of an interval, next for an open set, then for the characteristic function of a measurable set, then for a simple function.]

12. *(a)* Prove that a continuous function f is absolutely continuous if and only if $f(A)$ has Lebesgue measure 0 whenever A has Lebesgue measure 0.

(b) Show by example that we must assume that f is continuous.

4

Functional Analysis

One summer evening in 1916 while walking in the park
that surrounds the old centre of Cracow,
I overheard a conversation, or rather a couple of words;
the words "Lebesgue integral" were so unexpected
that I went over and made the speakers' acquaintance;
it was Stefan Banach and Otton Nikodym talking
about mathematics. H. Steinhaus

An important activity in modern analysis is the organization of classes of functions into spaces and the study of the spaces themselves from an axiomatic viewpoint. The area called linear functional analysis is an elegant example of this endeavor. The phrase "linear functional analysis" refers to the analysis of continuous linear operators between linear spaces. This chapter presents an introduction to four important classes of linear spaces: normed spaces, Banach spaces, Hilbert spaces, and locally convex spaces; the continuous linear operators between these spaces will also be discussed.

In Section 17 we define normed space and derive a few fundamental properties of these spaces and of continuous linear operators. The most important type of linear operator on a linear space is the linear functional, which maps the space into its associated scalar field. The *Hahn-Banach Theorem*, which we prove in Section 20, implies the existence of continuous linear functionals on various types of linear spaces. The space of continuous linear functionals on a normed space is itself a Banach space called the *dual space.*

Hilbert spaces, which we consider in Section 18, have more regular properties than Banach spaces. The topology of a Hilbert space is given by means of a norm determined by an inner product. This means that metric considerations play a larger role in Hilbert spaces than in Banach spaces. *The Metric Projection Theorem* asserts that for every point x not in a closed convex subset C of a Hilbert space there is a closest point in C to x. This result has valuable consequences for approximation and optimization theory. Another important property of a Hilbert space is that it represents its own dual space in a natural way just as finite dimensional spaces do.

The concept of orthogonality is older than that of Hilbert space. *Bessel's inequality* and *Parsival's identity* originated in the study of Fourier series. However, these ideas fit nicely into the Hilbert space framework as do facts about orthonormal sets in Hilbert space, found in Section 19.

The Baire Category Theorem has several useful consequences for Banach spaces such as the *Uniform Boundedness Principle* and the *Banach-Steinhaus Theorem*, which we present in Section 21, and the *Open Mapping* and *Closed Graph Theorems*, which we give in Section 22. The Uniform Boundedness Principle asserts that a sequence of continuous linear operators that is pointwise bounded on a Banach space is, in fact, uniformly bounded. The Banach-Steinhaus Theorem allows us to verify the pointwise convergence of a sequence of continuous linear operators by checking the convergence on a set (much smaller than the entire space). The Open Mapping and Closed Graph Theorems are equivalent results that allow us to assume the continuity of a linear operator by checking a formally less demanding hypothesis. The topological analogue of an algebraic basis for a linear space is the *Schauder* basis of a normed space, which we define in Section 20 and study further in Section 22. The Schauder basis can be used much like the basis of a finite dimensional space for representing linear operators and functionals on a normed space.

In Section 23 we present a few fundamental results about convex sets. First we give a separation theorem equivalent to the Hahn-Banach Theorem. Next we prove *Alaoglu's Theorem* on the compactness of the closed unit ball of a dual space with respect to the weak topology. Finally we introduce the concept of an extreme point and prove the *Krein-Mil'man Theorem*, that a compact convex subset of a locally convex space is the closed convex hull of its extreme points.

17

Normed Spaces

Between the two world wars there was a floodtide of mathematical creativity in the newly established nation of Poland. Much of this activity took place in the city of Lwów (formerly Lemberg in the Austrian Empire, now Lvov in the Soviet Ukraine.) The School of Lwów boasted numerous accomplished mathematicians such as Steinhaus, Ulam, Orlicz, Mazur, and Schauder. The leader of the School of Lwów was the eminent Stephen Banach (1892–1945) whose monograph *Théorie des Operations Linéaires*, 1932, is the source of much of the subject matter found in Sections 17, 20, 21, and 22.

In Section 8 you learned that a norm on a real or complex linear space E is a function $\| \ \|$ from E into the nonnegative reals that has the following properties.

(N1) For $x \in E$, $\|x\| = 0$ if and only if $x = 0$.

(N2) For $x \in E$ and a in the scalar field $\|ax\| = |a| \, \|x\|$.

(N3) For $x, y \in E$, $\|x + y\| \leqslant \|x\| + \|y\|$.

Theorem 8.4 shows that the function defined on $\mathbb{R}^n$ by

$$\|x\| = \left\{ \sum_{i=1}^{n} |x_i|^2 \right\}^{1/2}$$

satisfies the triangle inequality and is thus an example of a norm. On $\mathbb{C}^n$, the linear space of all n-tuples of complex numbers, a norm is defined by the same formula since we have taken the absolute value of each x_i. (Numerous examples of normed spaces were given in Section 8.)

Example 17.1:

Let $B(S)$ denote the space of all bounded functions on a set S. A norm on $B(S)$ is given by $\|f\| = \sup\{|f(s)|: s \in S\}$. If S is a compact metric space the same formula defines a norm on the linear space $C(S)$ of continuous scalar valued functions on S.

The space E with the topology defined by $\| \ \|$ is called a *normed space*. The next theorem is a basic characterization of continuous linear operators between normed spaces. Because of property (d) of the theorem, continuous linear operators are also called *bounded linear operators*. From this point on, whenever we speak of a linear operator from a linear space E to a linear space F we implicitly assume that E and F are over the same scalar field, that is either the real or complex numbers.

Theorem 17.2: Suppose T is a linear operator from the normed space E into the normed space F. The following statements are equivalent:

(a) T is continuous;

(b) T is continuous at one point;

(c) T is continuous at 0;

(d) there is $M > 0$ such that $\|Tx\| \leqslant M\|x\|$ for all $x \in E$.

Proof:

(a) $\Rightarrow$ (b), (c) is obvious.

(b) $\Rightarrow$ (c). Suppose T is continous at the point x_0. Let U be a nbd of 0 in F. Then U contains an open set of the form $B(0, \varepsilon) = \{x: \|x\| < \varepsilon\}$. The set

$$Tx_0 + B(0, \varepsilon) = \{Tx_0 + y : y \in B(0, \varepsilon)\} = B(Tx_0, \varepsilon)$$

is a nbd of Tx_0 so there is a nbd V of x_0 such that $V \subset T^{-1}(B(Tx_0, \varepsilon))$. This means that $\|Tx - Tx_0\| < \varepsilon$ for all x in V. Let $W = V - x_0$. The reader may verify that W is a nbd of 0 in E. If $x \in W$ then $x = u - x_0$ for some u in V so $\|Tx\| = \|T(u - x_0)\| = \|Tu - Tx_0\| < \varepsilon$. This means that $W \subset T^{-1}(B(0, \varepsilon)) \subset T^{-1}(U)$ so that T is continuous at 0 by definition.

(c) $\Rightarrow$ (d). Since T is continuous at 0 there is $\delta > 0$ such that $\|Tx\| < 1$ when $\|x\| < \delta$. If $M = 2/\delta$ then for all x in E, $\|\delta x/(2\|x\|)\| < \delta$ so that $\|T(\delta x/(2\|x\|))\| < 1$. Using the property of scalar homogeneity we conclude $\|Tx\| < (2/\delta)\|x\| = M\|x\|$.

(d) $\Rightarrow$ (a). Suppose $x_0 \in E$, and U is a nbd of Tx_0 in F. There is $\varepsilon > 0$ such that $B(Tx_0, \varepsilon) \subset U$. Let $V = B(x_0, \varepsilon/M)$. If $x \in V$, $\|x - x_0\| < \varepsilon/M$ so that

$$\|T(x) - Tx_0)\| = \|T(x - x_0)\| < M\|x - x_0\| < \varepsilon.$$

This shows that $T(x) \in U$ if $x \in V$. Therefore, T is continuous at each point x_0 of E. ■

Example 17.3:

Define S from $C[0, 1]$ into $C[0, 1]$ by the formula

$$(Sf)(t) = \int_0^t f(s)\, ds.$$

The function S is linear because of the linearity of the integral. It is bounded, hence continuous, because

$$|Sf(t)| \leqslant \int_0^t |f(s)|\, ds \leqslant \int_0^t \|f\|\, ds \leqslant \|f\|$$

from which it follows that $\|Sf\| \leqslant \|f\|$.

If T is a continuous linear operator from the normed space E to the normed space F we write $\|T\|$ for the greatest lower bound of all M such that $\|Tx\| \leqslant M\|x\|$ for all x in E. The following theorem justifies this notation.

Theorem 17.4:

(a) If T is a bounded linear operator from the normed space E to the normed space F then $\|T\|$ equals the following quantities

$$\|T\|_1 = \sup\{\|Tx\|: \|x\| = 1\}$$

$$\|T\|_2 = \sup\{\|Tx\|: \|x\| \leqslant 1\}$$

$$\|T\|_3 = \sup\{\|Tx\|/\|x\|: \|x\| \neq 0\}.$$

(b) The set $L(E, F)$ of all continuous linear operators from E into F forms a linear space on which $\|T\|$ is a norm.

Proof:

(a) It is obvious that $\|T\|_1 \leqslant \|T\|_2$ since for $\|T\|_2$ since for $\|T\|_2$ the least upper bound is taken over a larger set. If $\|x\| \leqslant 1$ and x is not 0, $\|Tx\| \leqslant \|Tx\|/\|x\|$ while if $x = 0$, $\|Tx\| = 0$ so $\|T\|_2 \leqslant \|T\|_3$. Finally, if $x \neq 0$, $\|Tx\|/\|x\| = \|T(x/\|x\|)\|$ and $\|(x/\|x\|)\| = 1$ so that $\|T\|_3 \leqslant \|T\|_1$. Since $\|Tx\| \leqslant \|T\|_3 \|x\|$ for all $x \in E$ it follows that $\|T\|_3 \geqslant \|T\|$. On the other hand, if $M > \|T\|$ we have $\|Tx\|/\|x\| < M$ whenever $x \neq 0$ so that $\|T\|_3 < M$. Therefore, we have $\|T\|_3 \leqslant \|T\|$.

(b) If T_1 and T_2 are bounded linear operators from E into F then for each x in E we have

$$\|(T_1 + T_2)x\| = \|T_1 x + T_2 x\| \leqslant \|T_1 x\| + \|T_2 x\| \leqslant (\|T_1\| + \|T_2\|)\|x\|.$$

This implies that $T_1 + T_2$ is continuous and $\|T_1 + T_2\| \leqslant \|T_1\| + \|T_2\|$.

If T is continuous and a is a scalar then for all x in E we have $\|aTx\| = |a|\,\|Tx\| \leqslant |a|\,\|T\|\,\|x\|$. Thus we have aT is continuous and $\|aT\| \leqslant |a|\,\|T\|$. If $a \neq 0$ we also have $\|T\| = \|(1/a)aT\| \leqslant |1/a|\,\|aT\|$ so that $|a|\,\|T\| \leqslant \|aT\|$. Consequently, we have $\|aT\| = |a|\,\|T\|$ for all a because the case $a = 0$ is obvious. ■

The topology on $L(E, F)$ determined by the norm $\| \ \|$ is called *the uniform topology of operators* and $\| \ \|$ is called the *uniform norm.*

Two normed spaces E and F are said to be *isomorphic* if there is a one-to-one continuous linear operator T from E onto F such that the inverse transformation T^{-1} from F onto E is also continuous. Our next result leads to the fact that all normed spaces of a given finite dimension over the same scalar field are isomorphic and thus they are equivalent as far as the theory of normed spaces is concerned.

Theorem 17.5: If $\| \ \|$ and $\| \ \|_*$ are two norms on a linear space E of finite dimension then $\| \ \|$ and $\| \ \|_*$ determine the same topology on E.

Proof: Suppose E has dimension n and $\{x_1, x_2, \ldots, x_n\}$ is a basis of E. Define a third norm on E as follows: if x in E has the unique expansion $x = \sum_{j=1}^n a_j x_j$, let $\|x\|_2 = \{\sum_{j=1}^n |a_j|^2\}^{1/2}$. You may verify that $\| \ \|_2$ is a norm, and that the function $Tx = (a_j)$ is an isometry from $(E, \| \ \|_2)$ onto $\mathbb{R}^n$ if E is a real space, and onto $\mathbb{C}^n$ if E is complex. We shall prove that the norm $\| \ \|$ is equivalent to the norm $\| \ \|_2$. A corresponding argument will then show that $\| \ \|_*$ is also equivalent to $\| \ \|$ so that all three norms $\| \ \|$, $\| \ \|_*$, and $\| \ \|_2$ determine the same topology on E. Using the Triangle Inequality for $\| \ \|$, and the Cauchy-Schwarz Inequality (Theorem 8.4) on the sum we find

(A) $\|x\| \leqslant \sum_{j=1}^n |a_j| \|x_j\| \leqslant \{\sum_{j=1}^n \|x_j\|^2\}^{1/2} \|x\|_2.$

This implies that the identity operator $Ix = x$ is continuous from $(E, \| \ \|_2)$ onto $(E, \| \ \|)$. Therefore, the topology determined by $\| \ \|_2$ is stronger than that determined by $\| \ \|$.

Next we show that the topology determined by $\| \ \|$ is stronger than that determined by $\| \ \|_2$. Let B denote the set of all x in E such that $\|x\|_2 = 1$. Since B is the image of a compact subset of $\mathbb{R}^n$ under the isometry from $\mathbb{R}^n$ onto E it follows that B is compact in E. From (A) we know there is $M > 0$ such that $\|x - y\| \leqslant M \|x - y\|_2$ so that the norm $\| \ \|$ is continuous on $(E, \| \ \|_2)$. Thus $\|x\|$ assumes its minimum, say m, on B. If $\|x\| < m$ we must conclude that $\|x\|_2 < 1$, for if $\|x\|_2$ were greater than one, we would have $(x/\|x\|_2)$ in B and $\|x/\|x\|_2\| < m$, in contradiction to the minimality of m. For every $\varepsilon > 0$ and every x in E, $\|mx/(1+\varepsilon)\|x\|\,\| < m$ so that $\|mx/(1+\varepsilon)\|x\|\,\|_2 < 1$. Using the homogeneity property of the norm we conclude that

$$\|x\|_2 < \frac{(1+\varepsilon)}{m} \|x\|$$

for all x in E. This proves that the identity mapping from $(E, \| \ \|)$ onto $(E, \| \ \|_2)$ is continuous so that the norm $\| \ \|$ determines a topology stronger than that determined by $\| \ \|_2$. Consequently, the two topologies coincide. ■

Corollary 17.6: Suppose E and F are normed spaces over the same scalar field. If E has finite dimension, then every linear operator from E into F is continuous.

Proof: If $\{x_1, x_2, \ldots, x_n\}$ is any basis for E then the function given by

$$\|x\|_1 = \sum_{j=1}^{n} |a_j| \quad \text{if } x = \sum_{j}^{n} a_j x_j$$

is a norm on E and thus equivalent to the original norm by Theorem 17.5. If T is a linear operator from E into F it suffices to prove T is continuous with respect to $\| \ \|_1$. Since

$$\left\| T\left(\sum_{j=1}^{n} a_j x_j\right) \right\| \leqslant \sum_{j=1}^{n} |a_j| \, \|Tx_j\| \leqslant \{\max \|Tx_j\|\} \|x\|_1$$

it follows from (d) of Theorem 17.2 that T is continuous with respect to $\| \ \|_1$. ■

Corollary 17.7:

(a) Every finite dimensional normed space is complete.

(b) Every finite dimensional subspace of a normed space is closed.

A complete normed space is called a *Banach space.* The corollary above shows that every finite dimensional normed space is a Banach space. It is not hard to see that $B(S)$ is a Banach space for every S; if S is infinite this provides an example of an infinite dimensional Banach space. Spaces of the form $C(S)$ where (S, ρ) is a metric space are also Banach spaces and are often of infinite dimension. We proved the completeness of $C(S)$ in Example 9.3 on page 61. It is often easier to verify completeness of a Banach space in terms of series rather than in terms of sequences. We say series $\sum_{j=1}^{\infty} x_j$ in a normed space E *converges to* x in E if for each $\varepsilon > 0$ there is K such that $\|x - \sum_{j=1}^{k} x_j\| < \varepsilon$ when $k > K$. The series $\sum_{j=1}^{\infty} x_j$ *converges absolutely* if $\sum_{j=1}^{\infty} \|x_j\| < \infty$.

Theorem 17.8: A normed space E is a Banach space if and only if every absolutely convergent series in E converges to a point x in E.

Proof: If $\sum_{j=1}^{\infty} x_j$ converges absolutely then the partial sums of the series form a Cauchy sequence. To verify this let $\varepsilon > 0$ be given. Let K be such that $\sum_{j=K}^{\infty} \|x_j\| < \varepsilon$. If $p > q > K$ then we have

$$\left\| \sum_{j=1}^{p} x_j - \sum_{j=1}^{q} x_j \right\| = \left\| \sum_{j=q}^{p} x_j \right\| \leqslant \sum_{j=q}^{p} \|x_j\| \leqslant \sum_{j=q}^{\infty} \|x_j\| < \varepsilon.$$

If E is complete then there is x in E to which the partial sums converge.

Suppose now that E is a normed space in which every absolutely convergent series converges. Let (x_n) be any Cauchy sequence in E. We define a subsequence (y_n) of (x_n) in the following way: let $y_i = x_n(i)$ where $n(i) > n(i-1)$ and if $p, q > n(i)$, $\|x_p - x_q\| < 2^{-i}$. Then the series $(y_2 - y_1) + (y_3 - y_2) \ldots$ converges absolutely since $\|y_{i+1} - y_i\| < 2^{1-i}$. The nth partial sum of this series is $y_{n+1} - y_1$ so if the series converges to y it follows that (y_n) converges to $y + y_1$. By Theorem 9.2 (b) on page 61 it follows that (x_n) converges. ■

In our first application of Theorem 17.8 we prove that if F is complete, then so is $L(E, F)$.

Theorem 17.9: If E is a normed space and F is a Banach space, then $L(E, F)$ is a Banach space.

Proof: Suppose (T_j) is a sequence in $L(E, F)$ with $\sum_j \|T_j\| < \infty$. For each x in E we have

$$\sum_j \|T_j x\| \leqslant \|x\| \sum_j \|T_j\| < \infty$$

so there is Tx in F with $\sum_j T_j x = Tx$. You may verify that the Tx is linear. The function T is bounded since $\|Tx\| \leqslant \|x\| \sum_j \|T_j\|$ for each x in E. The series $\sum_j T_j$ converges to T because if $\|x\| \leqslant 1$, then

$$\left\|\left(Tx - \sum_{j=1}^{k} T_j x\right)\right\| \leqslant \sum_{j=k+1}^{\infty} \|T_j\| \to 0. \blacksquare$$

A linear operator f from a normed space E into the scalar field is called a linear functional on E. The space of all continuous linear functionals on E is called the *dual space* of E, written E^*. Theorem 17.9 indicates that the dual space of a normed space is always a Banach space.

Suppose E is a normed space and F is a closed subspace of E. For each x in E we define the class $x + F$ to consist of all u in E such that $u = x + y$ for some y in F. The set of all classes $x + F$, $x \in E$ is called the *quotient space* of E by F and written E/F. For $u + F$ and $v + F$ in E/F and a in the scalar field we define $(u + F) + (v + F)$ to be $(u + v + F)$ and define $a(u + F)$ to be $au + F$. It is necessary to check that the operations are well defined; that is, we must show that if $u + F = u' + F$ and $v + F = v' + F$, then $(u + v) + F = (u' + v') + F$ and $au + F = au' + F$. With these operations E/F is a linear space in which $0 + F$ serves as the 0 element. A norm on E/F is given by the following formula as we shall verify in Theorem 17.10.

$$\|x + F\| = \inf\{\|x + y\|: y \in F\}. \tag{17–1}$$

Theorem 17.10: Suppose E is a normed space and F is a closed subspace of E. The function defined by (17–1) is a norm on E/F. If E is a Banach space so is E/F with the given norm.

Proof: The norm is well defined because if $x + F = x' + F$ then $x - x' \in F$ so

$$\begin{aligned}\inf\{\|x + y\|: y \in F\} &= \inf\{\|x' + (x - x') + y\|: y \in F\}\\ &= \inf\{\|x' + z\|: z \in F\}.\end{aligned}$$

It is not hard to see that $\| \;\; \|$ is scalar homogeneous. We verify the triangle inequality. Let x and y in E be given. For arbitrary $\varepsilon > 0$ let u and v in F be such that

$$\|x + u\| \leqslant \|x + F\| + \frac{\varepsilon}{2} \quad \text{and} \quad \|y + v\| \leqslant \|y + F\| + \frac{\varepsilon}{2}.$$

Then $u+v \in F$ so that

$$\|(x+y)+F\| \leqslant \|x+y+u+v\| \leqslant \|x+u\| + \|y+v\|$$
$$\leqslant \|x+F\| + \|y+F\| + \varepsilon.$$

Since $\varepsilon > 0$ was arbitrary, the triangle inequality is valid.

If $\|x+F\| = 0$, then for each $n = 1, 2, \ldots$ there is u_n in F with $\|x - u_n\| < 1/n$. This sequence converges to x in E so $x \in F$ since F is closed.

We use Theorem 17.8 to prove that if E is a Banach space then E/F is complete. Suppose $\sum_j (x_j + F)$ converges absolutely. For each j let v_j in F be such that $\|x_j + v_j\| < \|x_j + F\| + 2^{-j}$. Then the series $\sum_j (x_j + v_j)$ converges absolutely in E so there is x in E with $x = \sum_j (x_j + v_j)$. That $\sum_j (x_j + F)$ converges to $x + F$ follows since for each k we have $\|(x+F) - \sum_{j=1}^{k} (x_j + F)\| \leqslant \sum_{j=k+1}^{\infty} \|x_j + v_j\|$. ∎

Exercises

A

1. Show that if E and F are normed spaces then $E \times F$ is a normed space with the norm $\|(x, y)\|_\infty = \max\{\|x\|, \|y\|\}$, the norm $\|(x, y)\|_1 = \|x\| + \|y\|$, and the norm $\|(x, y)\|_2 = \{\|x\|^2 + \|y\|^2\}^{1/2}$.
2. Verify that the function on $\mathbb{C}^n$ defined by $\{\sum_{i=1}^{n} |x_i|^2\}^{1/2}$ is a norm.
3. Prove that the norm on a normed space is continuous with respect to the topology it determines.
4. Show that if E is a normed space and S is any set, then the space $B(S, E)$ of all bounded functions f from S into E is a normed space with the norm $\|f\| = \sup\{\|f(s)\| : s \in S\}$. If E is complete, so is $B(S, E)$.
5. Suppose E, F, and G are normed spaces, $T \in L(E, F)$ and $S \in L(F, G)$. Show that ST, the composition of S, with T is in $L(E, G)$ and $\|ST\| \leqslant \|S\| \, \|T\|$.
6. The space of bounded scalar sequences is denoted by l^∞. Let (a_j) be a sequence of scalars. Define the operator T from l^∞ into the space of all sequences by $Tx = (a_j x_j)$. Show that the range of T is contained in l^∞ if and only if (a_j) is bounded. In this case T is continuous and $\|T\| = \sup_j\{|a_j|\}$.
7. Prove that a linear operator is continuous, then it is uniformly continuous.
8. Verify that the relationship of two normed spaces being isomorphic is an equivalence relation.
9. Let e_j denote the sequence that has 1 in the jth place and 0's elsewhere. Show that no subsequence of $\{e_j\}$ converges in l^∞. Thus the closed unit ball of l^∞ is not compact. This means that the Bolzano-Weierstrass Theorem does not hold for l^∞. We shall shortly prove that it does not hold in any infinite-dimensional normed space.
10. Let T be a continuous linear operator from a normed space E to a normed space F.
    (a) Show that if $\{x_n\}$ is a Cauchy sequence in E then $\{Tx_n\}$ is a Cauchy sequence in F.
    (b) Prove that if F is isomorphic to E and E is complete, then F is also complete.

11. A subset C of a linear space is called *absolutely convex* if whenever x and y are in C and $|a| + |b| \leqslant 1$ then $ax + by \in C$. Show that the unit ball of every normed space is absolutely convex.

B

12. Let E and F be normed spaces.
    (a) Show that the three norms on $E \times F$ described in Exercise 1 are equivalent.
    (b) Show that if E and F are Banach spaces then so is $E \times F$ with the norm $\| \ \|_\infty$.
13. What is the norm $\|S\|$ for the operator S described in Example 17.3?
14. Show that $\|T\| = \sup\{\|Tx\| : \|x\| < 1\}$.
15. *(a)* Use the Baire Category Theorem to prove that no Banach space can have a countable (infinite) a-basis.
    (b) Thus there is no norm on the space ϕ of sequences, which are eventually 0, which makes ϕ a Banach space.
16. Show that the space l^1 of all sequences x such that $\|x\| = \sum_j |x_j| < \infty$ is a Banach space with the norm $\| \ \|$. It is separable because the set of all sequences (a_j) for which each a_j is rational and $a_j = 0$ for all but finitely many j is dense in l^1.

(In the following four problems refer to the list of spaces at the end of Section 8, page 52).

17. Prove that c is a closed subspace of l^∞ and thus a Banach space. Prove that c_0 is a closed subspace of c.
18. *(a)* Show that for each p with $1 \leqslant p < \infty$, l^p is a Banach space.
    (b) Show that each l^p is separable.
19. Show that bs is a Banach space and cs is a closed subspace of bs.
20. Show that bv is a Banach space and bv_0 is a closed subspace.
21. Suppose T is a bounded linear operator from a Banach space E into itself. Show that if a is a scalar with $|a| > \|T\|$ then $aI + T$ has a bounded inverse namely $a^{-1}\{I - T/a + T^2/a^2 - T^3/a^3 + \cdots\}$ where I is the identity operator $(Ix = x)$ for all x in E.
22. Prove that the operator $T(s_i) = (s_1, s_2 - s_1, s_3 - s_2, \ldots)$ is an isomorphism, in fact, an isometry from bv onto l^1.
23. Suppose T is a bounded linear operator from a normed space E into a normed space F. Let N be the set of all x in E such that $Tx = 0$.
    (a) Show N is a closed subspace of E.
    (b) Show that $T = \hat{T}Q$ where Q is the quotient operator defined from E onto E/N by $Qx = x + N$, and $\hat{T}$ is a one-to-one operator from E/N into F defined by $\hat{T}(x + N) = Tx$.
    (c) Verify that both $\hat{T}$ and Q are bounded linear operators.

C

24. *(a)* Prove that if $\mathscr{A}$ is an a-basis for a normed space E, then the set $\mathscr{B} = \{x/\|x\| : x \in \mathscr{A}\}$ is also an a-basis for E. An a-basis $\mathscr{B}$ in which each member has norm one is called a *normal a-basis for E*.

(b) Prove that if $\mathscr{B}$ is a normal a-basis for a normed space E and f is a continuous linear functional on E then the set $\{f(x): x \in \mathscr{B}\}$ is bounded.

(c) Use (b) to show that on every normed space of infinite dimension there exist linear functionals that are not continuous.

25. (a) Prove that if F is a proper closed subspace of a normed space E, then for each $t < 1$ there is a vector u in $E \backslash F$ such that $\|u - x\| > t$ for all x in F. (Let y be any vector in $E \backslash F$. Let v in F be such that $\|y - v\| \leqslant (1/t) \min\{\|y - w\|: w \in F\}$. Let $u = (y - v)/\|y - v\|$.)

(b) Prove that if E has infinite dimension then for each $t > 0$ there is a sequence $\{x_n: n = 1, 2, \ldots\}$ such that $\|x_n\| = 1$ for each n and $\|x_n - x_m\| > t$ for $n \neq m$.

(c) Prove that if E has infinite dimension, then the closed unit ball of E does not have the Bolzano Weierstrass Property and is thus not compact.

18

Hilbert Spaces

David Hilbert (1862–1943) is considered the foremost mathematician of the early twentieth century. Hilbert's development of infinite dimensional inner product spaces was the first step toward a theory of abstract spaces. The beginnings of this subject are found in a treatise published by Hilbert in 1901.

An *inner product* on a linear space E is a scalar valued function $\langle , \rangle$ on $E \times E$ that satisfies the following hypotheses.

(IP1) If x, y, and z are in E and a, b are scalars then

$$\langle ax + by, z\rangle = a\langle x, z\rangle + b\langle y, z\rangle.$$

(IP2) If x and y are in E then $\langle y, x\rangle$ is equal to the conjugate $\overline{\langle x, y\rangle}$ of $\langle x, y\rangle$.

(IP3) If $x \neq 0$ then $\langle x, x\rangle > 0$.

Condition (IP1) is referred to as *linearity in the first variable*, condition (IP2) as *conjugate symmetry*, and condition (IP3) as *positivity*. If the scalar field is $\mathbb{R}$ then (IP2) can be simplified to $\langle y, x\rangle = \langle x, y\rangle$ *symmetry*.

Example 18.1:

If $C[0, 1]$ denotes the space of all continuous complex valued functions on $[0, 1]$ the function

$$\langle f, g\rangle = \int_0^1 f(t)\overline{g(t)}\, dt$$

is easily seen to be an inner product on $C[0, 1]$.

Example 18.2:

The dot product on $\mathbb{R}^n$ given by

$$\langle x, y\rangle = \sum_{j=1}^{n} x_j y_j$$

where $x = (x_1, \ldots, x_n)$ and $y = (y_1, \ldots, y_n)$ is an inner product.

Two vectors x, y in a space E having an inner product $\langle , \rangle$ are said to be *orthogonal* if $\langle x, y\rangle = 0$. The analogy of this definition to geometric orthogonality in $\mathbb{R}^n$ is evident.

Part (a) of the following theorem is known as the Cauchy-Schwarz inequality. Compare it with (a) of Theorem 8.4 on page 49.

Theorem 18.3: If $\langle , \rangle$ is an inner product on a linear space E, then

(a) $|\langle x, y\rangle| \leqslant \sqrt{\langle x, x\rangle}\sqrt{\langle y, y\rangle}$ for all x, y in E;

(b) the function $\|x\| = \sqrt{\langle x, x\rangle}$ is a norm on E.

Equality holds in (a) if and only if x and y are proportional.

Proof:

(a) By (IP3) we know that $\langle x - cy, x - cy\rangle \geqslant 0$ for all c. If we expand $\langle x - cy, x - cy\rangle$ using (IP1) and (IP2) we obtain

$$\langle x - cy, x - cy\rangle = \langle x, x\rangle - c\overline{\langle x, y\rangle} - \bar{c}\langle x, y\rangle + |c|^2\langle y, y\rangle.$$

If $y \neq 0$ we take $c = \langle x, y\rangle/\langle y, y\rangle$ to obtain

$$\langle x, x\rangle - \frac{|\langle x, y\rangle|^2}{\langle y, y\rangle} - \frac{|\langle x, y\rangle|^2}{\langle y, y\rangle} + \frac{|\langle x, y\rangle|^2}{\langle y, y\rangle} = \langle x, x\rangle - \frac{|\langle x, y\rangle|^2}{\langle y, y\rangle} \geqslant 0,$$

which we can transform into the desired inequality. If $y = 0$ then both sides of the inequality are 0.

(b) It is clear that $\sqrt{\langle x, x\rangle}$ satisfies (N1) and (N2). To verify it also satisfies (N3) we calculate

$$\begin{aligned}\|x + y\|^2 &= \langle x + y, x + y\rangle = \langle x, x\rangle + 2Re(\langle x, y\rangle) + \langle y, y\rangle \\ &\leqslant \langle x, x\rangle + 2|\langle x, y\rangle| + \langle y, y\rangle \\ &\leqslant \langle x, x\rangle + 2\sqrt{\langle x, x\rangle}\sqrt{\langle y, y\rangle} + \langle y, y\rangle \\ &= (\|x| + \|y\|)^2.\end{aligned}$$

The proof of the last assertion is left as an exercise. ■

A linear space E with a topology derived from a norm $\| \ \|$ of the form $\|x\| = \sqrt{\langle x, x\rangle}$, where $\langle , \rangle$ is an inner product on E, is called an *inner product (ip) space*. If E is complete with this normed topology, E is called a *Hilbert space* (rather than a "Banach inner product space"). The Hilbert space concept is much more specific than that of the Banach space, and Hilbert spaces have more orderly properties.

The following lemma, the *Parallelogram Law*, is proved by direct calculation with inner products that we leave to you, the reader.

Lemma 18.4: If E is an ip space, then for all x, y in E we have

$$\|x+y\|^2+\|x-y\|^2=2\|x\|^2+2\|y\|^2.$$

The following theorem known as the *Projection Theorem* finds numerous applications in approximation and optimization theory.

Theorem 18.5: Suppose C is a closed convex subset of a Hilbert space H and u is a point not in C. There is a unique point $\pi_C(u)$ in C such that $\|u-\pi_C(u)\|$ is a minimum.

Proof: We may assume that $u=0$. Let $d=\inf\{\|x\|: x\in C\}$. For each $n=1, 2, \ldots$ there is a point x_n in C such that $\|x_n\|<d+1/n$. We show that the sequence (x_n) is Cauchy. By the Parallelogram Law we have for all m, n

$$\begin{aligned}\|x_m-x_n\|^2&=2\|x_m\|^2+2\|x_n\|^2-\|x_m+x_n\|^2\\&<2\left(d+\frac{1}{m}\right)^2+2\left(d+\frac{1}{n}\right)^2-\|x_m+x_n\|^2.\end{aligned}$$

Since C is convex the point $\frac{1}{2}x_m+\frac{1}{2}x_n$ is in C so that

$$\|x_m+x_n\|^2=4\|\tfrac{1}{2}x_m+\tfrac{1}{2}x_n\|^2\geqslant 4d^2.$$

Therefore, we have

$$\|x_m-x_n\|^2<2\left(d+\frac{1}{m}\right)^2+2\left(d+\frac{1}{n}\right)^2-4d^2$$

from which it follows that (x_n) is Cauchy.

If v denotes the limit of the sequence (x_n) then $\|v\|=\lim_n\|x_n\|=d$ so that $\|v\|$ is a minimum. If $\|w\|=d$ as well then

$$\|v-w\|^2=2\|v\|^2+2\|w\|^2-4\|\tfrac{1}{2}v+\tfrac{1}{2}w\|^2\leqslant 0$$

so that $\|v-w\|=0$. Therefore v is unique. ■

The *closest point function* $\pi_C(u)$ can be extended to all of H by setting $\pi_C(u)=u$ for u in C; the function π_C is often called the *metric projection* from H onto C. In general π_C is not a linear operator. But the following theorem, Theorem 18.7, shows that if set C is taken as a closed subspace of H, π_C is a continuous linear projection with the additional useful property that every x in the range of π_C is orthogonal to every y in the null space of π_C. If C is any subset of a Hilbert space H then $C^\perp$, the *annihilator* of C, is defined as the set of all y in H such that $\langle x, y\rangle=0$ for all x in C. The function $y\to\langle x, y\rangle$ is continuous because $|\langle x, y_1\rangle-\langle x, y_2\rangle|\leqslant\|x\|\,\|y_1-y_2\|$. Since $C^\perp=\bigcap_{x\in C}\{y: \langle x, y\rangle=0\}$ it follows that $C^\perp$ is a closed subset of H. It is also easy to see that $C^\perp$ is also a linear subspace of H.

Example 18.6:

As a simple example of the Projection Theorem we show the existence of a unique polynomial $p(x)$ of degree at most 4 such that

$$\int_0^1 |x^5 - p(x)|^2 \, dx$$

is a minimum. The space E of polynomials of degree at most 5 is a Hilbert space with the inner product given in Example 18.1. The space F of polynomials of degree at most 4 forms a closed subspace and is thus a closed convex subset. Thus by the Projection Theorem there is a unique p in F such that $\|x^5 - p\|$ is a minimum. Since the given integral is the square of the norm, p is the desired approximation. A similar argument shows that there is a unique polynomial q such that the degree of q is at most four,

$$\int_0^1 |x^5 - q(x)|^2 \, dx \quad \text{is a minimum} \quad \text{and} \quad \int_0^1 q(x)^2 \leqslant \tfrac{1}{2}.$$

The Projection Theorem does not help us to calculate the polynomials p and q, however. Some methods for doing this are developed in the next section. If F and G are two subspaces of a linear space H such that the only vector common to both F and G is zero, then the *direct sum* of F and G written $F \oplus G$ is defined to be the set of all vectors $x + y$ with x in F and y in G.

Theorem 18.7: Suppose G is a closed nonzero subspace of a Hilbert space H. Then

(a) $H = G \oplus G^{\perp}$.

(b) If $x = u + v$ with u in G and v in $G^{\perp}$ then $u = \pi_G(x)$.

(c) The function π_G is a continuous linear projection from H onto G with $\|\pi_G\| = 1$.

Proof: It is clear that $G \cap G^{\perp}$ contains only 0. If x is any vector in H we shall prove that $x - \pi_G(x)$ is in $G^{\perp}$. If u is in G then for all scalars a, $\pi_G(x) - au$ is in G so

$$\|x - \pi_G(x) + au\|^2 \geqslant \|x - \pi_G(x)\|^2.$$

If we expand the left-hand side of the above inequality we obtain

$$\|x - \pi_G(x) + au\|^2 = \|x - \pi_G(x)\|^2 + 2Re\langle au, x - \pi_G(x)\rangle + |a|^2 \|u\|^2$$

from which it follows that

$$2Re\langle au, x - \pi_G(x)\rangle \geqslant -|a|^2 \|u\|^2 \tag{18–1}$$

for all scalars a. If we take $a = 1/n$ in (18–1) we conclude that for all $n = 1, 2, \ldots$ we have

$$2Re\langle u, x - \pi_G(x)\rangle \geqslant -\frac{1}{n} \|u\|^2$$

from which it follows that $2Re\langle u, x-\pi_G(x)\rangle \geqslant 0$. If we take $a=-1/n$ we use (18–1) to show $2Re\langle u, x-\pi_G(x)\rangle \leqslant 0$. If we take $a=-i/n$ we have for $n=1, 2, \ldots$

$$2Im\langle u, x-\pi_G(x)\rangle = 2Re\langle iu, x-\pi_G(x)\rangle \geqslant -\frac{1}{n}\|u\|^2$$

so that $2Im\langle u, x-\pi_G(x)\rangle \geq 0$, and by taking $a=i/n$ that $2Im\langle u, x-\pi_G(x)\rangle \leqslant 0$. Therefore, $\langle u, x-\pi_G(x)\rangle$ must be 0. From this we conclude that (a) and (b) are valid and that π_G is a linear projection from H onto G.

For each x in H, $\|x\|^2 = \|x-\pi_G(x)\|^2 + \|\pi_G(x)\|^2$ because we have $\langle x-\pi_G(x), \pi_G(x)\rangle = 0$. Therefore, $\|\pi_G(x)\| \leqslant \|x\|$ from which it follows that $\|\pi_G\| \leqslant 1$. Since every nonzero projection has norm at least one, we conclude that $\|\pi_G\| = 1$ when G is a nontrivial subspace. ■

The space $G^\perp$ is called the *orthogonal complement* of the space G. The projection π_G is called the *orthogonal projection* from H onto G. Since in this case the metric projection is linear we shall denote it by P_G rather than by π_G.

The next theorem shows that, just as with the space $\mathbb{R}^n$, every Hilbert space represents its own dual space in a natural way by means of the inner product.

Theorem 18.8: For each x in a Hilbert space H the function f_x defined by $f_x(u) = \langle u, x\rangle$ is a continuous linear functional on H, and $\|f\| = \|x\|$. If f is any continuous linear functional on H, there is x in H such that $f(u) = \langle u, x\rangle$ for all u in H.

Proof: We leave to the reader the proof of the first assertion.

If f is a nonzero continuous linear functional on H, let G denote the null space of f, that is, the set of all u in H such that $f(u)=0$. Then G is a closed subspace of H; and if $f(y)$ is nonzero, it is not hard to see that $H = G \oplus [y]$. Therefore, $G^\perp$ must have dimension one. Thus there is a unique y in $G^\perp$ with $f(y)=1$. Since each u in H has a unique representation

$$u = v + f(u)y$$

where v is in G, $\langle u, y\rangle = \langle f(u)y, y\rangle = f(u)\|y\|^2$. If $x = y/\|y\|^2$ it follows that $\langle u, x\rangle = f(u)$ for all u in H.

Since $x = y/\|y\|^2$ we have $\|x\| = \|y\|/\|y\|^2 = 1/\|y\|$. If $\|u\| \leqslant 1$, we have $|f(u)| = |\langle u, x\rangle| \leqslant \|x\|$ so $\|f\| \leqslant \|x\|$. On the other hand, $\|y\|x$ has norm 1 and, $f(\|y\|x) = 1/\|y\| = \|x\|$ so $\|f\| \geqslant \|x\|$. ■

Exercises

A

1. Suppose $\langle , \rangle$ is an inner product on a linear space E. Show the following relations are valid
   (a) $\langle x, ay+bz\rangle = \bar{a}\langle x, y\rangle + \bar{b}\langle x, z\rangle$;
   (b) $\|x+y\|^2 = \|x\|^2 + 2Re\langle x, y\rangle + \|y\|^2$;
   (c) if x is orthogonal to y then $\|x+y\|^2 = \|x\|^2 + \|y\|^2$.
   (d) What theorem of geometry is equivalent to (c)?

2. Let E be an ip space. Prove that if $x_n \to x$ and $y_n \to y$ then $\langle x_n, y_n \rangle \to \langle x, y \rangle$. Thus the inner product is continuous from $E \times E$ into the scalar field.
3. Show that equality holds in 18.3 (a) if and only if x and y are proportional. (Reexamine the proof, and observe that equality holds if and only if there is c such that $\langle x - cy, x - cy \rangle = 0$.)
4. When does $\|x + y\| = \|x\| + \|y\|$? To what theorem of geometry does this correspond?
5. Prove that the function $\langle x, y \rangle = \sum_{i=1}^n x_i \bar{y}_i$ is an inner product on the space $\mathbb{C}^n$ of all n-tuples $x = (x_1, x_2, \ldots, x_n)$ of complex numbers.
6. Under the conditions of Theorem 18.5 show that $v = \pi_C(u)$ if and only if for all w in C we have $Re\langle u - v, v - w \rangle \geqslant 0$. Show
$$\|u - w\|^2 - \|u - v\|^2 \geqslant 2Re\langle u - v, v - w \rangle.$$
7. Prove the first assertion of Theorem 18.8.
8. Suppose (a_{ij}) is an $m \times n$ matrix of positive numbers and $M_i > 0$ for $i = 1, 2, \ldots, m$. Show that there exists a unique vector x in R^n such that $\sum_{j=1}^n x_j^2$ is a minimum and $\sum_{j=1}^n a_{ij} x_j \leqslant M_i$ for $i \leqslant m$.
9. Suppose p is a polynomial of degree n. Prove that if $m < n$ there is a unique polynomial q of degree $\leqslant m$ such that $\int_0^1 |p(t) - q(t)|^2 \, dt$ is a minimum.

B

10. Prove that if S is any subset of a Hilbert space E then $(S^\perp)^\perp$ is the closed linear span of S.
11. Show that if $\| \;\; \|$ is a norm on a linear space that satisfies the parallelogram law then there is an inner product $\langle , \rangle$ on E such that $\|x\| = \sqrt{\langle x, x \rangle}$ for all x in E. [*Hint*: Use
$$\langle x, y \rangle = \tfrac{1}{2}\{\|x + y\|^2 - \|x - y\|^2 + \|x + iy\|^2 - \|x - iy\|^2\}.]$$
12. Prove that the space l^2 of all sequences $x = (x_n)$ such that $\sum_n |x_n|^2 < \infty$ forms a Hilbert space with inner product $\langle x, y \rangle = \sum_n x_n \bar{y}_n$.
13. A sequence (x_n) in a Hilbert space E is said to converge weakly to x if $\lim_n \langle x_n, y \rangle = \langle x, y \rangle$ for all y in E.
    (a) Show that the sequence (e_n) in l^2 converges weakly to 0.
    (b) Show that if (x_n) converges weakly to x in a finite dimensional ip space then (x_n) converges to x in norm. (Use a basis.)
    (c) Show that if (x_n) converges weakly to x and $\|x_n\| \to \|x\|$ then (x_n) converges to x in norm.
    (d) Show that if (x_n) is a bounded sequence in a separable Hilbert space then there is a subsequence of (x_n) that converges weakly.
14. If C is a closed and bounded convex subset of a Hilbert space and u is not in C show that $\pi(u)$ is on the boundary of C, that is, $\pi(u) \in C \setminus \text{interior}(C)$.
15. Suppose f is any continuous function on $[a, b]$. Show there is a unique polynomial q of degree $\leqslant m$ such that $\int_a^b |f(t) - q(t)|^2 \, dt$ is a minimum.
16. *(a)* Show that the uniform norm on l^∞ does not satisfy the Parallelogram Law.

(b) Let x denote the sequence $(1, -1, 1, -1, \ldots)$. Show that $\|x-y\| \geqslant 1$ for each sequence y in the space c of convergent sequences.
(c) If $u=(1, 1, 1, \ldots)$ and $v=(-1, -1, \ldots)$ then $\|x-u\|$ and $\|x-v\|$ are both 1. Thus l^{∞} does not have the unique approximation property.
(d) Show there are infinitely many z in C^2 such that $\|x-z\|=1$.
(e) Let $w_{2n-1}=1-1/n$, $w_{2n}=1/n-1$ for $n=1, 2, \ldots$. Show there is no best approximation u in c to $w=(w_n)$.

19

Orthogonality

As we stated at the beginning of this chapter orthogonality of functions arose in connection with the Fourier series. The fact that this concept can be incorporated into the framework of Hilbert spaces is perhaps the main reason that Hilbert space is important. A subset S of nonzero elements in an ip space E is called *orthogonal* if $\langle x, y\rangle=0$ for all x, y in S with $x \neq y$. The set S is called *orthonormal* (ON) if S is orthogonal and $\|x\|=1$ for all x in S. An orthogonal set S is called *complete* if there is no nonzero vector y in E such that $\langle x, y\rangle=0$ for all x in S. A complete orthonormal set is referred to as a CON set.

Our first result is to use a straightforward maximality argument to show that every inner product space contains a complete orthonormal set.

Theorem 19.1: Every inner product space E that contains a nonzero member contains a CON set.

Proof: Denote by $\mathscr{U}$ the collection of all ON sets in S. Since $\mathscr{U}$ contains all singletons $\{x\}$ with $\|x\|=1$ it follows that $\mathscr{U}$ is not empty. It is easy to see that the chain union of ON sets is also an ON set so that by Postulate 3.6 on page 15, $\mathscr{U}$ contains a maximal member S. The set S is complete because if $y \neq 0$ is such that $\langle x, y\rangle=0$ for all x in S, the set $S \cup \{y/\|y\|\}$ is an ON set that properly contains S in contradiction to the maximality of S. ■

Since every orthogonal set is li, it follows that a CON set in a finite dimensional space F is a basis for F. It is easy to see that there are also bases for a finite dimensional space F that are not orthogonal. The following key theorem characterizes the point in the linear span of a finite ON set that is closest to a given point.

Theorem 19.2: Suppose $\{x_1, x_2, \ldots, x_n\}$ is an ON set in an ip space E.

(a) For each x in F the linear span of $\{x_1, x_2, \ldots, x_n\}$ we have

$$x=\sum_{i=1}^{n} \langle x, x_i\rangle x_i.$$

(b) If x is in $E \backslash F$ the point $u=\sum_{i=1}^{n} \langle x, x_i\rangle x_i$ is the closest point in F to x;
(c) $\|x\|^2 \geqslant \sum_{i=1}^{n} |\langle x, x_i\rangle|^2$ for all x in E.

Proof: We leave the proof of (a) to you.

(b) Let b_i denote $\langle x, x_i \rangle$ for $i = 1, 2, \ldots, n$; and let v be any point of F. Since $\{x_1, x_2, \ldots, x_n\}$ is a basis for F, v has an expansion $v = \sum_{i=1}^{n} a_i x_i$. Thus we have

$$\|x - v\|^2 = \left\| x - \sum_{i=1}^{n} a_i x_i \right\|^2 = \|x\|^2 + \sum_{i=1}^{n} (|a_i|^2 - a_i \bar{b}_i - \bar{a}_i b_i)$$

$$= \|x\|^2 - \sum_{i=1}^{n} |b_i|^2 + \sum_{i=1}^{n} (|a_i|^2 - a_i \bar{b}_i - \bar{a}_i b_i + |b_i|^2)$$

$$= \|x\|^2 - \sum_{i=1}^{n} |b_i|^2 + \sum_{i=1}^{n} |a_i - b_i|^2.$$

The term $\|x\|^2 - \sum_{i=1}^{n} |b_i|^2$ depends only on x, and the term $\sum_{i=1}^{n} |a_i - b_i|^2$ is minimized when $a_i = b_i$ for each i. Therefore, $\|x - v\|$ is a minimum when $v = u$. In this case,

$$\|x\|^2 - \sum_{i=1}^{n} |b_i|^2 = \|x - u\|^2 \geq 0$$

so that (c) is valid. ■

In order to study expansions with respect to arbitrary ON sets we must extend the idea of unconditional and absolute convergence to series having more than countably many terms. If f is a nonnegative function on an arbitrary set A we denote by $\sum_{s \in A} f(s)$ the supremum of all numbers $\sum_{s \in D} f(s)$ as D ranges over all finite subsets of A. It is not hard to show that this is the usual sum of nonnegative numbers if A is finite or countable. If A is uncountable and $\sum_{s \in A} f(s) < \infty$ then the set of all s such that $f(s) > 1/n$ must be finite for all $n = 1, 2, \ldots$. Hence, the set of all s for which $f(s) > 0$ must be countable. If U is a function from A into a normed space E we say that the series $\sum_{s \in A} U(s)$ *converges unconditionally* to the point x in E if for each $\varepsilon > 0$ there is a finite subset D_ε of A such that whenever D is a finite subset that contains D_ε we have $\|x - \sum_{s \in D} U(s)\| < \varepsilon$. This concept corresponds to the usual sum of a series if A is finite. If A is countable and E is a finite dimensional space, it corresponds to unconditional convergence, which is equivalent to absolute convergence in this case.

Theorem 19.3: Suppose S is an ON set in a Hilbert space E.

(a) For each x in E we have $\sum_{u \in S} |\langle x, u \rangle|^2 \leq \|x\|^2$ (Bessel's inequality).

(b) The series $\sum_{u \in S} \langle x, u \rangle u$ converges unconditionally in E to the closest point v to x in the closed linear span of S in E.

(c) The set S is a CON set if and only if $\sum_{u \in S} |\langle x, u \rangle|^2 = \|x\|^2$ for all x in E.

Proof:

(a) This is essentially a restatement of Theorem 19.2 (c), which implies that $\sum_{u \in D} |\langle x, u \rangle|^2 < \|x\|^2$ for every finite subset D of S.

(b) Denote by F the closed linear span of S in E. By Theorem 18.5 on page 146 there is a point v in F that is closest to x. By (a),

$$\sum_{u \in S} |\langle v, u\rangle|^2 \leqslant \|v\|^2$$

so that $\langle v, u\rangle = 0$ except for a countable set $u_1, u_2, \ldots$. We shall treat the case where the countable set is infinite; the case where it is finite is easier and it is mostly handled in Theorem 19.2. If $v_n = \sum_{i=1}^{n} \langle v, u_i\rangle u_i$ for $n = 1, 2, \ldots$, then (v_n) is a Cauchy sequence in E because if $p < q$, $\|v_q - v_p\|^2 = \sum_{i=p+1}^{q} |\langle v, u_i\rangle|^2$. Since E is complete this sequence has a limit v_0. Since $\langle v_n, u_m\rangle = \langle v, u_m\rangle$ for $n \geqslant m$ and $\lim_n \langle v_n, u_m\rangle = \langle v_0, u_m\rangle$ it follows that $\langle v_0, u_m\rangle = \langle v, u_m\rangle$ for all m. This implies that $\|(v - v_0) - \sum_{i=1}^{m} a_i u_i\|^2 = \|v - v_0\|^2 + \sum_{i=1}^{m} |a_i|^2$ for all linear combinations $\sum_{i=1}^{m} a_i u_i$. Since v is known to be in the closure of $[u_i]$ it follows that $v = v_0$. The series $\sum_{i=1}^{\infty} \langle v, u_i\rangle u_i$ converges unconditionally to v because if $\|\sum_{i=1}^{m} \langle v, u_i\rangle u_i - v\| < \varepsilon$ and D contains $\{1, 2, \ldots, m\}$ then $\sum_{i \in D} \langle v, u_i\rangle u_i$ is the closest point in $[u_i]_{i \in D}$ so it has distance less than ε from v. That $\langle v, u_i\rangle = \langle x, u_i\rangle$ for each i now follows by an argument used to establish (c) of Theorem 19.2 employing infinite rather than finite series. That $\langle x, u\rangle = 0$ for each u in $S \backslash \{u_i\}$ follows since if $\langle x, u\rangle \neq 0$ we would have $v + \langle x, u\rangle u$ strictly closer to x than v is. Lastly we note that since $\sum_{i=1}^{\infty} \langle v, u_i\rangle u_i$ converges to v it follows that

$$\|v\|^2 = \lim_n \left\| \sum_{i=1}^{n} \langle v, u_i\rangle u_i \right\|^2 = \sum_{i=1}^{\infty} |\langle v, u_i\rangle|^2.$$

(c) If there is x in E such that $\sum_{u \in S} |\langle x, u\rangle|^2 < \|x\|^2$, then x is not in the closed linear span F of S so that if v is the closest point to x in F we have $(x - v)$ is nonzero and $\langle (x - v), u\rangle = 0$ for all u in S. Therefore, S is not complete. If S is not complete, there is x in E such that all $\langle x, u\rangle = 0$ but $x \neq 0$. It is obvious that in this case we have

$$\sum_{u \in S} |\langle x, u\rangle|^2 < \|x\|^2. \blacksquare$$

Corollary 19.4: For S an ON set in a Hilbert space E the following statements are equivalent:

(a) S is complete;

(b) the closure of the linear span of S is dense in E;

(c) every member x of E has a unique expansion $x = \sum_{u \in S} a_u u$;

(d) for each x in E the series $\sum_{u \in S} \langle x, u\rangle u$ converges unconditionally to x;

(e) for each x in E, $\sum_{u \in S} |\langle x, u\rangle|^2 = \|x\|^2$.

In view of the corollary above a CON set in a Hilbert space E is often called an *orthonormal basis* for E. The construction found in the theorem below is called the *Gram-Schmidt process*. It provides an algorithm for constructing an orthonormal sequence from a given countable li set.

Theorem 19.5: Suppose $\{y_1, y_2, \ldots\}$ is a li subset of an inner product space E. There is an ON set $\{x_1, x_2, \ldots\}$ such that $(*)[x_1, x_2, \ldots, x_n] = [y_1, y_2, \ldots, y_n]$ for all $n = 1, 2, \ldots$. The set $\{x_i\}$ is constructed by the following algorithm: $x_1 = y_1/\|y_1\|$; if $x_1, x_2, \ldots, x_{m-1}$ have been constructed such that (*) holds for $n \leqslant m-1$, let $u_m = y_m - \sum_{j=1}^{m-1} \langle y_m, x_j \rangle x_j$ and let $x_m = u_m/\|u_m\|$.

Proof: For each m, $u_m \neq 0$ since y_m is not in $[x_1, \ldots, x_{m-1}] = [y_1, \ldots, y_{m-1}]$. It can be calculated directly that u_m is orthogonal to each x_i for $i < m$. Thus $\{x_1, \ldots, x_m\}$ is ON. ■

Example 19.6:

The functional defined on $C[-1, 1] \times C[-1, 1]$ by the formula $\langle f, g \rangle = \int_{-1}^{1} f(x)g(x)\,dx$ is an inner product on the space $C[-1, 1]$. If we apply the Gram-Schmidt process to the li sequence $\{1, t, t^2, \ldots\}$ the first few terms are as follows:

$$x_0 = \frac{1}{\sqrt{2}}, \qquad x_1 = \sqrt{\tfrac{2}{3}}\,t, \qquad x_2 = \sqrt{\tfrac{5}{2}}\left(\tfrac{3}{2}t^2 - \tfrac{1}{2}\right).$$

In general, the function u_m determined by the Gram-Schmidt process will be $t^m - q_{m-1}$ where q_{m-1} is the polynomial in $[1, t, t^2, \ldots, t^{m-1}]$ that is closest to t^m. Thus u_m will be the unique polynomial of degree m, which is orthogonal to $1, t, t^2, \ldots, t^{m-1}$ and has lead coefficient one. The polynomial $d^m/dt^m(t^2-1)^m$ obviously has degree m. To show it is orthogonal to $1, t, \ldots, t^{m-1}$ let $k < m$. Then we have

$$\int_{-1}^{1} \frac{d^m}{dt^m}(t^2-1)^m t^k\,dt = \left[\frac{d^{m-1}}{dt^{m-1}}(t^2-1)^m t^k\right]_{-1}^{1} - \int_{-1}^{1} \frac{d^{m-1}}{dt^{m-1}}(t^2-1)^m k t^{k-1}\,dt.$$

The first summand on the right-hand side is zero because (t^2-1) is a factor of the resulting polynomial. Thus we can reduce the integral on the left to

$$(-1)^k k! \int_{-1}^{1} \frac{d^{m-k}}{dt^{m-k}}(t^2-1)^m\,dt = \left[\frac{d^{m-k-1}}{dt^{m-k-1}}(t^2-1)^m\right]_{-1}^{1} = 0.$$

Therefore, u_m is a multiple of $d^m/dt^m(t^2-1)^m$. Since the lead coefficient of the latter polynomial is $(2m)!/m!$, it follows that $u_m = m!/(2m)!\,d^m/dt^m(t^2-1)^m$. In order to calculate x_m we must find $\|u_m\|$; we have

$$\|u_m\|^2 = \int_{-1}^{1} u_m^2\,dt = \int_{-1}^{1} u_m(t)t^m\,dt$$

because u_m is orthogonal to powers of $t < m$. Therefore, we have

$$\frac{(2m)!}{m!}\|u_m\|^2 = \int_{-1}^{1} \frac{d^m}{dt^m}(t^2-1)^m t^m\,dt$$

$$= \left[\frac{d^{m-1}}{dt^{m-1}}(t^2-1)^m t^m\right]_{-1}^{1} - m\int_{-1}^{1} \frac{d^{m-1}}{dt^{m-1}}(t^2-1)^m t^m\,dt,$$

which can be reduced in m steps to $(-1)^m m! \int_{-1}^{1} (t^2-1)^m\, dt$. Again applying the process of integration by parts we can show that

$$\int_{-1}^{1} (t^2-1)^m\, dt = (-1)^m \frac{2^{m+1} m!}{(1 \cdot 3 \cdot \cdots \cdot \{2m+1\})}.$$

Solving for $\|u_m\|$ we finally calculate

$$x_m = \frac{\sqrt{m+1/2}}{(2^m m!)} \frac{d^m}{dt^m} (t^2-1)^m.$$

The polynomial $P_m(t) = 1/(2^m m!) d^m/dt^m (t^2-1)m$ is known as the mth *Legendre polynomial.*

Exercises

A

1. Find three CON sets in $\mathbb{R}^3$ with no common vector.
2. Show that if T is an ON set in an ip space E there is a CON set S in E that contains T.
3. Establish (a) of Theorem 19.2.
4. Find the polynomial $p(t)$ of degree at most 4 such that $\int_0^1 |t^6 - p(t)|^2\, dt$ is a minimum.
5. *(a)* Use the trigonometric formulas for the sine and cosine of the sums of angles to show the following

$$\sin A \cos B = \tfrac{1}{2}\{\sin(A+B) + \sin(A-B)\}$$
$$\cos A \cos B = \tfrac{1}{2}\{\cos(A+B) + \cos(A-B)\}$$
$$\sin A \sin B = \tfrac{1}{2}\{\cos(A-B) - \cos(A+B)\}$$

   (b) Show that the set $\{1, \cos nt, \sin nt\colon n = 1, 2, \ldots\}$ is an orthogonal set on $C[0, 2\pi]$ with inner product $\langle f, g\rangle = \int_0^{2\pi} f(t)g(t)\, dt$.
6. Prove that for every continuous function f on an interval $[a, b]$ there is a unique polynomial p_n of degree $\leqslant n$ such that $\int_a^b |f(t) - p_n(t)|^2\, dt$ is minimum.
7. Prove that if F and G are closed subspaces of an ip space E such that $\langle x, y\rangle = 0$ for all x in F and y in G then the direct sum $F \oplus G$ of F and G is closed.
8. If E and F are inner product spaces, show $E \times F$ is an inner product space with the inner product $\langle (u, v), (x, y)\rangle = \langle u, x\rangle_E + \langle v, y\rangle_F$. If E and F are Hilbert spaces so is $E \times F$ with the inner product described.

B

9. *(a)* Prove that if $\{x_1, x_2, \ldots\}$ is an ON sequence in an inner product space then x_n converges weakly to 0.
   (b) Prove that for every continuous function f on $[0, 2\pi]$, $\int_0^{2\pi} f(t)\cos(nt)\, dt$ and $\int_0^{2\pi} f(t)\sin(nt)\, dt$ converge to 0.
10. Show that a generalized series $\sum_{s \in A} U(s)$ of reals converges unconditionally if and only if $\sum_{s \in A} |U(s)|$ converges. Do the same for a complex series.

11. Show that the sequence $\{(2\pi)^{-1/2}e^{int}: n=0, \pm 1, \pm 2, \ldots\}$ is an ON set in the space $C[0, 2\pi]$ of continuous complex valued functions on $[0, 2\pi]$ with the inner product $\langle f, g\rangle = \int_0^{2\pi} f(t)\overline{g(t)}\, dt$.
12. Show that the sequence $\{(n/\pi)^{1/2}z^{n-1}: n=1, 2, \ldots\}$ is an ON sequence in the space $C(D)$ of all continuous complex valued functions f on the unit disk $\{z: |z| \leqslant 1\}$ with the inner product

$$\langle f, g\rangle = \iint_{|z|<1} f(z)\overline{g(z)}\, dx\, dy, \quad \text{where } z = x+iy.$$

13. Let $\| \ \|$ denote the usual norm on $C[0, 1]$ and let $\| \ \|_2$ denote the norm determined by the inner product $\langle f, g\rangle = \int_0^1 f(t)\overline{g(t)}\, dt$. If F is a subspace of $C[0, 1]$ for which there is $M > 0$ such that $\|f\| \leqslant M\|f\|_2$ for all f in F, then for any finite ON set $\{u_1, u_2, \ldots, u_n\}$ in F, we have $\sum_{i=1}^n |f_i(x)|^2 \leqslant M^2$ for all x in $[0, 1]$. Therefore, the dimension of F cannot exceed M. From this conclude that the norms $\| \ \|$ and $\| \ \|_2$ do not induce equivalent topologies on $C[0, 1]$.
14. Use the Gram-Schmidt process to construct an orthonormal sequence in $C[0, 1]$ with the inner product $\int_0^1 f(t)g(t)\, dt$ from the sequence $1, t, t^2, \ldots$.
15. For every set S let $l^2(S)$ consist of all scalar functions f on S such that $\sum_{s\in S} |f(s)|^2 < \infty$.
    (a) Show $l^2(S)$ is a Hilbert space with inner product $\langle f, g\rangle = \sum_{s\in S} f(s)g(s)$ where the series converges unconditionally.
    (b) Show that if S and T are sets having the same cardinality then $l^2(S)$ is isometric to $l^2(T)$.
    (c) Show every Hilbert space is isometric to $l^2(S)$ for some set S. (Let $\{x_\alpha: \alpha \in A\}$ be a CON set in E; show E is isometric to $l^2(A)$.)

20

The Hahn-Banach Theorem

The *Hahn-Banach Theorem* can be interpreted analytically as an existence theorem for continuous linear functionals on a normed space, or geometrically as a theorem about the separation of convex sets by closed hyperplanes. In this section we shall consider some of the analytic forms of this theorem.

A *sublinear form* is a real valued function p on a real linear space E such that $p(ax) = ap(x)$ if $a > 0$ and $p(x+y) \leqslant p(x) + p(y)$ for all x, y in E. A norm on E and a linear functional on E are examples of sublinear forms.

The following theorem is known as the analytic form of the Hahn-Banach Theorem.

Theorem 20.1: Suppose p is a sublinear form on a real linear space E, and f is a linear functional defined on a subspace F of E such that $f(x) \leqslant p(x)$ for all x in F. Then there exists a linear functional $\hat{f}$ on E such that

(HB1) $\hat{f}$ is an extension of f, that is, $\hat{f}(x)=f(x)$ for x in F

(HB2) $\hat{f}(x) \leqslant p(x)$ for all x in E.

Proof: We shall first use Postulate 3.6 on page 15 to obtain a maximal extension of f. Then we shall show that this extension has domain all of E for otherwise f could be extended.

Let $\mathscr{A}$ consist of all subsets A of $E \times \mathbb{R}$ that have the form $[G, g]=\{(x, g(x)): x \in G\}$ where G is a subspace of E containing F and g is a linear functional on G that satisfies (HB1) and the inequality (HB2) for all x in G. $[G, g]$ is known as the *graph* of g. At present we know that $\mathscr{A}$ contains at least the set $[F, f]$. Suppose $\mathscr{B}$ is a chain in $\mathscr{A}$. If $[G_\alpha, g_\alpha]$ and $[G_\beta, g_\beta]$ are in this chain, then one set must be contained in the other. If $[G_\alpha, g_\alpha]$ is contained in $[G_\beta, g_\beta]$, then for each x in G_α, $(x, g_\alpha(x)) \in [G_\beta, g_\beta]$ so that $x \in G_\beta$ and $g_\beta(x)=g_\alpha(x)$. In other words G_α is a subspace of G_β and g_β is an extension of g_α to G_β. A corresponding statement holds if $[G_\beta, g_\beta]$ is contained in $[G_\alpha, g_\alpha]$. The reader may show that $G=\bigcup_\alpha G_\alpha$ is a subspace of E. If $x \in G$, there is α such that $x \in G_\alpha$; if x is also in G_β we know that $g_\alpha(x)=g_\beta(x)$. Define g on G as the common value. Then g is a linear functional on G that satisfies (HB1) and (HB2). The reader may show that $[G, g]=\bigcup_\alpha [G_\alpha, g_\alpha]$. Thus $\mathscr{A}$ satisfies the hypothesis of Postulate 3.6 so that it has a maximal element that we denote by $[E_1, \hat{f}]$.

It remains to show that E_1 must equal E. Suppose for the sake of obtaining a contradiction that there is a vector y in $E \backslash E_1$. For each $s>0$ and x in E we have

$$2\hat{f}\left(\frac{x}{s}\right) \leqslant 2p\left(\frac{x}{s}\right) \leqslant p\left(\frac{x}{s}+y\right)+p\left(\frac{x}{s}-y\right)$$

so that

$$\hat{f}\left(\frac{x}{s}\right)-p\left(\frac{x}{s}-y\right) \leqslant p\left(\frac{x}{s}+y\right)-\hat{f}\left(\frac{x}{s}\right).$$

Therefore, if we let $A=\inf\{p(x/s+y)-\hat{f}(x/s): s>0\}$ and $a=\sup\{\hat{f}(x/s)-p(x/s-y)\}$ we shall have $a \leqslant A$. Let c be any real number with $a \leqslant c \leqslant A$. Define $f^\#$ on E_2 the linear span of E_1 and y by

$$f^\#(x+ty)=\hat{f}(x)+ct.$$

The linear functional $f^\#$ satisfies (HB1) because it is an extension of $\hat{f}$. To show that $f^\#$ satisfies (HB2) suppose first that $t>0$. Since $c \leqslant p(x/t+y)-\hat{f}(x/t)$, we have $f^\#(x+ty)=\hat{f}(x)+ct \leqslant p(x+ty)$. If $t<0$, let $s=-t$. Since $c \geqslant -p(x/s-y)+\hat{f}(x/s)$, it follows that $f^\#(x+ty)=\hat{f}(x)+ct=\hat{f}(x)-cs \leqslant p(x-sy)=p(x+ty)$. Since $f^\#$ satisfies (HB1) and (HB2) it follows that the set $[E_2, f^\#]$ properly contains $[E_1, \hat{f}]$, contradicting the fact that $[E_1, \hat{f}]$ is maximal. ■

The next result is a normed space form of the Hahn-Banach Theorem. The extension to the complex case is due to Bohnenblust and Sobczyk.

Theorem 20.2: Suppose F is a subspace of a normed space E and f is a bounded linear functional defined on F. There is a bounded linear functional $\hat{f}$ on E such that $\|\hat{f}\|=\|f\|$ and $\hat{f}(x)=f(x)$ for all x in F.

Proof:

(Case 1) E a real normed space. Let $M = \|f\|$. Since $p(x) = M\|\ \|$ is a sublinear form on E and $|f(x)| \leqslant M\|x\|$ for x in F it follows that there is an extension $\hat{f}$ of f to E such that $|\hat{f}(x)| \leqslant M\|x\|$ for x in E.

(Case 2) E is a complex normed space. We can consider E a linear space over the reals by restricting multiplication to real numbers. If f is any linear functional on E we can decompose $f(x)$ into its real and complex parts

$$f(x) = f_1(x) + if_2(x)$$

where f_1 and f_2 are real functions of x. You may show that both f_1 and f_2 are linear functionals on E considered as a real space. Since $f(ix)$ is equal to both $f_1(ix) + if_2(ix)$ and $if(x) = if_1(x) - f_2(x)$, it follows by equating real and complex parts that $f_1(ix) = -f_2(x)$ and $f_2(ix) = f_1(x)$. Thus we may rewrite the decomposition as

$$f(x) = f_1(x) - if_1(ix).$$

Suppose f is a complex linear functional defined on a subspace F of E with $\|f\| = M$. If f_1 is the real part of f in the above decomposition, then for each x in F we have

$$|f_1(x)| = |Re(f(x))| \leqslant |f(x)| \leqslant M\|x\|$$

so $\|f_1\| \leqslant M$. Let $\hat{f}_1$ be an extension of f_1 to E such that $\|\hat{f}_1\| \leqslant M$, and let f be defined by

$$\hat{f}(x) = \hat{f}_1(x) - i\hat{f}_1(ix).$$

If $\|x\| \leqslant 1$ let c be a complex number such that $c\hat{f}(x) = |\hat{f}(x)|$ and $|c| = 1$. Then we shall have

$$|\hat{f}(x)| = \hat{f}(cx) = \hat{f}_1(cx) \leqslant M\|cx\| = M\|x\|$$

because $\hat{f}(cx)$ is real and $|c| = 1$. This proves that $|\hat{f}(x)| \leqslant M\|x\|$ for x in E. Since $\|f\| = M$ as a functional on F we must also have $\|f\| = M$. ■

Corollary 20.3: For each $x \neq 0$ in a normed space E there is a continuous linear functional f on E such that $\|f\| = 1$ and $f(x) = \|x\|$.

Proof: Let F be the span of x in E, that is, $F = \{ax\colon a \text{ is a scalar}\}$. Define f_1 on F by $f_1(ax) = a\|x\|$. Since $\|ax\| = |a|\,\|x\| = |f_1(x)|$ for all ax in F it follows that $\|f_1\| = 1$. Use Theorem 20.2 to extend f_1 to a function f defined on all of E such that $\|f\| = 1$. Then $f(x) = f_1(x) = \|x\|$. ■

Corollary 20.4: Suppose F is a closed subspace of a normed space E and x is a point in $E \backslash F$. There is a continuous linear functional f on E such that $\|f\| = 1$, $f(y) = 0$ for $y \in F$ and

$$f(x) = \operatorname{dist}(x, F) = \inf\{\|x - y\|\colon y \in F\}.$$

Proof: Define f_1 on $\text{span}\{x, F\} = \{ax + y: y \in F\}$ by the formula $f_1(ax + y) = ad$ where $d = \text{dist}(x, F)$. If $\|ax + y\| \leqslant 1$ and $a \neq 0$ then $d \leqslant \|x + y/a\| \leqslant 1/|a|$ so that $|f_1(ax + y)| = d|a| \leqslant 1$. To see that $\|f_1\| = 1$ let $\varepsilon > 0$ be given. Let $y \in F$ be such that $\|x - y\| < d + \varepsilon$. If $u = (x - y)/\|x - y\|$ then $\|u\| = 1$ and $f_1(u) = d/\|x - y\| \geqslant d/(d + \varepsilon)$. Therefore, we have $\|f_1\| \geqslant d/(d + \varepsilon)$. Since this inequality is true for each $\varepsilon > 0$ it follows that $\|f_1\| \geqslant 1$. By Theorem 20.2 the functional f_1 can be extended to a functional f on all of E with the required properties. ■

Corollary 20.3 shows that the dual space E^* of a nontrivial normed space contains nonzero members. The process of taking duals can be continued to form the *second dual* $E^{**} = (E^*)^*$. For each x in E there is $\hat{x}$ in E^{**} defined by the formula

$$\hat{x}(f) = f(x) \quad \text{for all } f \text{ in } E^*.$$

If $\|f\| \leqslant 1$ then $|\hat{x}(f)| = |f(x)| \leqslant \|x\|$ so that $\|\hat{x}\| \leqslant \|x\|$ for all x in E. The mapping from x to x is often called the *James operator* and written $J(x) = \hat{x}$. If the range of J is all of E^{**} then E is said to be *reflexive.*

Theorem 20.5: The mapping J from E into E^{**} is a linear isometry.

Proof: We shall let the reader verify that J is linear. We have already seen that $\|J\| \leqslant 1$. That $\|\hat{x}\| = \|x\|$ for all x in E follows from Corollary 20.3. Given x in E let f in E^* be such that $f(x) = \|x\|$ and $\|f\| = 1$. Then $\|\hat{x}\| \geqslant |\hat{x}(f)| = |f(x)| = \|x\|$. ■

A useful device for representing the dual space of a normed space as well as for other purposes is the *Schauder basis* (or *s-basis.*) A sequence (x_n) in a normed space E of infinite dimension is called an *s-basis* if it satisfies the following two conditions.

(SB1) Each x in E has a unique expansion $x = \sum_{n=1}^{\infty} a_n x_n$.

(SB2) For each x in E the mapping $f_n(x) = a_n$ is continuous.

An s-basis differs from an a-basis in that the s-basis admits infinite expansions and the coefficient functionals are required to be continuous. An s-basis is called an *unconditional* Schauder basis if convergence in (SB1) is unconditional, that is, if for each x in E and each $\varepsilon > 0$ there is a finite subset B of $\mathbb{N}$ such that if C is a finite subset of $\mathbb{N}$ that contains B we have $\|x - \sum_{n \in C} a_n x_n\| < \varepsilon$. If (x_n) is a Schauder basis for a normed space E and $f \in E^*$ then for each x in E we must have

$$f(x) = \sum_{n=1}^{\infty} f(x_n) f_n(x) = \sum_{n=1}^{\infty} fn(x) f(x_n)$$

where $x = \sum_{n=1}^{\infty} a_n x_n$ is the expansion given by (SB1) and (f_n) is the sequence of functionals given by (SB2). If (x_n) is an unconditional basis, then the series $\sum_{n=1}^{\infty} f_n(x) f(x_n)$ converges unconditionally, thus absolutely.

Example 20.6:

Let e_n be the sequence that has a 1 in the nth coordinate and 0's elsewhere. Then (e_n) is an unconditional basis for the space c_0. For suppose $x = (x_n)$ is in c_0 and $\varepsilon > 0$ is given. There is M such that $|x_n| < \varepsilon$ if $n > M$. If B is any finite subset of $\mathbb{N}$ that contains $\{1, 2, \ldots, M\}$ then $\|x - \sum_{n \in B} x_n e_n\| < \varepsilon$ because if n is not in B, $n > M$. Thus if f is in c_0^* it follows that $\sum_{n=1}^{\infty} |f(e_n)x_n| < \infty$ for each x in c_0. For each k let $u^{(k)}$ be the sequence for which $u_i^{(k)} = |f(e_i)|/f(e_i)$ if $i \leqslant k$ and $f(e_i) \neq 0$ and $u_i^{(k)} = 0$ otherwise. Then each $u^{(k)}$ is in c_0 and $\|u^{(k)}\| \leqslant 1$ so that we have

$$f(u^{(k)}) = \sum_{n=1}^{k} |f(e_i)| \leqslant \|f\|.$$

This implies that if f is in c_0^* we have $\sum_{n=1}^{\infty} |f(e_i)| \leqslant \|f\|$. On the other hand if $\sum_{n=1}^{\infty} |c_n| < \infty$, the series $\sum_{n=1}^{\infty} c_n x_n$ converges for each x in c_0 and defines a linear functional f on c_0 with $\|f\| \leqslant \sum_{n=1}^{\infty} |c_n|$ because $|\sum_{n=1}^{\infty} c_n x_n| \leqslant \sup_n |x_n| \sum_{n=1}^{\infty} |c_n|$. We express this by saying that the space l^1 of all absolutely convergent series is an isometric representation of the dual space of c_0.

The sequence (e_n) is also an unconditional s-basis for l^1. Proceeding as above we can show that f is in $(l^1)^*$ if and only if there is a bounded sequence (u_n) such that $f(e_n) = u_n$ for each n, and that for each bounded sequence (u_n) the functional defined on l^1 by $f(x) = \sum_{n=1}^{\infty} u_n x_n$ is continuous on l^1 and that $\|f\| = \sup_n |u_n| = \|u\|$. Thus the space l^∞ of bounded sequences is an isometric representation for the dual space of l^1. The mapping J with respect to these representations is the inclusion map from c_0 into l^∞. Therefore, the space c_0 is not reflexive.

If g is any continuous function from a topological space S into $\mathbb{R}$ then $g^{-1}(0)$ is a closed subset of S because (0) is closed in $\mathbb{R}$. The following theorem characterizes continuous linear functionals on a normed space by means of this property.

Theorem 20.7: A linear functional f on a normed space E is continuous if and only if $f^{-1}(0)$ is a closed subspace of E. If f is not continuous then $f^{-1}(0)$ is dense in E.

Proof: We have already noted that f continuous implies $f^{-1}(0)$ closed. Suppose $f^{-1}(0) = F$ is closed. If $F = E$ then we are finished since f is the zero functional. Otherwise there is x in E such that $f(x) = 1$. If u is any other vector in E then $u - f(u)x$ is in F. Thus each u in E has a unique expansion $u = f(u)x + z$ where z is in F. To show f is continuous suppose $u_n \to u$ in E; let $u_n = f(u_n)x + z_n$. The sequence $(f(u_n))$ is bounded because if we could find a subsequence $(f(u_{k(n)}))$ with $|f(u_{k(n)})| \to \infty$, we would then have

$$\frac{u_{k(n)}}{f(u_{k(n)})} = x + \frac{z_{k(n)}}{f(u_{k(n)})}.$$

Since $(u_{k(n)})$ converges to u we must have $(u_{k(n)}/f(u_{k(n)}))$ converges to 0 so that

$z_{k(n)}(f(u_{k(n)}) \to -x$. But each vector $z_{k(n)}(f(u_{k(n)})$ is in F and F is closed so $-x$ is in F, which is a contradiction. Thus $(f(u_n))$ must be bounded.

If $(f(u_{k(n)}))$ is any convergent subsequence of $(f(u_n))$, then since $u_{k(n)} = f(u_{k(n)})x + z_{k(n)}$ it follows that $(z_{k(n)})$ converges to some point z in F. Thus we shall have $u = (\lim_n f(u_{k(n)}))x + z$ from which we conclude $f(u) = \lim_n f(u_{k(n)})$. Since every convergent subsequence of $(f(u_n))$ converges to $f(u)$ it follows that the sequence itself converges to $f(u)$.

If f is not continuous then $f^{-1}(0)$ is not closed so the closure F of $f^{-1}(0)$ properly contains $f^{-1}(0)$. The reader may verify that the closure of a linear subspace is itself a linear subspace so F is a subspace of E. But every u in E can be written in the form $u = ax + z$ where $f(x) \neq 0$ and $f(z) = 0$. Therefore, F must be all of E. ■

Exercises

A

1. Show that if p is a sublinear form and $a < 0$ then $p(ax) = -ap(-x)$.
2. Show that if $\{p_\alpha\}$ is a set of sublinear forms on a linear space E and $\sup_\alpha p_\alpha(x) = p(x) < \infty$ for all x in E then p is a sublinear form.
3. Is the sum of two sublinear forms a sublinear form?
4. Let $E = \mathbb{R}^2$; let F consist of all (x, y) in $\mathbb{R}^2$ for which $y = 0$. Define f on F by $f(x, y) = x$.
   (a) Show there is only one extension g of f to all of E such that $|g(x, y)| \leqslant \|(x, y)\|$ for all (x, y) in E when $\mathbb{R}^2$ is given the Euclidean norm.
   (b) Show there are infinitely many extensions g of f to all E such that $|g(x, y)| \leqslant 1.01\,\|(x, y)\|$ for all (x, y) in E.
   (c) If E is given the norm $\|(x, y)\| = |x| + |y|$ show there are infinitely many extensions of g to E with $|g(x, y)| \leqslant \|(x, y)\|$ for all (x, y) in E.
5. Prove that if $\{E_\alpha\}$ is a chain of linear subspaces of a linear space E then $\bigcup_\alpha E_\alpha$ is a linear subspace of E.
6. Show that if f is a complex linear functional on a complex linear space E and $f = f_1 + if_2$, where f_1 and f_2 are real functions, then f_1 and f_2 are linear over $\mathbb{R}$.
7. Prove that every finite dimensional normed space is reflexive.
8. Prove that if $\{x_1, x_2, \ldots, x_k\}$ is an a-basis for a finite dimensional normed space, then it satisfies (SB1) and (SB2) with k replacing ∞ in (SB1).
9. Show that (e_n) is an unconditional basis for l^1 and l^∞ is a representation for the dual space.
10. Show that the closure of a linear subspace F of a normed space E is itself a linear subspace. You must show that if $x_n \to x$ and $y_n \to y$ then $x_n + y_n \to x + y$ and $ax_n \to ax$.

B

11. Prove the converse of Theorem 17.9. (Suppose F is not complete. Let (x_n) be a Cauchy sequence in F that has no limit in F. Let f be any continuous linear

functional on E. How do you know there is one? Consider the sequence (T_n) in $L(E, F)$ defined by $T_n x = f(x)x_n$.)

12. A *projection* is a linear operator P such that $P^2 = P$.

(a) If $\{x_1, x_2, \ldots, x_n\}$ is a finite li subset of a normed space E show there are continuous linear functionals $f_1, f_2, \ldots, f_n$ on E such that $f_i(x_j) = 1$ if $i = j$ and 0 if $i \neq j$.

(b) Use (a) to show that if F is a subspace of E having finite dimension there is a continuous projection P from E with range F.

13. Suppose F is a closed subspace of a normed space E having finite *codimension*. This means there is a finite set $\{x_1, x_2, \ldots, x_n\}$ such that E is the span of $F \cup \{x_1, x_2, \ldots, x_n\}$. Show there is a projection P from E onto F.

14. Prove that a Banach space E is reflexive if and only if E^* is reflexive.

15. Suppose E is a normed space and E^{***} is the third dual, that is, the dual of E^{**}. For each functional θ in E^{***} let $T\theta$ be the restriction of θ to E^*.

(a) Show that T is a continuous linear operator from E^{***} onto E^* with $\|T\| \leqslant 1$.

(b) If J is the James mapping from E^* into E^{***} then $TJf = f$ for all f in E^*.

(c) Define P from E^{***} into itself by $P\theta = JTf$. Show P is a continuous linear operator having norm one with range equal to all of $J(E^*)$.

(d) If θ is in $J(E^*)$ then $P\theta = \theta$. The results of (c) and (d) are expressed by saying that P is a continuous projection from E^{***} onto E^*.

(e) Show that $E^{***} = J(E^*) \oplus N$ where $N = \{\theta : \theta \in E^{***},\ P\theta = 0\}$, and $\oplus$ means direct sum.

16. Let cs denote the space of all scalar sequences $x = (x_n)$ such that $\sum_{n=1}^{\infty} x_n$ converges.

(a) Show that cs is a Banach space with the norm $\|x\| = \sup_k |\sum_{n=1}^{k} x_n|$.

(b) Show that $\{e_n\}$ is a Schauder basis for cs that is not unconditional.

17. Let bv denote the space of all scalar sequences $y = (y_n)$ with $\sum_{n=1}^{\infty} |y_{n+1} - y_n| < \infty$.

(c) Show that bv is a Banach space with the norm $\|y\| = |y_1| + \sum_{n=1}^{\infty} |y_{n+1} - y_n|$.

(d) Show that bv represents the dual space of cs. (For this you will need the inequality

$$\left|\sum_{j=1}^{k} x_j y_j\right| \leqslant \left(\sup_{n<k+1} \sum_{j=1}^{n} x_j\right)\left(|y_1| + \sum_{j=2}^{k} |x_j - x_{j-1}|\right).$$

18. *(a)* Prove that the space c of convergent sequences is equal to $c_0 \oplus [e]$ where e is the sequence $(1, 1, \ldots)$.

(b) Prove that $e, e_1, e_2, \ldots$ is an s-basis of c.

(c) Prove that $a_0 e + \sum_{n=1}^{\infty} a_n e_n$ converges to a point in c if and only if (a_n) converges to 0.

(d) Prove that c is isomorphic to c_0.

(e) Represent the dual space of c as l^1.

C

19. (a) Show that the function defined on l^∞ by $p(x) = \limsup 1/n \sum_{j=1}^{n} x_j$ is a sublinear form.
   (b) Define λ on the space c of convergent sequences by $\lambda(x) = \lim_n x_n$. Then $\lambda(x) \leqslant p(x)$ for x in c.
   (c) Show that there is a linear functional Λ defined on all of l^∞ such that $\Lambda(x) = \lambda(x)$ for x in c and $\Lambda(x) \leqslant p(x)$ for x in l^∞.
   (d) Prove that if $\Lambda(x) \leqslant p(x)$, then $\Lambda(x) \geqslant \liminf x_n$. ($\Lambda(x) = \Lambda(-x) \leqslant \limsup -x_n = -\liminf x_n$.)
   (e) Show that every such function Λ determined by (c) satisfies the following properties (i) $\Lambda(x) = \lim_n x_n$ if x converges; (ii) if $x_i \geqslant 0$ for all i then $\Lambda(x) \geqslant 0$; (iii) for each x in l^∞, $\Lambda(x) = \Lambda(x_2, x_3, \ldots)$. (A functional Λ that satisfies (i), (ii), and (iii) is called a Banach limit.)
20. Show that in every infinite dimensional normed space E there are two convex subsets C and D such that
   (a) C and D are disjoint;
   (b) both C and D are dense in E;
   (c) $C \cup D = E$. [*Hint*: Use the fact that there exists a discontinuous linear functional on E.]

21

Uniform Boundedness Principle

A direct consequence of the Baire Category Theorem is the *Uniform Boundedness Principle*, which asserts that if a sequence of continuous linear operators on a Banach space is pointwise bounded then it is uniformly bounded. By combining the Uniform Boundedness Principle with the uniform continuity of continuous linear operators we obtain the *Banach-Steinhaus Theorem*. This theorem asserts that pointwise boundedness and convergence on a fundamental subset (a subset with dense span) implies pointwise convergence on the entire space. The following theorem is a precise statement of the Uniform Boundedness Principle.

Theorem 21.1: If (T_n) is a sequence of continuous linear operators from a Banach space E into a normed space F such that $\sup_n \|T_n x\| < \infty$ for each x in E then $\sup_n \|T_n\| < \infty$.

Proof: For each pair of positive integers m, n let $C_{m,n}$ denote the set of all x in E such that $\|T_m x\| \leqslant n$. Since each operator T_m is continuous, each set $C_{m,n}$ is closed. Hence, the set $C_n = \bigcap_{m=1}^{\infty} C_{m,n}$ is also closed. Since $\sup_n \|T_n x\| < \infty$ for each x in E it follows that $E = \bigcup_{n=1}^{\infty} C_n$. By the Baire Category Theorem, Theorem 9.7 on page 64, one of the sets C_k has a nonempty interior. If x_0 is an interior point of C_k, then there is $\varepsilon > 0$ such that $B(x_0, \varepsilon) = \{x\colon \|x - x_0\| < \varepsilon\}$ is contained in C_k. This means that if $\|x - x_0\| < \varepsilon$, $\|T_n x\| \leqslant k$ for all n. Thus, if $\|y\| < \varepsilon$ we have

$$\|T_n y\| = \|T_n(y + x_0) - T_n(x_0)\| \leqslant \|T_n(y + x_0)\| + \|T_n x_0\| \leqslant k + \|T_n x_0\|.$$

If $\|y\| < 1$ it follows that for each n we have

$$\|T_n y\| = \left(\frac{1}{\varepsilon}\right) \|\varepsilon T y\| \leqslant \frac{k + \|T_n x_0\|}{\varepsilon}$$

so that

$$\sup_n \|T_n\| \leqslant \frac{k + \|T_n x_0\|}{\varepsilon} < \infty. \blacksquare$$

Theorem 21.2: Suppose T is a continuous linear operator from a dense subspace F of a normed space E into a Banach space G. Then there exists a unique continuous linear extension $\hat{T}$ from E into G. (See Problem 19 of Section 8 on page 58.)

Proof: Let x be a point in E. Since F is dense in E there is a sequence (x_n) in F that converges to x. The sequence (Tx_n) is a Cauchy sequence in G because $\|Tx_m - Tx_n\| \leqslant \|T\| \|x_m - x_n\|$ for all m, n. Thus there is a point y in G such that $\lim_n Tx_n = y$. If (u_n) is any other sequence that converges to x then (Tu_n) converges to y because $(u_1, x_1, u_2, x_2 \ldots)$ converges to x so that $(Tu_1, Tx_1, Tu_2, Tx_2, \ldots)$ is a Cauchy sequence in G that has a subsequence converging to y. Thus the function $\hat{T}x = \lim_n Tx_n$ where (x_n) is any sequence in F converging to x is a well defined extension of T.

To verify that $\hat{T}$ is linear, suppose u, v are in E and a, b are scalars. Let (u_n) in F converge to u and (v_n) in F converge to v. Then we have

$$\begin{aligned}
\hat{T}(au + bv) &= \lim_n T(au_n + bv_n) \\
&= \lim_n aTu_n + \lim bTv_n \\
&= a \lim_n Tu_n + b \lim_n Tv_n = a\hat{T}u + b\hat{T}v.
\end{aligned}$$

To verify that $\hat{T}$ is bounded suppose x is in E and (x_n) in F converges to x. Then we have

$$\|\hat{T}x\| = \lim_n \|Tx_n\| \leqslant \|T\| \lim_n \|x_n\| = \|T\| \|x\|.$$

To show that $\hat{T}$ is the unique continuous extension of T suppose f is a continuous function from E into G such that $f(u) = T(u)$ for u in F. If x is in E there is a sequence (x_n) in F that converges to x. By the continuity of f we have $f(x) = \lim_n f(x_n) = \lim_n T(x_n) = \hat{T}(x)$. $\blacksquare$

A subset D of a normed space E is said to be *fundamental in* E if $[D]$ the linear span of D is dense in E. The following result is known as the *Banach-Steinhaus Theorem.*

Theorem 21.3: Suppose $\{T_n\}$ is a sequence of continuous linear operators from a Banach space E into a Banach space F such that

(a) $\sup_n \|T_n x\| < \infty$ for all x in E and

(b) $\lim_n T_n x$ exists for each x in a fundamental subset D of E.

Then there is a continuous linear operator T from E into F such that $\lim_n T_n x = Tx$ for all x in E.

Proof: Because of (a) and the Uniform Boundedness Principle it follows that $\sup_n \|T_n\| = M < \infty$.

Since $(T_n x)$ converges for each x in D and each T_n is a linear operator it follows that $(T_n u)$ converges for each u in the linear span $[D]$ of D. To show that $(T_n x)$ converges for each x in E we show that this sequence is a Cauchy sequence in the Banach space F. Let x in E and $\varepsilon > 0$ be given. Since $[D]$ is dense in E there is u in $[D]$ such that $\|x-u\| < \varepsilon/3M$. Since $(T_n u)$ converges there is K such that $\|T_i u - T_k u\| < \varepsilon/3$ when i and k are greater than K. Thus, for $i, k > K$ we shall have

$$\begin{aligned} \|T_i x - T_k x\| &\leqslant \|T_i(x-u)\| + \|T_i u - T_k u\| + \|T_k(u-x)\| \\ &\leqslant M\|x-u\| + \frac{\varepsilon}{3} + M\|x-u\| < \varepsilon. \end{aligned}$$

If we denote $\lim_n T_n x$ by Tx, then T is linear because

$$\begin{aligned} T(au+bv) &= \lim_n T_n(au+bv) \\ &= \lim_n aT_n u + \lim_n bT_n v \\ &= a \lim_n T_n u + b \lim_n T_n v = aTu + bTv. \end{aligned}$$

for all scalars a and b and all u, v in E. Since $\|T_n x\| \leqslant M\|x\|$ for all x in E it follows that $\|Tx\| \leqslant M\|x\|$ for all x in E so that T is continuous. ∎

As an application of the Banach-Steinhaus Theorem we derive the following useful criterion for a sequence in a Banach space to be an s-basis.

Theorem 21.4: Suppose (x_n) is a fundamental li sequence in a Banach space E. Then (x_n) is an s-basis for E if

(*) there is $M > 0$ such that for each $\sum_{n=1}^m a_n x_n$ in the linear span $[x_n]$ of (x_n) we have $\|\sum_{n=1}^k a_n x_n\| \leqslant M \|\sum_{n=1}^m a_n x_n\|$ for all $k < m$.

Proof: Denote $[x_n]$ by F. For each $k = 1, 2, \ldots$ let F_k be the finite dimensional subspace $[x_1, x_2, \ldots, x_k]$. The function P_k defined from F into F_k by $P_k(\sum_{n=1}^m a_n x_n) = \sum_{n=1}^{\min(k,m)} a_n x_n$ is linear and because of (*) it follows that $\|P_k\| \leqslant M$. By Theorem 21.2 there is a unique continuous extension $\hat{P}_k$ of P_k to all of E, with $\|\hat{P}_k\| \leqslant M$. Since

$$\lim_k \hat{P}_k x_n = \lim_k P_k x_n = x_n \quad \text{for all } n,$$

it follows from the Banach-Steinhaus Theorem that $\lim_k \hat{P}_k x = x$ for each x in E. For each m and k we have $P_m P_k = P_{\min(m,k)}$ from which we conclude that $\hat{P}_m \hat{P}_k = \hat{P}_{\min(m,k)}$. Therefore, if $\hat{P}_k x = \sum_{n=1}^k a_n x_n$, and $\hat{P}_m x = \sum_{n=1}^m b_n x_n$, with $k < m$ we have $a_n = b_n$ for $n = 1, 2, \ldots, k$. This means that for each x in E the sequence $(\hat{P}_k)$ of operators determines a unique sequence of scalars (a_n) such that

$\hat{P}_k x = \sum_{n=1}^{k} a_n x_n$ for each k. Since $(\hat{P}_k x)$ converges to x it follows that x has the unique expansion

(EXP) $x = \sum_{n=1}^{\infty} a_n x_n$.

This proves that the sequence (x_n) satisfies (SB1).

In order to prove that (x_n) satisfies (SB2) suppose $f_k(x) = a_k$ where x has the expansion (EXP). By going to the limit in (*) we find

$$\left\| \sum_{j=1}^{k} a_j x_j \right\| \leqslant M \|x\| \quad \text{for all } k.$$

Thus we shall have

$$\| f_k(x) x_k \| \leqslant \left\| \sum_{j=1}^{k} a_j x_j \right\| + \left\| \sum_{j=1}^{k-1} a_j x_j \right\| \leqslant 2M \|x\|$$

from which we conclude $\|f_k\| \leqslant 2M / \|x_k\|$. ■

The converse of Theorem 21.4 will be established in the following section by means of the Closed Graph Theorem.

Example 21.5:

In the Banach sequence spaces c_0 and l^p, $1 \leqslant p \leqslant \infty$, let e_n denote the sequence with 1 in the nth coordinate and 0's elsewhere. The norm on each of these spaces has the property that $\|\sum_{n=1}^{k} x_n e_n\| \leqslant \|x\|$ for all x in the space. In order to show that $\{e_n\}$ is a basis for each of the spaces it suffices by the previous theorem to show $[e_n]$ is dense in the space.

A subset B of a normed space E is called *bounded* if there is $M > 0$ such that $\|x\| < M$ for each x in B. The set B is called *weakly bounded* if for each $f \in E^*$ we have $\sup\{|f(x)|: x \in B\} < \infty$. If B is bounded then B is weakly bounded because $|f(x)| \leqslant \|f\| \, \|x\|$ for all x in B. We can use the Hahn-Banach Theorem and the Uniform Boundedness Principle to derive the converse of this fact.

Theorem 21.6: A weakly bounded subset B of a normed space E is bounded.

Proof: Suppose B is a subset of a normed space E that is not bounded. We shall verify that B is not weakly bounded. For each n let x_n in B be such that $\|x_n\| > n$. Define $\hat{x}_n$ on E^*, the dual space of E, by $\hat{x}_n(f) = f(x_n)$. If $\sup_n |\hat{x}_n(f)|$ were $< \infty$ for each f in E^* it would follow from the Uniform Boundedness Principle that $\sup_n \|\hat{x}_n\| = \sup_n \|x_n\| < \infty$. But this is not the case so there is f in E^* such that $\sup_n |f(x_n)| = \infty$. ■

Exercises

A

1. A function p from a linear space E into the nonnegative reals is called a *seminorm* if $p(x+y) \leqslant p(x) + p(y)$ and $p(ax) = |a| p(x)$ for all x, y in E and a in the scalar field.

(a) Show that if T is a linear operator from a normed space E into a normed space F then the function defined on E by $p(x) = \|Tx\|$ is a seminorm on E.

(b) Show that a seminorm p on E is continuous if and only if there is $M > 0$ such that $p(x) \leq M\|x\|$ for all x in E.

(c) Imitate the proof of Theorem 21.1 to show that if $\{p_n\}$ is a sequence of continuous seminorms on a Banach space E such that $p(x) = \sup_n p_n(x) < \infty$ for all x in E, then p is a continuous seminorm on E.

(d) Show Theorem 21.1 is a special case of this result.

2. Prove that a sequence (T_n) of continuous linear operators from a Banach space E into a normed space F is uniformly bounded if and only if for each x in E and f in F^* we have $\sup_n |f(T_n x)| < \infty$.

3. Put the norm $\|x\| = \sum_n |x_n|$ on the space ϕ of finitely nonzero sequences. For each n define T_n from ϕ into ϕ by $T_n x = x + (n-1)x_n e_n$. Show that each T_n is continuous from ϕ into ϕ with this norm and that $\sup_n \|T_n x\| < \infty$ for all x in ϕ, but that $\sup_n \|T_n\| = \infty$. This illustrates the necessity of completeness in Theorem 21.1.

4. Let $\tilde{E}$ denote the completion of the normed space E. Show that the restriction operator $Tf = g$ where $g(x) = f(x)$ for x in E is a linear isometry from $\tilde{E}^*$ onto E^*. Thus for all practical purposes $\tilde{E}^*$ can be identified with E^*.

5. *(a)* If $\{T_n\}$ is a sequence of continuous linear operators from a normed space E into a normed space F such that $\lim_n T_n x = Tx$ for all x in E show that T is bounded and $\|T\| \leq \liminf_n \|T_n\|$.

   (b) Define T_n from l^1 into l^1 by $T_n x = (x_n, x_{n+1}, \ldots)$ for each n. Then $\|T_n\| = 1$ for each n but $\lim_n T_n x = 0$ for all x.

   (c) Prove that if E is finite dimensional then $\|T - T_n\| \to 0$.

6. Use Theorem 21.4 to show that $\{e_n\}$ is a basis for cs.

7. *(a)* Prove that a subspace F of a normed space E is dense in E if and only if whenever f in E^* is 0 on F it is 0 on all of E.

   (b) Prove that a set B is fundamental in E if and only if there is no nonzero f in E^* that is 0 on B.

8. Suppose (x_n) is a sequence in l^1, (that is, a sequence of sequences $x_n = (x_{nj})$) such that $\sup_n |\sum_{j=1}^{\infty} x_{nj} y_j| < \infty$ for all (y_j) in c_0. Use the Uniform Boundedness Principle and the fact that l^1 is the dual space of c_0 to prove that $\sup_n \sum_{j=1}^{\infty} |x_{nj}| < \infty$.

B

9. Let E and F be linear spaces. A function U from $E \times F$ into the scalar field is called *bilinear* if for each fixed y in F, scalars a, b and u, v in E we have $U(au + bv, y) = aU(u, y) + bU(v, y)$ and for each fixed x in E, scalars a, b and y, z in F we have $U(x, ay + bz) = aU(x, y) + bU(x, z)$.

   (a) Show that if E and F are Banach spaces then a bilinear form U on $E \times F$ is continuous if and only if there is $M > 0$ such that $|U(x, y)| \leq M\|x\|\|y\|$ for all x in E and y in F.

   (b) Show that if T is a continuous linear operator from E into F^* the function $U(x, y) = (Ty)(x)$ is a bilinear form on $E \times F$.

(c) Prove that every continuous bilinear form on $E \times F$ has the form $U(x, y) = (Ty)(x)$ for some T in $L(E, F^*)$.

10. Let (x_n) be a fundamental li sequence in a Banach space E.

(a) Prove that the space S of all sequences (a_n) such that $\sum_{n=1}^{\infty} a_n x_n$ converges in E is a Banach space with the norm $\|(a_n)\| = \sup_k \|\sum_{n=1}^{k} a_n x_n\|$.

(b) Prove that the sequence (e_n) is an s-basis of S.

11. A *biorthogonal sequence* in a Banach space E is a double sequence (x_n, f_n) with each x_n in E and each f_n in E^* such that $f_n(x_m) = 1$ if $n = m$ and 0 if $n \neq m$. Show that if (x_n, f_n) is a biorthogonal sequence in a Banach space E such that $\sup_k |\sum_{n=1}^{k} f_n(x) f(x_n)| < \infty$ for each x in E and each f in E^* then (x_n) is an s-basis for its closed linear span in E and (f_n) is an s-basis for its closed linear span in E^*.

C

12. Let D be a countable dense subset of the interval $[0, 1]$.

(a) Show that for each continuous real function f on $[0, 1]$ and each $\varepsilon > 0$ there are points $t_1 < t_2 < \cdots < t_n$ in D such that for the function g whose graph consists of the line segments joining the points $(0, f(0))$, $(t_1, f(t_1))$, $(t_2, f(t_2))$, $\ldots$, $(t_n, f(t_n))$, $(1, f(1))$ we have $|f(t) - g(t)| < \varepsilon$ for all t in $[0, 1]$. The function g is called the *polygonal approximation* of f with nodes $t_1, t_2, \ldots, t_n$.

(b) Suppose $\{t_1, t_2, \ldots\}$ is any ennumeration of $D \backslash \{0, 1\}$. For each n, and each f in $C[0, 1]$ let $P_n(f)$ denote the polygonal approximation of f with nodes $t_1, t_2, \ldots, t_n$. Show that each P_n is a linear operator from $C[0, 1]$ into $C[0, 1]$ with $\|P_n\| \leqslant 1$. Show that each P_n is a projection, and $\lim_n P_n(f) = f$ for all f in $C\{0, 1]$.

(c) Define $u_{-1}(t) = t$, $u_0(t) = 1 - t$. For $n = 1, 2, \ldots$, let r_1 be the largest t_i less than t_n, or 0 if there is no such t_i; let r_2 be the smallest t_i greater than t_n or 1 if there is no such t_i. Define u_n to be the function whose graph consists of the line segments joining $(0, 0)$ to $(r_1, 0)$ to $(t_n, 1)$ to $(r_2, 0)$ to $(1, 0)$. Show that $\{u_{-1}, u_0, \ldots, u_n\}$ is a basis for the range of P_n for each n. Show that $\{u_{-1}, u_0, u_1, \ldots\}$ is an s-basis for $C[0, 1]$.

22

The Open Mapping and Closed Graph Theorems

A consequence of the Baire Category Theorem, less immediate than the Uniform Boundedness Principle, is the *Open Mapping Theorem,* which states that the image of an open set under a continuous onto linear operator is open. The *Closed Graph Theorem,* which is equivalent to the Open Mapping Theorem, gives a condition for the continuity of a linear operator; this condition is often much easier to verify than continuity.

Theorem 22.1: Suppose T is a continuous linear operator from a Banach space E onto a Banach space F and that O is an open subset of E. Then $T(O)$, the image of O under T, is an open subset of F.

Proof: The first step is to show that for each $\varepsilon > 0$, the zero point of F is interior to the closure $\overline{T(B(0, \varepsilon))}$ of the image of the open ball with center 0 and radius ε. Since T is onto we know that $F = \bigcup_{n=1}^{\infty} n\overline{T(B(0, \varepsilon))}$ so that one of the sets $n\overline{T(B(0, \varepsilon))}$ must have an interior point y because of the Baire Category Theorem. If $B(y, \delta)$ is contained in $n\overline{T(B(0, \varepsilon))}$, then so is $B(-y, \delta)$ for if z is in $B(-y, \delta)$ then $-z$ is in $B(y, \delta)$ so there is a sequence (v_j) in $B(0, \varepsilon)$ such that (nTv_j) converges to $-z$. The sequence $(nT(-v_j))$ thus converges to z. The set

$$U = \tfrac{1}{2}B(y, \delta) + \tfrac{1}{2}B(-y, \delta) = \{\tfrac{1}{2}u + \tfrac{1}{2}v\colon u \in B(y, \delta),\ v \in B(-y, \delta)\}$$

is an open subset of F, contains 0, and is contained in $\overline{TB(0, \varepsilon)}$ because the latter set is convex. Therefore, we conclude that 0 is an interior point of $n\overline{TB(0, \varepsilon)}$. If η is such that $B(0, \eta) \subset n\overline{TB(0, \varepsilon)}$, it follows that $B(0, \eta/n) \subset \overline{TB(0, \varepsilon)}$. Thus 0 is an interior point of $\overline{TB(0, \varepsilon)}$.

The next step is to show that for each $\varepsilon > 0$, $\overline{TB(0, \varepsilon)}$ is contained in $TB(0, 2\varepsilon)$. This implies that 0 is an interior point of $TB(0, \delta)$ for all $\delta > 0$. Suppose y is a point in $\overline{TB(0, \varepsilon)}$. Let (δ_n) be a sequence of positive reals that converges to 0. Since 0 is an interior point of $\overline{TB(0, \varepsilon/2)}$ and y is in the closure of $TB(0, \varepsilon)$ there is a point u_1 in $B(0, \varepsilon)$ such that $y - Tu_1$ is in $\overline{TB(0, \varepsilon/2)}$ and $\|y - Tu_1\| < \delta_1$. If we have defined u_n in $B(0, \varepsilon/2^n)$ such that $y - Tu_1 - Tu_2 - \cdots - Tu_n$ is in $\overline{TB(0, \varepsilon/2^n)}$ and has norm less than δ_n, let u_{n+1} in $B(0, \varepsilon/2^n)$ be such that $y - Tu_1 - \cdots - Tu_n - Tu_{n+1}$ is in $\overline{TB(0, \varepsilon/2^{n+1})}$ and has norm less than δ_{n+1}. In this way we define a sequence (u_n) in E such that $\|u_j\| < \varepsilon/2^{j-1}$ for each j. Since E is a Banach space the series $\sum_{n=1}^{\infty} u_n$ converges to a point x that is in $B(0, 2\varepsilon)$ because

$$\sum_{n=1}^{\infty} \|u_n\| < \sum_{n=1}^{\infty} \frac{\varepsilon}{2^{j-1}} = 2\varepsilon.$$

Since $\|y - Tu_1 - \cdots - Tu_n\| < \delta_n$ for all n it follows that $Tx = y$. Therefore, we conclude that y is in $TB(0, 2\varepsilon)$.

To complete the proof of the theorem suppose y is in $T(O)$ where O is open in E. If $Tx = y$ there is $\varepsilon > 0$ such that $B(x, \varepsilon) \subset O$. Suppose $\delta > 0$ is such that $B_F(0, \delta) \subset TB(0, \varepsilon)$. Here $B_F(0, \delta)$ is the open ball in F with center 0 and radius δ. Then $B_F(y, \delta) \subset T(O)$ because if $\|z - y\| < \delta$, $y - z \in B_F(0, \delta)$ so $y - z \in TB(0, \varepsilon)$. Hence, there is u with $\|u\| < \varepsilon$ such that $Tu = y - z = Tx - z$. But $\|(x + u) - u\| = \|u\| < \varepsilon$ so $x + u \in O$. Consequently, we have $T(x + u) = z \in T(O)$. ∎

Corollary 22.2: Suppose E is a linear space on which two norms $\|\ \|_1$ and $\|\ \|_2$ are defined such that $(E, \|\ \|_1)$ and $(E, \|\ \|_2)$ are both Banach spaces. If the identity mapping from $(E, \|\ \|_1)$ onto $(E, \|\ \|_2)$ is continuous, then both norms determine the same topology.

Proof: Since the identity mapping from $(E, \| \ \|_1)$ onto $(E, \| \ \|_2)$ is continuous and onto, it carries open sets to open sets. Therefore, it is a homeomorphism so the two topologies coincide. ■

Our next goal is the Closed Graph Theorem. The *graph* of a function f from a topological space R into a topological space S is the subset $\Gamma(f)$ of $R \times S$ given by the formula

$$\Gamma(f) = \{(r, f(r)) \colon r \text{ is in the domain of } f\}.$$

The graph $\Gamma(f)$ is called *closed* if $\Gamma(f)$ is closed in the product topology on $R \times S$. We first prove a preliminary result characterizing the linear operators between normed spaces that have a closed graph.

Theorem 22.3: Suppose T is a linear operator from a subspace of a normed space E into a normed space F. Then $\Gamma(T)$ is closed in $E \times F$ if and only if the following condition holds.

(CG) Whenever (x_n) converges to x in E and Tx_n converges to y in F it follows that x is in the domain of T and $Tx = y$.

Proof: Necessity of **(CG).** We recall that $E \times F$ is a normed space and that a sequence (x_n, y_n) converges to (x, y) if and only if $x_n \to x$ in E and $y_n \to y$ in F. If $x_n \to x$ and $Tx_n \to y$ it follows that (x_n, Tx_n) is in $\Gamma(T)$, and that (x_n, Tx_n) converges to (x, y) in $E \times F$. Thus if $\Gamma(T)$ is closed we must have (x, y) in $\Gamma(T)$ from which it follows that Tx is defined and equals y.

We leave the proof of the sufficiency of (CG) to you, the reader. ■

It is easy to see that a continuous linear operator satisfies condition (CG). But this condition is formally weaker than continuity because it allows you to assume that (Tx_n) converges to some member of F, whereas with continuity you must verify that (Tx_n) converges.

The next result is the Closed Graph Theorem.

Theorem 22.4: Suppose T is a linear operator whose domain is a Banach space E and whose range is a subspace of a Banach space F. If the graph $\Gamma(T)$ of T is closed in $E \times F$ then the operator T is continuous.

Proof: The product space $E \times F$ is a Banach space with the norm $\|(x, y)\| = \sup\{\|x\|, \|y\|\}$. Thus $\Gamma(T)$, being a closed subspace of $E \times F$, is also a Banach space. The function U defined from $\Gamma(T)$ into F by $U(x, Tx) = Tx$ is continuous since $\|U(x, Tx)\| \leqslant \|(x, Tx)\|$ for each member of $\Gamma(T)$. The function V defined from $\Gamma(T)$ onto E by $V(x, Tx) = x$ is likewise continuous. By the Open Mapping Theorem, V sends open sets into open sets. Since V is also one-to-one it follows that the inverse V^{-1} of V is continuous from E onto $\Gamma(T)$. Since the composition of continuous operators is continuous and since for all x in E we have $Tx = UV^{-1}x$, it follows that T is continuous. ■

As a second application of the Open Mapping Theorem we shall prove the converse of Theorem 21.4.

Theorem 22.5: A fundamental li sequence (x_n) in a Banach space E is an s-basis for E if and only if

(*) there is $M>0$ such that for each $\sum_{n=1}^{m} a_n x_n$ in the linear span $[x_n]$ of (x_n) we have $\|\sum_{n=1}^{k} a_n x_n\| \leqslant M \|\sum_{n=1}^{m} a_n x_n\|$ for all $k<m$.

Proof: The sufficiency of (*) was given as Theorem 21.4.

In order to prove that (*) is necessary, suppose the sequence (x_n) is an s-basis for the Banach space E. Thus each x in E has a unique expansion

$$x = \sum_{n=1}^{\infty} a_n x_n. \tag{22–1}$$

Since the series in (22–1) converges it follows that the partial sums $\{\sum_{n=1}^{m} a_n x_n\colon m=1, 2, \ldots\}$ form a bounded set in E. Thus for each x in E we have

$$\|x\|_* = \sup_m \left\| \sum_{n=1}^{m} a_n x_n \right\| < \infty.$$

It is easy to see that the real valued function $\|x\|_*$ is a norm on E and that $\|x\| \leqslant \|x\|_*$ for each x in E. Thus the topology on E determined by the norm $\|\ \|_*$ is stronger than that determined by $\|\ \|$.

We shall now prove that $(E, \|\ \|_*)$ is complete. Suppose (y_k) is a Cauchy sequence in $(E, \|\ \|)$. Then not only is (y_k) a Cauchy sequence in $(E, \|\ \|)$ but if $y_k = \sum_{n=1}^{\infty} a_{kn} x_n$, then for each $m=1, 2, \ldots$ $(\sum_{n=1}^{m} a_{kn} x_n)$ is also a Cauchy sequence in $(E, \|\ \|)$. Moreover, since $a_{km} x_m = \sum_{n=1}^{m} a_{kn} x_n - \sum_{n=1}^{m=1} a_{kn} x_n$ the sequence $(a_{km} x_m)$ is a Cauchy sequence in $(E, \|\ \|)$ for each m so that (a_{km}) is a Cauchy sequence of scalars. If $\lim_k a_{km} = a_m$ for each m, it follows that the Cauchy sequence $(\sum_{n=1}^{m} a_{kn} x_n)$ converges to $\sum_{n=1}^{m} a_n x_n$ in $(E, \|\ \|)$. If (y_k) converges to y, we have reason to suspect that $y = \sum_{n=1}^{\infty} a_n x_n$. To see this is the case let $\varepsilon>0$ be given. Let M be such that $\|y_p - y_q\|_* < \varepsilon/3$ when $p,\ q \geqslant M$. Let K be such that $\|y_M - \sum_{n=1}^{m} a_{Mn} x_n\| < \varepsilon/3$ when $m \geqslant K$. Then if $m \geqslant K$ we shall have

$$\left\| y - \sum_{n=1}^{m} a_n x_n \right\| \leqslant \|y - y_M\| + \left\| y_M - \sum_{n=1}^{m} a_{Mn} x_n \right\| + \left\| \sum_{n=1}^{m} a_{Mn} x_n - \sum_{n=1}^{m} a_n x_n \right\|$$

$$< \frac{\varepsilon}{3} + \frac{\varepsilon}{3} + \frac{\varepsilon}{3}.$$

Finally we shall prove that (y_k) converges to y in $(E, \|\ \|_*)$. Given $\varepsilon>0$, let M be such that $\|y_p - y_q\|_* < \varepsilon$ when $p,\ q > M$. For each $p,\ q>M$ and each m we have $\|\sum_{n=1}^{m} a_{pn} x_n - \sum_{n=1}^{m} a_{qn} x_n\| < \varepsilon$, so that if we let $q \to \infty$ we will have $\|\sum_{n=1}^{m} a_{pn} x_n - \sum_{n=1}^{m} a_n x_n\| \leqslant \varepsilon$. If we take the supremum over all m we obtain $\|y_p - y\|_* \leqslant \varepsilon$ for all $p > M$. Therefore, (y_k) converges to y in $(E, \|\ \|_*)$.

Since $\|x\| \leqslant \|x\|_*$ for all x in E it follows that the identity map from $(E, \|\ \|_*)$ into $(E, \|\ \|)$ is continuous. Since both norms make E a Banach space it follows from Corollary 22.2 that the two topologies coincide, and that the identity map is a homeomorphism. If M denotes the norm of the identity mapping from $(E, \|\ \|)$ onto $(E, \|\ \|_*)$ we have $\|x\|_* \leqslant M \|x\|$ for all x in E so that if $x = \sum_{n=1}^{m} a_n x_n$ is in

$[x_n]$ and $k < m$ we have

$$\left\| \sum_{n=1}^{k} a_n x_n \right\| \leqslant \left\| \sum_{n=1}^{m} a_n x_n \right\|_* \leqslant M \left\| \sum_{n=1}^{m} a_n x_n \right\| . \blacksquare$$

The proof of the following result is left to you. It shows that when E is a Banach space, condition (SB1) implies (SB2).

Corollary 22.6: If (x_n) is a sequence in a Banach space E that satisfies (SB1), then it also satisfies (SB2).

Exercises

A

1. If U and V are open subsets of a normed space show that $U + V = \{x + y: x \in U, y \in V\}$ is also open.
2. Suppose E is a linear space and $\| \ \|_1$ and $\| \ \|_2$ are two norms on E, which make E a Banach space.
   (*a*) Prove that if there is $K > 0$ such that $\|x\|_1 \leqslant K\|x\|_2$ for all x in E then there is $M > 0$ such that $\|x\|_2 \leqslant M\|x\|_1$ for all x in E and the two norms give the same topology on E.
   (*b*) Prove that if E is a Banach space with the norm $\|x\| = \|x\|_1 + \|x\|_2$ then the three norms determine the same topology on E.
3. A projection is a linear operator P from a linear space into itself such that $P^2 = P$.
   (*a*) Show that if $P \neq 0$ is a continuous projection from a normed space into itself then $\|P\| \geqslant 1$.
   (*b*) Suppose F and G are subspaces of a linear space E such that $E = F \oplus G$; this means that each x in E has a unique representation $x = y + z$ with y in F and z in G. Show that the function $P(x) = y$ is a projection from E into E with range F and null space G. Show how every projection can be determined in this way.
   (*c*) Prove that if P is a projection from a Banach space E into itself then the P is continuous if and only if the range and null space of P are both closed. (Use the Closed Graph Theorem.)
4. A *BH* space is a Banach space E that is a linear subspace of a linear space $\mathscr{X}$ with a Hausdorff topology such that the inclusion operator from E into $\mathscr{X}$ is continuous.
   (*a*) Show that if E and F are *BH* spaces with respect to the same containing space $\mathscr{X}$ and if $E \subset F$ then the inclusion mapping is continuous.
   (*b*) Show that if there is a *BH* topology on a subspace E of $\mathscr{X}$, then that topology is unique.
5. Suppose T is a one-to-one continuous linear operator from a normed space E into itself with range F. Show that the linear operator T^{-1} from F into E has a closed graph.
6. Let $C^1[0, 1]$ be the subspace of $C[0, 1]$ that consists of all functions with continuous first derivatives. Show the linear operator D with domain $C^1[0, 1]$

is a closed linear operator in $C[0, 1]$ that is not continuous. (Show that D is the inverse of $Sf(t) = \int_0^t f(s)\, ds$.)

7. Establish the sufficiency of condition (CG) in Theorem 22.3.
8. Let T be a continuous linear operator from a Banach space E into a Banach space F such that the null space of T is Z and the range space of T is G. Show that G is isomorphic to E/Z if and only if G is closed in F.
9. Suppose E, F, and G are Banach spaces, S is a one-to-one continuous linear operator from F into G, T is a linear operator from E into F, and $S \circ T$ is a continuous linear operator from E into G. Show that T is continuous.
10. Show that if (x_n) is an s-basis for a Banach space E then the double sequence (x_n, f_n) where each f_n is given by Corollary 22.6 is a biorthogonal sequence in E. (See Exercise 11 of Section 21.)

B

11. *(a)* Suppose F and G are closed subspaces of a Banach space E. Then $F + G = \{y + z \colon y \in F,\ z \in G\}$ is closed if and only if there is $M > 0$ such that for all x in $F + G$ we have

$$\inf\{\|u\| + \|v\| \colon u \in F,\ v \in G,\ u + v = x\} \leqslant M\|x\|.$$

[*Hint*: Define T from $F \times G$ onto $F + G$ by the formula $T(u, v) = u + v$. Show that T is continuous if $F \times G$ is given the norm $\|(u, v)\| = \|u\| + \|v\|$. Let Z be the null space of T. Show that the operator $\hat{T}$ from $F \times G/Z$ onto $F + G$ given by $\hat{T}((u, v) + Z) = u + v$ is one-to-one and continuous. If $F + G$ is closed then $\hat{T}$ is an isomorphism by the Open Mapping Theorem.]

(b) Show that if F, G, and H are closed subspaces such that $F + G + H$ is closed, then $F + G$ is closed.

12. Let $P[0, 1]$ denote the space of polynomial functions f on $[0, 1]$ with the norm $\|f\| = \sup\{|f(t)| \colon 0 \leqslant t \leqslant 1\}$. Show that the function $Df = f'$ from $P[0, 1]$ into $P[0, 1]$ has a closed graph but is not continuous.
13. If (x_n) is an s-basis for a Banach space E and (x_{n_k}) is a subsequence, then (x_{n_k}) is an s-basis for its closed linear span $\overline{[x_{n_k}]}$.
14. Show that if T is a continuous linear operator from a Banach space E into a Banach space F such that $T(E)$ is of the second category in F, then $T(E) = F$.
15. Let E and F be Banach spaces and let T be a continuous linear operator from E onto F. Show there exists $M > 0$ such that if (y_n) is a convergent sequence in F, there is a convergent sequence (x_n) in E such that $\|x_n\| \leqslant M\|y_n\|$ and $Tx_n = y_n$ for each n.
16. A BK-space is a Banach space of sequences on which each linear functional E_n defined by $E_n(x_i) = x_n$ is continuous.

(a) Show that the following sequence spaces are BK-spaces: l^1, c_0, c, l^p, cs, bs, bv, cs, bv_0.

(b) Show that if E and F are BK-spaces with $E \subset F$ then the inclusion from E into F is continuous.

(c) Show that if E and F are BK-spaces and that (r_j) is a sequence such that $(r_j x_j)$ is in F for (x_j) in E then the operator T defined by $T(x_j) = (r_j x_j)$ is continuous from E into F.

(*d*) Denote by $M(E, F)$ the set of all sequences (r_j) determined in (c). Show that $M(E, F)$ is a BK-space with the norm $\|(r_j)\| = \sup\{\|(r_j x_j)\|_F: (x_j) \in E, \|(x_j)\|_E \leqslant 1\}$. [*Hint*: Show that $M(E, F)$ is a closed subspace of $L(E, F)$.]

17. Suppose E and F are linear spaces. Find necessary and sufficient conditions on a subset $\mathscr{G}$ of $E \times F$ for $\mathscr{G}$ to be (a) the graph of a linear operator from a subspace D of E into F; (b) the graph of a continuous linear operator from E into F.

18. Suppose T is a bounded linear operator from the Banach space E into the Banach space G, and U is a bounded linear operator from the Banach F into G. Suppose that for each x in F there is a unique y in F such that $Tx = Uy$. Show that the function $Vx = y$ is a continuous linear operator from E into F.

23

Convexity

A *seminorm* is a function p on a linear space E that satisfies conditions (N2) and (N3) of Section 17, but does not necessarily satisfy (N1). A *locally convex space* is a linear space E upon which there is a topology determined by a collection $\mathscr{R}$ of seminorms in the following way.

(*LC1*) A subset U of E is open if and only if for each $x \in U$ there is a finite subset $\{p_1, p_2, \ldots, p_n\}$ of $\mathscr{R}$ and positive numbers $\varepsilon_1, \varepsilon_2, \ldots, \varepsilon_n$ such that the set of all y where $p_i(x-y) < \varepsilon_i$ for $i = 1, 2, \ldots, n$ is a subset of U.

(*LC2*) If x in E is nonzero, there is p in $\mathscr{R}$ with $p(x) > 0$.

Example 23.1:

For each $\alpha \in \Lambda$, let E_α be a normed space with norm $\|\ \|_\alpha$; let $E = \prod_{\alpha \in \Lambda} E_\alpha$. The product topology on E is a locally convex topology determined by the collection $\{p_\alpha: \alpha \in \Lambda\}$ of seminorms defined by $p_\alpha(f) = \|f(i)\|_\alpha$. To verify this suppose U is open in the product topology. For each f in U there is a basic open set V such that $f \in V \subset U$. The set V has the form $\prod_{\alpha \in \Lambda} V_\alpha$ where $V_\alpha = E_\alpha$ for all but finitely many α and V_α is open in E_α for all α. Suppose $V_\alpha = E_\alpha$ for all α except for $\alpha_1, \alpha_2, \ldots, \alpha_n$. Since each V_{α_i} is open there is $\varepsilon_i > 0$ such that V_{α_i} contains all y with $\|y - f(\alpha_i)\|_{\alpha_i} < \varepsilon_i$. This implies that V contains all g in E with $p_{\alpha_i}(f-g) < \varepsilon_i$ for $i = 1, 2, \ldots, n$. Therefore, U is open in the sense of the definition of locally convex space. A corresponding argument shows that if U is open in the sense of this definition, then every point in U is contained in a basic open set that is contained in U.

Important examples of locally convex spaces are the weak topology on a normed space and the weak* topology on the dual space of a normed space. Suppose E is a normed space with dual space E^*. A subset U of E is called *weakly open* or *open in the weak topology* if for each x in U there are $f_1, f_2 \ldots, f_n$ in E^* and positive numbers $\varepsilon_1, \varepsilon_2, \ldots, \varepsilon_n$ such that y is in U whenever $|f_i(y-x)| < \varepsilon_i$ for

$i = 1, 2, \ldots, n$. A subset U of E^* is called *weakly open* or *open in the weak* topology* if for each f in U there are $x_1, x_2, \ldots, x_n$ in E and positive numbers $\varepsilon_1, \varepsilon_2, \ldots, \varepsilon_n$ such that g is in U whenever $|(g - f)(x_i)| < \varepsilon_i$ for $i = 1, 2, \ldots, n$. Thus the seminorms that determine the weak topology on E have the form $p_f(x) = \|f(x)\|$ as f ranges over E^*, and the seminorms that determine the weak* topology on E^* have the form $p_x(f) = \|f(x)\|$ as x ranges over E. If U is a subset of E, which is open with respect to the weak topology, then U is open in the normed topology. For if $u \in U$ there are $f_1, f_2, \ldots, f_n$ in E^* and $\varepsilon_1, \varepsilon_2, \ldots, \varepsilon_n > 0$ such that the set V defined by

$$V = \{x\colon |f_i(u - x)| < \varepsilon_i \quad \text{for } i = 1, 2, \ldots, n\}$$

is contained in U. Since each f_i is continuous, the set V is open in E. Therefore, U is open in E. A similar proof shows that every subset of E^* that is open in the weak* topology of E^* is open in E^*.

Theorem 23.2: Suppose E is a linear space having a locally convex topology determined by a collection $\mathscr{R}$ of seminorms. A seminorm p on E is continuous if and only if there is a finite subset $q_1, q_2, \ldots, q_n$ of $\mathscr{R}$ and $M > 0$ such that

$$p(x) < M \max\{q_i(x)\colon 1 \leqslant i \leqslant n\}$$

for all x in E.

Proof: Suppose p is continuous on E. Then the set U of all x in E such that $p(x) < 1$ is open in E. Since 0 is in U there are $q_1, \ldots, q_n$ in $\mathscr{R}$ and $\varepsilon_1, \ldots, \varepsilon_n > 0$ such that the set V of all x in E such that $q_i(x) < \varepsilon_i$ for all i is contained in U. Let $M = \max\{1/\varepsilon_i\colon 1 \leqslant i \leqslant n\}$. For each x in E, let $L = M \max\{q_i(x)\colon 1 \leqslant i \leqslant n\}$. Since $q_i(x/L) \leqslant \varepsilon_i$ it follows that $p(x/L) \leqslant 1$. Therefore, $p(x) \leqslant L$, which was to be shown.

Suppose the seminorm p on E satisfies the given condition. Then p is continuous at x for all x in E because for each $\varepsilon > 0$ we have $p(x - y) < \varepsilon$ when $q_i(x - y) < \varepsilon/M$ for all i. ■

The following result is known as *Alaoglu's Theorem.*

Theorem 23.3: If U is the closed unit ball of a dual space E^* of a normed space E then U is compact in the weak* topology.

Proof: We carry out the proof for a real normed space E, and leave the case of E, a complex space to you. The closed unit ball U of E^* is the set of functions f on E such that $-\|x\| \leqslant f(x) \leqslant \|x\|$ for all x in E. Thus, U is a subset of the product space $\prod = \prod_{x \in E} [-\|x\|, \|x\|]$. By Tychonoff's Theorem (Theorem 10.12), $\prod$ is compact in the product topology. Hence, it suffices to prove that (a) U is a closed subset of $\prod$ and (b) the weak* topology on U coincides with the relative product topology.

Statement (b) is valid since the definition of weak* open sets is essentially the same as sets open in the product topology. To show (a) suppose g is in the closure of U in $\prod$. We must show that g is linear and $g(x) \in [-\|x\|, \|x\|]$ for all x in E. Suppose x, y are in E and a, b are scalars. Given $\varepsilon > 0$ there is f in U with

$$|f(x) - g(x)| < \varepsilon, \ |f(y) - g(y)| < \varepsilon \quad \text{and} \quad |f(ax + by) - g(ax + by)| < \varepsilon$$

because the set of all f satisfying these three conditions is a (basic) nbd of g in the product topology. Since f is linear we shall have

$$\begin{aligned}
&|g(ax+by)-ag(x)-bg(y)| \\
&\qquad = |g(ax+by)-ag(x)-bg(y)-f(ax+by)+af(x)+af(y)| \\
&\qquad \leqslant |g(ax+by)-f(ax+by)|+|a||g(x)-f(x)|+|b||g(y)-f(y)| \\
&\qquad < (1+|a|+|b|)\varepsilon.
\end{aligned}$$

Since $\varepsilon>0$ is arbitrary it follows that g is linear. ■

The Hahn-Banach Theorem may be interpreted in terms of the separation of convex sets in a locally convex space. The following theorem is the most elementary of these. A point x of a subset S of a locally convex space E is called an *internal point* of S if for each y in E there is $\varepsilon>0$ such that $x+ty\in S$ for all t with $|t|<\varepsilon$.

Theorem 23.4: Suppose U and C are disjoint convex subsets of a real linear space E such that U contains an internal point. There is a nonzero linear functional f on E such that $f(x)\leqslant f(y)$ for all x in U and y in C.

Proof: First suppose 0 is the internal point of U. Then every point of $C-U=\{y-x\colon y\in C,\ x\in U\}$ is an internal point of $C-U$. Let x_0 be any selected point of $C-U$, and let $V=C-U-x_0$. It is not hard to see that V is a convex set and 0 is an internal point of V. Define a function q on E by

$$q(u)=\inf\left\{t\colon t>0,\quad \frac{u}{t}\in V\right\}.$$

For each u in E, $q(u)<\infty$ because 0 is an internal point of V. We shall prove that q is a sublinear functional on E. If $a>0$ we have

$$\begin{aligned}
q(au) &= \inf\left\{t\colon t>0,\quad \frac{au}{t}\in V\right\} \\
&= \inf\left\{as\colon s>0,\quad \frac{u}{s}\in V\right\} \quad \text{(Substitute } t=as.) \\
&= aq(u).
\end{aligned}$$

Suppose u and v are points in E. Given $\varepsilon>0$ there are t_1 and $t_2>0$ such that

$$q(u)\geqslant t_1-\frac{\varepsilon}{2},\quad \frac{u}{t_1}\in V \quad\text{and}\quad q(v)\geqslant t_2-\frac{\varepsilon}{2},\quad \frac{v}{t_2}\in V.$$

Since V is convex the point

$$(t_1+t_2)^{-1}(u+v)=(t_1+t_2)^{-1}\left(t_1\left(\frac{u}{t_1}\right)+t_2\left(\frac{u}{t_2}\right)\right)$$

is in V so that

$$q(u+v)\leqslant t_1+t_2\leqslant q(u)+q(v)+\varepsilon.$$

Since the inequality above holds for each $\varepsilon>0$ it follows that q is subadditive.

Since C and U are disjoint, 0 is not in $C-U$ so that $-x_0$ is not in V. This implies that $q(-x_0) \geqslant 1$. Define a linear functional $\hat{f}$ on the span $[x_0]$ of x_0 by $\hat{f}(ax_0) = -aq(-x_0)$. Then $\hat{f}(y) \leqslant q(y)$ for all y in $[x_0]$ because if $t \geqslant 0$, $\hat{f}(tx_0) \leqslant 0 \leqslant q(tx_0)$ while if $t<0$, $\hat{f}(tx_0) = -tq(-x_0) = q(tx_0)$. By the Hahn-Banach Theorem (Theorem 20.1) there is an extension f of $\hat{f}$ to all of E. If x is in U and y is in C, then $x-y-x_0$ is in V so $q(x-y-x_0) \leqslant 1$. Therefore, we have

$$f(x-y) \leqslant 1 + f(x_0) = 1 - q(-x_0) \leqslant 0.$$

If the internal point u_0 of U is different from 0, we can apply the argument above to find a functional f on E such that $f(u) \leqslant f(v)$ for each u in $U - u_0$ and v in $C - u_0$. This functional will satisfy the conclusion of the theorem. ■

The following result is obtained by applying Theorem 23.4 to a complex space E regarded as a space over the real numbers.

Corollary 23.5: Suppose U and C are convex subsets of a complex locally convex space E, and U has an internal point. Then there is a continuous linear functional f on E such that $Re(f(x)) \leqslant Re(f(y))$ for all x in U and y in C.

Theorem 23.6: Suppose C is a closed subset of a locally convex space E, and y is a point not in C. Then there is a continuous linear functional f on C such that $Re(f(y)) < \inf\{Re(f(x): x \in C\}$.

Proof: We assume first that $y=0$. There is a continuous seminorm p on E and $\varepsilon > 0$ such that the set $U = \{u: p(u) < \varepsilon\}$ is disjoint from C. By Theorem 23.4 or its corollary there is a nonzero linear functional f on E such that for each u in U and v in C we have $Re(f(u)) \leqslant Re(f(v))$. Since f is nonzero there is x_0 in E such that $Re(f(x_0)) > 0$. Since $tx_0 \in U$ for all $t < \varepsilon/p(x_0)$ it follows that $Re(f(x)) > 0$ for all x in C. To show f is continuous note that if $M = f(x)$ for some x in C we have $|f(u)| \leqslant |M/\varepsilon| p(u)$ for all u in E.

If $y \neq 0$, we apply the above argument to $C - y$. The resulting functional satisfies $Re(f(y)) < \inf\{Re(f(x)): x \in C\}$. ■

Corollary 23.7: A convex subset C of a normed space E is closed if and only if it is closed in the weak topology of E.

Proof: Since the weak topology on E is weaker than the normed topology it follows that every weakly closed subset of E is closed.

Suppose C is closed in the normed topology and convex. For each $y \in C$ there is a continuous linear functional f such that $Re(f(y)) < \inf\{Re(f(x)): x \in C\}$. If $c = \inf\{Re(f(x)): x \in C\}$ then the set $\{z: Re(f(z)) < c\}$ is a weakly open subset of E that contains y and has empty intersection with C. Therefore, the complement of C is weakly open. ■

A point x of a convex set C is called an *extreme point* if it is not an internal point of a line segment contained in C. That is, if $x = tu + (1-t)v$ with u, v in C and $0 \leqslant t \leqslant 1$ it necessarily follows that $u = v = x$. A subset S of C is called *extremal* if whenever x is in S and $x = tu + (1-t)v$ with u, v in C, and $0 \leqslant t \leqslant 1$ it necessarily follows that u and v are in S. Thus a point x is an extreme point if and only if the singleton $\{x\}$ is extremal. The vertices of a polyhedron are its only extreme points while the faces and edges are extremal sets. Every point of the surface of a sphere is an extreme point. See Figure 23.1. The *closed convex hull* of a set of points S is the smallest closed convex set C that contains S, that is, $C = \cap\{D: D \text{ is closed and convex and } D \supset S\}$. The following result is known as the *Krein-Mil'man Theorem.*

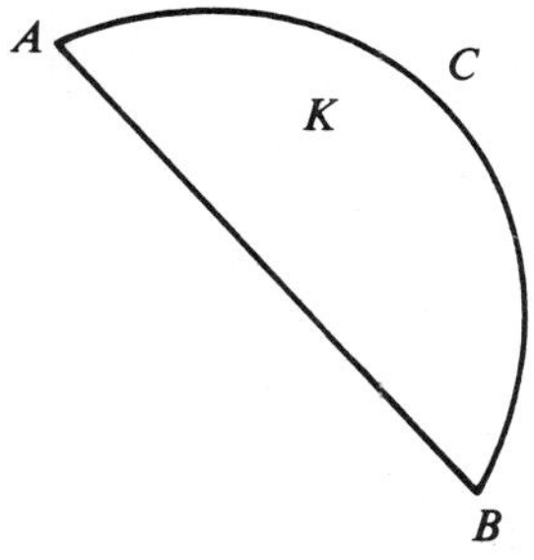

Each point on curve C is an extreme point of K as are A and B but no point on the line joining A to B is an extreme point.

Figure 23.1

Theorem 23.8: Every compact convex subset C of a locally convex space is the closed convex hull of its set of extreme points.

Proof: We shall regard the locally convex space as a real space. The main task is to show that the set S of extreme points of C is nonempty. Let $\mathscr{M}$ denote the collection of all compact, convex extremal subsets of C; then certainly C is in $\mathscr{M}$. If $\mathscr{N}$ is a chain in C then $\mathscr{N}$ has the fip so $D = \cap \mathscr{N}$ is nonempty, compact, and convex. We verify that D is an extremal set of C. If x is in D and $x = tu + (1-t)v$ with u, v in C then each B in $\mathscr{N}$ contains u and v since B is extremal. Therefore, u and v are in D.

Using Postulate 3.8 on page 15 we can show the collection $\mathscr{M}$ has a minimal set D. We verify that D must be a singleton. Suppose x_1 and x_2 are distinct points in D. By Theorem 23.6 there is a continuous linear functional f on E such that $f(x_1) < f(x_2)$. Let b denote the minimum of f on D, and let B consist of all x in D such that $f(x) = b$. The set B is nonempty since D is compact. The reader may verify that B is also convex and compact. The set B is extremal because if $x \in B$ and $x = tu + (1-t)v$ with u, v in C it follows that u, v are also in D because D is extremal. Since b is the minimum of f on D we conclude $f(u)$ and $f(v) \geqslant b$. But if either are greater than b, we must have $f(x) > b$, which is impossible. Therefore, both u and v are in B. However, this contradicts the minimality of D since B does not contain the point x_2 of D. Therefore, our assumption that D is not a singleton leads to a contradiction. Since D is an extremal singleton, the member of D is an extreme point.

Let C_1 denote the closed convex hull of the extreme points of C. If there is a point x in $C \backslash C_1$ then there is a continuous linear functional f on E such that $f(x) \leqslant \inf\{f(y): y \in C_1\}$ because of Theorem 23.6. Let b denote the minimum of f on C and let B denote the set of all u in C such that $f(u) = b$. Then B is a compact, convex, extremal subset of C so it must contain an extreme point x_0 of C. This contradicts the fact that all of the extreme points of C are contained in C_1 from which B is disjoint. ■

Exercises

A

1. Show that if C and D are convex subsets of a linear space E, so is $C + D$.
2. Show that the interior and closure of convex subsets of a locally convex space E are convex.
3. Prove that if T is a linear operator from a linear space E into a linear space F and C is a convex subset of E, then $T(C)$ is a convex subset of F.
4. *(a)* Prove that if p and q are seminorms defined on a linear space E, then so are the functions r and s defined as follows

$$r(x) = \max\{p(x), q(x)\}; \qquad s(x) = \inf\{p(u) + q(v): u + v = x\}.$$

   (b) Show r is the smallest seminorm greater than p and q, and that s is the largest seminorm smaller than p and q.
5. Prove that the collection of open sets given in the definition of locally convex space satisfies the definition for a topology.
6. Show that (e_n) converges to 0 in the weak topology of c_0, the weak* topology of l^1, but not in the weak topology of l^1 or in any of the Banach space topologies. Does (e_n) converge to 0 in the weak* topology on l^∞?
7. *(a)* Show that a linear functional on a normed space E is continuous if and only if it is weakly continuous.
   (b) Show that a linear functional F on a dual space E^* is weak* continuous if and only if there is x in E such that $F(f) = f(x)$ for all f in E^*.
   (c) Conclude from this that the weak and weak* topologies on E^* coincide if and only if E is reflexive.
8. Suppose $\mathscr{T}$ is a locally convex topology on a space E and $\mathscr{T}$ is determined by a collection $\mathscr{R}$ of seminorms. Show there is a maximal collection $\hat{\mathscr{R}}$ of seminorms that determines $\mathscr{T}$, and characterize $\hat{\mathscr{R}}$.
9. *(a)* Show that if p is a seminorm on a linear space E then the set U of all x in E with $p(x) \leqslant 1$ has the following two properties: (b1) 0 is an internal point of U; (b2) if u and v are in U and a, b are scalars with $|a| + |b| \leqslant 1$ then $au + bv$ are in U.
   (b) Show that if U is a subset of E having properties (b1) and (b2) then the function p defined by $p(x) = \inf\{t: t > 0, x/t \in U\}$ is a seminorm on E and

$$\{x: p(x) < 1\} \subset U \subset \{x: p(x) \leqslant 1\}.$$

B

10. *(a)* Prove that a sequence (x_n) in a normed space E converges to a point x in the weak topology if and only if $\lim_n f(x_n) = f(x)$ for all x in E^*.
    (b) Prove that if (x_n) is an ON sequence in a Hilbert space then (x_n) converges weakly to 0.

(c) Prove that if (x_n) is a sequence in a Hilbert space that converges weakly to x then (x_n) converges strongly to x if and only if $\lim_n \|x_n\| = \|x\|$.

11. *(a)* Prove that if in Example 23.1 the set Λ is countable, then $\prod_{\alpha \in \Lambda} E_\alpha$ is a complete metric space.

(b) Show that a sequence (f_n) in the product space converges to f if and only if $(f_n(\alpha))$ converges to $f(\alpha)$ for all α in Λ.

(c) Prove that if Λ is countably infinite then $\prod_{\alpha \in \Lambda} E_\alpha$ is a complete metric locally convex space that is not a Banach space.

12. Prove that every weak* closed and bounded subset of the dual space E^* of a normed space E is compact in the weak* topology.

13. *(a)* Prove that for each u in the closed unit ball U of c_0 there are x, and y in U with $x \neq y$ and $0 < t < 1$ such that $u = tx + (1 - t)y$.

(b) Show that there is no normed space E such that c_0 is isomorphic to E^*.

14. *(a)* Show that the closed unit ball U of the real linear space $C[0, 1]$ contains only two extreme points.

(b) Show that $C[0, 1]$ is not isomorphic to the dual space of a normed space.

5

Spaces of Functions and Sequences

Nature is not embarrassed by difficulties of analysis.
A. Fresnel
Mathematical Analysis is as extensive as Nature, Herself.
J. Fourier

Much of analysis can be interpreted as an application of the ideas described in preceding chapters to specific spaces of functions or sequences. In Chapter 5, we will give you a brief introduction to this vast area and we shall describe various spaces of continuous functions, integrable functions, differentiable functions, and sequences. We will also present a few of the most important results concerning these spaces.

Section 24 concerns the space $C(S)$ of continuous functions on a compact metric space S. The first result in this section is the important Stone-Weierstrass Theorem that describes conditions under which a subalgebra of $C(S)$ is dense. The density of a subspace is best understood in terms of approximation: namely that every member of the space can be approximated as closely as desired by members of the subspace. Next we prove the surprising theorem of Korovkin on the convergence of positive operators on $C[a, b]$. This theorem leads to a different proof of the classical Weierstrass Polynomial Approximation Theorem via the Bernstein Polynomial. Finally we present the Arzela-Ascoli Theorem, which characterizes compact subsets of $C(S)$. The space $BV(I)$ of functions of bounded variation on an interval is related to $C(I)$ through the Riemann-Stieltjes integral, a special case of the Lebesgue-Stieltjes integral. We discuss these matters in Section 25, where the main result is the Riesz Representation Theorem for the dual of $C(I)$.

Section 26 is an introduction to the space of measurable functions. This is an example of an important function space that cannot be regarded as a Banach space.

In Section 27 we consider the Lebesgue spaces $L^p(S, \mathscr{A}, \mu)$ where $(S, \mathscr{A}, \mu)$ is a measure space. After establishing the important inequalities of Hölder and Minkowski, we prove that the Lebesgue spaces are complete. This is a generalization of the classical theorem of Riesz-Fischer. The final result in this section is a representation theorem for the dual space of L^p.

We will study spaces of sequences and matrix operators between them in Section 28. Numerous sequence spaces are described, all of which belong to the class of BK-spaces. The most important result is that matrix operators between BK-spaces are always continuous. When the coordinate vectors form a basis for a BK-space E (we express this by saying E has AK), then we can represent the dual of E in a very nice way. In Section 29 we introduce several classes of spaces of differentiable functions such as $C^{(n)}(I)$ and Sobolev spaces and we prove a few results regarding integral operators between them.

24

Spaces of Continuous Functions

If S is a compact Hausdorff topological space, then the space $C(S)$ of continuous scalar valued functions on S is a Banach space with the uniform norm $\|f\| = \sup\{|f(t)|: t \in S\}$. If f and g are in $C(S)$ the product function fg is also in $C(S)$ and we have $\|fg\| \leqslant \|f\| \|g\|$. A Banach space with such a multiplication is called a *Banach algebra.* Moreover, $C(S)$ is closed under the finite lattice operations $\max\{f, g\}$ and $\min\{f, g\}$. See Theorem 7.5 on page 41.

If S is a compact interval of the real line $[a, b]$ then the space $P[a, b]$ of polynomials is a subalgebra of $C[a, b]$ since the sum and product of polynomials are polynomials. Moreover, for all s, t in $[a, b]$ with $s \neq t$ there is p in $P[a, b]$ (of degree one, in fact) such that $p(s) \neq p(t)$. Thus we say that $P[a, b]$ separates the points of $[a, b]$. In the nineteenth century, Karl Weierstrass (1815–1897) proved that for $C[a, b]$ over the reals, $P[a, b]$ is dense in $C[a, b]$. That is, for each continuous real valued function f on $[a, b]$ and each $\varepsilon > 0$ there is a polynomial g on $[a, b]$ such that $|f(s) - g(s)| < \varepsilon$ for all s in $[a, b]$. In the twentieth century, Marshall Stone (1901–1989), son of Supreme Court Justice Harlan Stone, proved the following far reaching generalization of the Weierstrass Polynomial Approximation Theorem, which is now called the *Stone-Weierstrass Theorem.*

Theorem 24.1: Suppose $\mathscr{A}$ is a subalgebra of the space $C(S)$ of continuous real valued functions on a compact Hausdorff space S that separates points of S and contains all constant functions. Then $\mathscr{A}$ is dense in $C(S)$.

To prepare for the proof of this theorem we shall first prove the following lemma, which is clearly a special case of the Weierstrass Theorem.

Lemma 24.2: For each $\varepsilon > 0$ there is a polynomial g such that $||s| - g(s)| < \varepsilon$ for each s in $[-1, 1]$.

Proof: Let $\sum_{n=0}^{\infty} c_n t^n$ be the Taylor series for $(1 - t)^{1/2}$. Then for each n we have

$$c_n = 1 \cdot 3 \cdot \cdots \cdot \frac{2n - 3}{2^n n!}$$

so that $\sum_{n=0}^{\infty} c_n$ converges by Raabe's test. Therefore, $\sum_{n=0}^{\infty} c_n t^n$ converges uniformly on $[0, 1]$. This means that for any $\varepsilon > 0$ there is a polynomial h, namely, a partial sum of the Taylor series, such that $|(1-t)^{1/2} - h(t)| < \varepsilon$ for all t in $[0, 1]$. If $g(t) = h(1-t^2)$, then g is a polynomial and for all t in $[-1, 1]$ we have

$$||t| - g(t)| = ||t| - h(1-t^2)| = |(1-u)^{1/2} - h(u)| < \varepsilon,$$

where $u = 1 - t^2$. ■

Proof of Theorem 24.1: Let $\bar{\mathscr{A}}$ denote the closure of $\mathscr{A}$ in $C(S)$. We leave the reader the task of showing that $\bar{\mathscr{A}}$ is also a subalgebra of $C(S)$.

If f is in $\bar{\mathscr{A}}$, we prove that $|f|$ is also in $\bar{\mathscr{A}}$. If $\|f\| \leqslant 1$, let $\varepsilon > 0$, and let q be a polynomial such that $||t| - q(t)| < \varepsilon$ for all t in $[-1, 1]$. If $q(t) = a_0 + a_1 t + \cdots + a_n t^n$ then $q(f) = a_0 + a_1 f + \cdots + a_n (f)^n$ is in $\bar{\mathscr{A}}$ because $\bar{\mathscr{A}}$ is an algebra. Since $-1 \leqslant f(s) \leqslant 1$ for all s in S, it follows that

$$||f(s)| - q(f)(s)| < \varepsilon$$

for all s in S. This shows that $|f|$ is in the closure of $\bar{\mathscr{A}}$ so that $|f|$ is in $\bar{\mathscr{A}}$. If $\|f\| > 1$, apply the previous argument to show that $|f|/\|f\|$ is in $\bar{\mathscr{A}}$ from which we conclude that $|f|$ is in $\bar{\mathscr{A}}$

Since we have

$$\inf(f, g) = \tfrac{1}{2}(f+g) - \tfrac{1}{2}|f-g| \quad \text{and} \quad \sup(f, g) = \tfrac{1}{2}(f+g) + \tfrac{1}{2}|f-g|$$

it follows that $\bar{\mathscr{A}}$ is closed under these finite lattice operations.

Let f be an arbitrary function in $C(S)$, and let $\varepsilon > 0$ be given. We shall construct an ε-approximation of f in three steps. First we show that for each pair of points s, t in S there is a function $f_{s,t}$ in $\bar{\mathscr{A}}$ such that $f_{s,t}(s) = f(s)$ and $f_{s,t}(t) = f(t)$. Let $g(u) = f(s)$ for all u in S. Then g is in $\bar{\mathscr{A}}$ because $\mathscr{A}$ contains constant functions. Let h in $\mathscr{A}$ be such that $h(s) \neq h(t)$. Then the function

$$f_{s,t}(u) = g(u) + (f(t) - f(s)) \frac{h(u) - h(s)}{h(t) - h(s)}$$

is in $\bar{\mathscr{A}}$ and has the desired properties.

Next we show that for each s in S there is a function f_s in $\bar{\mathscr{A}}$ such that $f_s(s) = f(s)$ and $f_s(u) < f(u) + \varepsilon$ for all u in S. For each t in S there is a nbd U_t of t such that $f_{s,t}(u) < f(u) + \varepsilon$ for all u in U_t. Since S is compact there is a finite subset F of S such that $\{U_t : t \in F\}$ covers S. If

$$f_s = \inf\{f_{s\,t} : t \in F\}$$

then f_s is in $\bar{\mathscr{A}}$ and has the desired properties.

Finally, for every $s \in S$, there is a nbd V_s such that for each u in V_s we have

$$f_s(u) > f(u) - \varepsilon.$$

Since the collection $\{V_s : s \in S\}$ is an open covering of S there is a finite subset G of S such that $\{V_s : s \in G\}$ also covers S. Let

$$f_\varepsilon = \sup\{f_s : s \in G\}.$$

Then f_ε is in $\mathscr{A}$ and $-\varepsilon < f(u) - f_\varepsilon(u) < \varepsilon$ for all u in S. Thus f_ε is the desired approximation to f. ■

Since the collection of polynomials on any interval satisfies the hypothesis of Theorem 24.1 we obtain the Weierstrass Theorem as an immediate consequence. However, the following example shows the possibility of even wider applications. See the exercises for others.

Example 24.3:

Denote by $EP[0, a]$ the space of even polynomials on the interval $[0, a]$; that is, polynomials having the form $f(x) = g(x^2)$ where g is a polynomial. In other words, f has no terms of odd degree. Since $u^2 \neq v^2$ if $u \neq v$, the space $EP[0, a]$ separates points. That $EP[0, a]$ contains constants is obvious. Since $EP[0, a]$ is an algebra it must be dense in $C[0, a]$. The space $EP[-a, a]$ is not dense in $C[-a, a]$, however, because it does not separate points; for example, $f(-u) = f(u)$ for each u in $[-a, a]$.

The Stone-Weierstrass Theorem is not valid for complex spaces $C(S)$. For example, if S is the unit disk $\{u + vi: u^2 + v^2 \leqslant 1\}$, any function that is in the uniform closure of the space of polynomials must be analytic on the interior of S. But there are many continuous functions on S that are not analytic on the interior of S.

Another approach to the Weierstrass Theorem is the *Theorem of Korovkin* on the convergence of positive operators. Although less sweeping than the Stone-Weierstrass Theorem, Korovkin's Theorem is equally spectacular. We restrict our attention to the real Banach space $C[0, 1]$, but our discussion can be extended without modification to spaces of the form $C[a, b]$. A linear operator T from $C[0, 1]$ into itself is called *positive* if whenever $f(t) \geqslant 0$ for each t in $[0, 1]$, we then have $(Tf)(t) \geqslant 0$ for all t in $[0, 1]$. We define three functions f_0, f_1, and f_2 on $[0, 1]$ by $f_0(t) = 1$ for all t in $[0, 1]$; $f_1(t) = t$ for all t in $[0, 1]$; and $f_2(t) = t^2$ for all t in $[0, 1]$. Thus f_0 is the constant 1, f_1 is the identity, and f_2 the square function.

Theorem 24.4: Let (T_n) be a sequence of positive linear operators from the real space $C[0, 1]$ into itself such that $\lim_n T_n f_j = f_j$ for $j = 0, 1, 2$. Then we have $\lim_n T_n f = f$ for all f in $C[0, 1]$.

Proof: For each t in $[0, 1]$ let K_t be the function in $C[0, 1]$ defined by the formula

$$K_t(s) = (s - t)^2 = f_2(s) - 2tf_1(s) + t^2 f_0(s).$$

Given $\varepsilon > 0$ let n_0 be such that

$$\| T_n f_2 - f_2 \| + 2 \| T_n f_1 - f_1 \| + \| T_n f_0 - f_0 \| < \varepsilon$$

when $n > n_0$. Since $K_t(t) = 0$, we have for $n > n_0$ and all t in $[0, 1]$

$$|(T_n K_t)(t)| = |(T_n K_t - K_t)(t)| \leqslant \| T_n K_t - K_t \| < \varepsilon.$$

Let f in $C[0, 1]$ and $\varepsilon > 0$ be given. Since f is uniformly continuous there is $\delta > 0$ such that $|f(s) - f(t)| < \varepsilon$ when $|s - t| < \delta$. Let $M = \|f\|$. If $|s - t| < \delta$ then $-\varepsilon \leqslant f(s) - f(t) \leqslant \varepsilon$ whereas if $|s - t| > \delta$ we have

$$-2M \frac{(s - t)^2}{\delta^2} \leqslant -2M \leqslant f(s) - f(t) \leqslant 2M \leqslant 2M \frac{(s - t)^2}{\delta^2}.$$

Combining these two observations we conclude that

$$-\varepsilon - 2M\frac{(s-t)^2}{\delta^2} \leqslant f(s) - f(t) \leqslant \varepsilon + 2M\frac{(s-t)^2}{\delta^2}$$

for all s, t in $[0, 1]$. For a fixed t in $[0, 1]$ we can rewrite this inequality

$$-\varepsilon f_0 - \frac{2M}{\delta^2}K_t \leqslant f - f(t)f_0 \leqslant \varepsilon f_0 + \frac{2M}{\delta^2}K_t.$$

Since each T_n is positive we shall have

$$-\varepsilon T_n f_0 - \frac{2M}{\delta^2}T_n K_t \leqslant Tf - f(t)T_n f_0 \leqslant \varepsilon T_n f_0 + \frac{2M}{\delta^2}T_n K_t.$$

If n_1 is such that $\|T_n f_0 - f_0\| < \varepsilon/M$ and $|T_n K_t(t)| < \delta^2\varepsilon/2M$ for all t in $[0, 1]$ when $n > n_1$ we shall then also have

$$\begin{aligned}-\varepsilon\{f_0 + \|T_n f_0 - f_0\|\} - \varepsilon &\leqslant T_r f - f(t)f_0 + f(t)f_0 - f(t)T_n f_0\\ &\leqslant T_r f - f(t)f_0 + M\|f_0 - T_n f_0\|\\ &\leqslant T_r f - f(t) + \varepsilon.\end{aligned}$$

From this we may conclude that

$$-3\varepsilon - \frac{\varepsilon^2}{M} \leqslant T_n f - f(t)$$

for all t in $[0, 1]$. Similarly we find that

$$T_n f - f(t) \leqslant 3\varepsilon + \frac{\varepsilon^2}{M}$$

for all t in $[0, 1]$. Since $\varepsilon > 0$ is arbitrary it follows that $T_n f$ converges uniformly to f. ■

If f is a real function on $[0, 1]$ the nth *Bernstein Polynomial Approximation* (BPA) of f is the polynomial $B_n(f)$ defined by the formula

$$\{B_n(f)\}(t) = \sum_{k=0}^{n} \binom{n}{k} f\left(\frac{k}{n}\right) t^k(1-t)^{n-k}.$$

Theorem 24.5: For each f in $C[0, 1]$ the sequence $(B_n(f))$ converges uniformly to f.

Proof: We leave to the reader the task of showing that B_n is a positive linear operator from $C[0, 1]$ into $C[0, 1]$. In order to establish the theorem it suffices to prove that $B_n(f_i)$ converges uniformly to f_i for $i = 0$, 1, and 2. For all $n = 1, 2, \ldots$ we have $B_n(f_0) = f_0$ and $B_n(f_1) = 1$ because

$$B_n(f_0) = \sum_{k=0}^{n} \binom{n}{k} t^k(1-t)^{n-k} = (t + (1-t))^n = 1$$

and

$$B_n(f_1) = \sum_{k=1}^{n} \binom{n}{k}\left(\frac{k}{n}\right) t^k(1-t)^k = \sum_{k=1}^{n} \binom{n-1}{k-1} t^k(1-t)^{n-k}$$
$$= t \sum_{k=0}^{n-1} \binom{n-1}{k} t^{k+1}(1-t)^{n-1-k} = t.$$

The situation for $B_n(f_2)$ is more complicated. We have

$$\begin{aligned}
B_n(f_2) &= \sum_{k=0}^{n} \binom{n}{k}\left(\frac{k}{n}\right)^2 t^k(1-t)^{n-k} \\
&= \sum_{k=1}^{n} \binom{n-1}{k-1}\left(\frac{k}{n}\right) t^k(1-t)^{n-k} \\
&= \sum_{k=0}^{n-1} \binom{n-1}{k}\left(\frac{k+1}{n}\right) t^{k+1}(1-t)^{n-1-k} \\
&= \sum_{k=0}^{n-1} \binom{n-1}{k}\left(\frac{k}{n-1}\right) t^{k+1}(1-t)^{n-1-k} \\
&\quad + \sum_{k=0}^{n-1} \binom{n-1}{k}\left(\frac{n-k-1}{n(n-1)}\right) t^{k+1}(1-t)^{n-1-k} \\
&= t^2 + c_n(t)
\end{aligned}$$

where

$$c_n(t) = \sum_{k=0}^{n-1} \binom{n-1}{k}\left(\frac{n-k-1}{n(n-1)}\right) t^{k+1}(1-t)^{n-1-k}.$$

Since

$$0 \leqslant c_n(t) \leqslant \frac{t}{n} \sum_{k=0}^{n-1} \binom{n-1}{k} t^k(1-t)^{n-1-k} = \frac{t}{n}$$

for $n > 1$ it follows that $B_n(f_2)$ converges uniformly to f_2. Therefore, we can apply Korovkin's Theorem to conclude that $B_n(f)$ converges uniformly to f for all f in $C[0, 1]$. ■

The Weierstrass Polynomial Approximation Theorem is, of course, an immediate consequence of Theorem 24.5.

Our final goal in this section is to characterize the compact subsets of the space $C(S)$ for S, a compact metric space. A set F of scalar functions on a metric space (S, ρ) is called *equicontinuous* if for each $\varepsilon > 0$ there is $\delta > 0$ such that $|f(x) - f(y)| < \varepsilon$ for each f in F whenever $\rho(x, y) < \delta$. Thus every function in F is uniformly continuous and F behaves in a regular way. The following "formula" is a useful aid to remembering the meaning of equicontinuity. Of course, you have to interpret the symbolism in the proper way:

$$|F(x) - F(y)| < \varepsilon \quad \text{whenever } \rho(x, y) < \delta.$$

The following result is known as the *Arzela-Ascoli Theorem.*

Theorem 24.6: Suppose S is a compact metric space. A bounded and closed subset F of $C(S)$ is compact if and only if F is equicontinuous.

Proof: Necessity of equicontinuity. Let $\varepsilon > 0$ be given. By Theorem 8.9 on page 52, each f in F is uniformly continuous so there is δ_f such that $|f(x)-f(y)| < \varepsilon$ when $\rho(x, y) < \delta_f$. For each $n = 1, 2, \ldots,$ let $U_n = \{g: |g(x)-g(y)| < \varepsilon$ when $\rho(x, y) \leqslant 1/n\}$. Each U_n is open in $C(S)$. For suppose $g_0 \in U_n$. Let D denote the set of all pairs (x, y) in $S \times S$ such that $\rho(x, y) \leqslant 1/n$. Since the function $h(x, y) = \varepsilon - |g_0(x) - g_0(y)|$ is continuous on $S \times S$, it attains its minimum δ on the compact subset D; see Theorem 5.8 on page 27 and Theorem 7.11 on page 44. This implies that $\delta > 0$. If $\|g_0 - f\| < \delta/2$, then for each x, y in S we have

$$\begin{aligned}|f(x)-f(y)| &\leqslant |g_0(x)-g_0(y)| + |g_0(x)-f(x)| + |g_0(y)-f(y)| \\ &< |g_0(x)-g_0(y)| + \delta/2 + \delta/2 \leqslant \varepsilon.\end{aligned}$$

Since for each f in F there is n such that $1/n < \delta_f$ it follows that the collection $\{U_n\}$ is an open covering of F. Since F is compact and $U_1 \subset U_2 \subset \ldots$ it follows that there is n such that $F \subset U_n$. This implies that for all f in F we have $|f(x)-f(y)| < \varepsilon$ when $\rho(x, y) < 1/n$.

Sufficiency of equicontinuity. We shall prove that every sequence in F has a convergent subsequence. The compactness of F then follows from Theorem 8.8 (page 50). Suppose $\{f_n\}$ is any sequence in F. Since S is compact and metric it is separable by the proof of (b) $\Rightarrow$ (a) of Theorem 8.8. Suppose $D = \{s_1, s_2, \ldots\}$ is a countable dense subset of S. Since F is bounded so is the sequence of scalars $\{f_n(s_1)\}$. Therefore, we can find a convergent subsequence $\{f_{n(1)}(s_1)\}$. The sequence of scalars $\{f_{n(1)}(s_2)\}$ is also bounded so we can find a convergent subsequence $\{f_{n(2)}(s_2)\}$. Proceeding inductively we can for each j find a subsequence $\{f_{n(j)}\}$ of $\{f_{n(j-1)}\}$ such that $\{f_{n(j)}(s_i)\}$ converges for all $i \leqslant j$. The sequence $\{f_{n(n)}\}$ becomes a subsequence of $\{f_{n(j)}\}$ when $n \geqslant j$. Thus $\{f_{n(n)}(s_i)\}$ converges for all i. This technique is called the *diagonal procedure* or *Cantor's diagonal procedure.* For convenience let $g_n = f_{n(n)}$. Define a scalar function g on D by the formula $g(s_i) = \lim_n g_n(s_i)$. To verify that the function g is uniformly continuous on D let $\varepsilon > 0$ be given. Let $\delta > 0$ be such that $|f(x)-f(y)| < \varepsilon/2$ for all f in F when $\rho(x, y) < \delta$. If t and u are in D and $\rho(t, u) < \delta$, then $|g_n(t) - g_n(u)| < \varepsilon/2$ for each n so that $|g(t) - g(u)| = \lim_n |g_n(t) - g_n(u)| \leqslant \varepsilon/2 < \varepsilon$. Since g is uniformly continuous on D it has a unique continuous extension to all of S, which we also denote by g. See Exercise 19 (a) on page 58 of Section 8.

It remains only to prove that $\{g_n\}$ converges uniformly to g. To do this let $\varepsilon > 0$ be given. Let δ be such that $|f(x)-f(y)| < \varepsilon/3$ for all f in F when $\rho(x, y) < \delta$. We have seen that this implies $|g(x) - g(y)| \leqslant \varepsilon/3$. Let B be a finite subset of D such that for each x in S there is t in B with $\rho(x, t) < \delta$. Since B is finite there is M such that for all t in B we have $|g_n(t) - g(t)| < \varepsilon/3$ when $n > M$. For any x in S, let t_x in B be such that $\rho(x, t_x) < \delta$. If $n > M$ we shall have

$$\begin{aligned}|g_n(x) - g(x)| &\leqslant |g_n(x) - g_n(t_x)| + |g_n(t_x) - g(t_x)| + |g(t_x) - g(x)| \\ &< \frac{\varepsilon}{3} + \frac{\varepsilon}{3} + \frac{\varepsilon}{3} = \varepsilon. \quad \blacksquare\end{aligned}$$

Example 24.7:

The unit ball of $C[0, 1]$, $B = \{f: f \in C[0, 1], \|f\| \leqslant 1\}$ is not equicontinuous. For if $f_n(t) = t^n$ for $n = 1, 2, \ldots$ then there is no $\delta > 0$ such that $|f_n(t) - 1| < \varepsilon$ for each n when $|1 - t| < \delta$. Another way to see that B is not equicontinuous is to observe that no subsequence of $\{f_n\}$ converges.

On the other hand, if we take the set C of all f in B that satisfies a Lipschitz condition $|f(s) - f(t)| \leqslant M|s - t|$ for some $M > 0$ then C is clearly equicontinuous and thus compact since it is closed and bounded.

Exercises

A

1. Show that if f and g are in $C(S)$ then $fg \in C(S)$ and $\|fg\| \leqslant \|f\| \|g\|$.
2. Prove that the sequence spaces c_0 and l^∞ are Banach algebras with the uniform norm and coordinatewise multiplication of sequences.
3. Prove that the closure of a subalgebra $\mathscr{A}$ of $C(S)$ is also a subalgebra.
4. A linear sublattice Λ of $C(S)$ is a subspace of $C(S)$, which is closed under the maximum and minimum of pairs of functions. Prove that if Λ is a sublattice of $C(S)$, which separates points and contains the constant functions then Λ is dense in $C(S)$. [*Hint*: Examine the proof of Theorem 24.1, and verify that it was the lattice properties of $\mathscr{A}$ that we used to show $\mathscr{A} = C(S)$.]
5. Suppose T is a positive linear operator from $C[0, 1]$ into itself such that $Tf_j = f_j$ for $j = 0, 1, 2$. Show that T must be the identity operator. Find an example of a positive operator T such that $Tf_0 = f_0$, $Tf_1 = f_1$, but $Tf_2 \neq f_2$.
6. Show that every positive operator T on $C[0, 1]$ is continuous. [*Hint*: If $\|f\| \leqslant 1$ then $-f_0 \leqslant f \leqslant f_0$.]
7. What is the probabilistic interpretation of the nth BPA of a function f?
8. Prove that if f is in $C[0, 1]$ and $\int_0^1 f(t)t_n \, dt = 0$ for all n, then f is identically 0.
9. For $M > 0$ let U_M denote the set of all f in $C[0, 1]$ such that $|f(x) - f(y)| \leqslant M|x - y|$ for all x, y in $[0, 1]$.
   (a) Show that U_M is closed and equicontinuous but not bounded.
   (b) Let V_M consist of all f in U_M such that $\|f\| \leqslant M$. Show that V_M is compact in $C[0, 1]$.
10. Suppose $\{f_n\}$ is a bounded sequence of scalar functions on a set S, and T is a countable subset of S. Show there is a subsequence $\{f_m\}$ of $\{f_n\}$ such that $\{f_m(t)\}$ converges for each t in T. [*Hint*: Use the Cantor diagonal argument.]

B

11. Suppose S and T are compact Hausdorff spaces. Show that for each h in the real space $C(S \times T)$ and each $\varepsilon > 0$ there are functions $f_1, f_2, \ldots, f_n$ in $C(S)$ and $g_1, g_2, \ldots, g_n$ in $C(T)$ such that for all s in S and t in T we have $|h(s, t) - \sum_{j=1}^n f(s)g(t)| < \varepsilon$.

12. Suppose F is a collection of real valued functions on a compact Hausdorff space S that separates the points of S. Prove that each continuous real valued function on S can be uniformly approximated by a function of the form $q(f_1(s), f_2(s), \ldots, f_n(s))$ where q is a polynomial in a finite number of variables and $f_1, \ldots, f_n$ are functions in F.
13. Let T denote the unit circle, $\{(x, y): x^2 + y^2 = 1\}$. Show that the space $C(T)$ can be identified with the closed subspace of $C[0, 2\pi]$ consisting of all f such that $f(0) = f(2\pi)$.
    (a) Show that the space TP of trigonometric polynomials $p(t) = a_0/2 + \sum_{n=1}^{m} (a_n \cos nt + b_n \sin nt)$, is a subalgebra of $C[T]$.
    (b) Prove that TP is dense in $C[T]$.
    (c) Let $P(\mathbb{R})$ denote the space of all continuous functions f on the real line $\mathbb{R}$ with period 2π, that is, $f(t + 2\pi) = f(t)$ for all t in $\mathbb{R}$. Show that for each f in $P(\mathbb{R})$ and each $\varepsilon > 0$ there is a trigonometric polynomial p such that $|f(t) - p(t)| < \varepsilon$.
14. Suppose $\mathscr{A}$ is a subalgebra of the complex space $C(S)$ where S is compact and Hausdorff. If $\bar{f}$, the conjugate of f, is in $\mathscr{A}$ for all f in $\mathscr{A}$ show the closure of $\mathscr{A}$ is all of $C(S)$.
15. For each $n = 1, 2, \ldots$ let $p_n(t) = c_n(1 - t^2)^n$ where c_n is chosen to make the integral $\int_{-1}^{1} p_n(t)\, dt = 1$. Let f be a continuous function on $\mathbb{R}$ such that $f(t) = 0$ for $t \leqslant 0$ and $t \geqslant 1$. Show that the sequence of polynomials

$$q_n(t) = \int_{-1}^{1} p_n(t + s) f(s)\, ds$$

    converges uniformly to f on $[0, 1]$.
16. Suppose $\mathscr{A}$ is a subalgebra of the complex space $C[a, b]$ that has the property that $\bar{f}$, the conjugate function of f, is in $\mathscr{A}$ whenever f is in $\mathscr{A}$. Prove that $\mathscr{A}$ is dense in $C[a, b]$.

C

17. Let (T_n) be a sequence of positive linear operators from the space $C[0, 2\pi]$ into itself.
    (a) Show that if $\lim_n T_n f = f$ for $f = f_0$, $f(t) = \sin t$, and $f(t) = \cos t$, then $\lim_n T_n f = f$ for all f in $C[0, 2\pi]$.
    (b) Show that condition (a) is not satisfied for $F_n f =$ the nth partial sum of the Fourier series for f.
    (c) Show that condition (a) is satisfied for $\phi_n(f) = 1/n \sum_{j=1}^{n} F_n f$ where $F_n f$ is the nth partial sum of the Fourier series for f.

25

The Dual of C[a, b]

The purpose of this section is to represent the dual of the space of continuous functions on an interval as a space of functions on that interval.

In Section 14 we defined the variation $V(f, a, b)$ of a real valued function f on an interval $[a, b]$ to be the supremum of all numbers

$$\sum_{i=1}^{n} |f(u_i)-f(u_{i-1})|$$

as $\{u_0, \ldots, u_n\}$ ranges over all finite sets of reals with $a \leqslant u_0 \leqslant u_1 \leqslant \ldots \leqslant u_n \leqslant b$, and defined f to be of bounded variation on $[a, b]$ if $V(f, a, b) < \infty$. The set of all functions of bounded variation on an interval $I=[a, b]$ is denoted by $BV([a, b])$ or $BV(I)$. It is easy to verify that $BV(I)$ forms a linear space.

Theorem 25.1: For each interval, $I=[a, b]$, the space $BV(I)$ is a Banach space with the norm $\|f\| = |f(a)| + V(f; a, b)$. For each $t \in I$ the evaluation mapping $F_t(f)=f(t)$ is a continuous linear functional on $BV(I)$ with norm 1.

Proof: Suppose $\{f_n\}$ is a sequence in $BV(I)$ with $\sum_n \|f_n\| < \infty$. For each t in I and each f in $BV(I)$ we have

$$|f(t)| \leqslant |f(a)| + |f(t)-f(a)| \leqslant V(f, a, t) + |f(a)| \leqslant \|f\|.$$

Therefore, $\sum_n f_n(t)$ converges absolutely for each t in I to a value $f(t)$. The function f is in $BV(I)$ because for all partitions $\prod = \{t_0, t_1, \ldots, t_m\}$ of I we have

$$V(f, \textstyle\prod) \leqslant \displaystyle\sum_{i=1}^{m} \sum_{n=1}^{\infty} |f_n(t_i)-f_n(t_{i-1})| \leqslant \sum_{n=1}^{\infty} V(f_n; a, b).$$

The series $\sum_{n=1}^{\infty} f_n$ converges to f because for each $k=1, 2, \ldots$ and all $\prod$ we have as above

$$\left\| f - \sum_{n=1}^{k} f_n \right\| \leqslant \sum_{n=k+1}^{\infty} \|f_n\|.$$

We have already noted that $|f(t)| \leqslant \|f\|$ for all f in $BV(I)$ so that $\|F_t\| \leqslant 1$ for all t. That $\|F_t\| \geqslant 1$ follows since the function f_t defined by

$$f_t(s)=1 \quad \text{for } a \leqslant s \leqslant t, \qquad f_t(s)=0 \quad \text{for } t < s < 1.$$

has norm 1 and $F_t(f_t)=1$. ■

We leave the proof of the following theorem to you. It is the fundamental theorem on the existence of the Riemann-Stieltjes integral. If g is Riemann-Stieltjes integrable with respect to f on an interval I then g is also Lebesgue-Stieltjes integrable with respect to f. See Section 14.

Theorem 25.2: Suppose f is in $BV(I)$ and g is in $C(I)$ where $I=[a, b]$. There is a number A such that for each $\varepsilon > 0$ there is a finite partition $\prod_\varepsilon$ of I such that whenever $\prod = \{s_0, s_1, \ldots, s_n\}$ is a partition of I, which contains $\prod_\varepsilon$ we have

$$\left| A - \sum_{j=1}^{n} g(t_j)\{f(s_j)-f(s_{j-1})\} \right| < \varepsilon$$

where t_j is any number in the interval (s_{j-1}, s_j).

The value A is called the *Riemann-Stieltjes integral* of g with respect to f and we write

$$A = \int_a^b g(t)df(t).$$

Some properties of the Riemann-Stieltjes integral are summarized in the following theorem. We denote by $\|g\|$ the norm of g in $C(I)$. ■

Theorem 25.3: Suppose f_1 and f_2 are in $C(I)$, g_1 and g_2 are in $BV(I)$ where $I = [a, b]$, and c_1, c_2 are in R. The following formulas are valid for the Riemann-Stieltjes integral.

(a) $\int_a^b (c_1 f_1 + c_2 f_2)\, dg_1 = c_1 \int_a^b f_1 dg_1 + c_2 \int_a^b f_2 dg_1$.

(b) $\int_a^b f_1 d(c_1 g_1 + c_2 g_2) = c_1 \int_a^b f_1 dg_1 + c_2 \int_a^b f_2 dg_2$.

(c) $|\int_a^b f_1 dg_1| \leqslant \|f_1\| V(g_1; a, b)$.

(In (c) $\|f_1\|$ denotes the norm of f_1 in $C(I)$.)

Proof: We omit the proof of (a) and (b).

(c) If $\prod = \{s_0, s_1, \ldots, s_n\}$ is any partition of I, and t_i is in (s_{i-1}, s_i) for $i = 1, 2, \ldots, n$, then we have

$$\left| \sum_{j=1}^{n} f_1(t_j)\{g_1(s_j) - g_1(s_{j-1})\} \right| \leqslant \sum_{j=1}^{n} |f_1(t_j)|\, |g_1(s_j) - g_1(s_{j-1})|$$

$$\leqslant \|f_1\| V(g_1, \textstyle\prod) \leqslant \|g_1\| V(g_1; a, b).$$

By passing to the limit in this inequality we obtain the desired relation. ■

One implication of Theorem 25.3 is that each member f of $BV(I)$ determines a continuous linear functional F_f on $C(I)$ by the formula

$$F_f(g) = \int_a^b g df.$$

Different members f_1 and f_2 of $BV(I)$ can determine the same functional on $C(I)$ if $f_1(t+) - f_2(t+)$ is a constant where

$$f(t+) = \lim_{s \downarrow t} f(s).$$

Let $NBV(I)$ consist of all f in $BV(I)$ that are 0 at a and continuous from the right, that is, such that $f(t+) = f(t)$ for all t in $[a, b)$. For each f in $BV(I)$ define $\bar{f}$ by $\bar{f}(t) = f(t+) - f(a+)$. Then $\bar{f}$ is in NBV and the following lemma holds.

Lemma 25.4: For each f in $BV(I)$, $\bar{f}$ is the unique member of NBV such that $\int_a^b g\, d\bar{f} = \int_a^b g\, df$ for all g in $C(I)$.

The next result is known as the *Riesz Representation Theorem for the Dual of* $C(I)$. It asserts that every continuous linear functional F on $C(I)$ can be represented uniquely by a member of $NBV(I)$ and that the correspondence is isometric.

Theorem 25.5: For each F in the dual space $C(I)^*$ of the Banach space $C(I)$ there is a unique function g_F in $NBV(I)$ such that

$$F(f) = \int_a^b f\, dg_F \quad \text{for all } f \in C(I)$$

and $\|F\| = \|g_F\|$ where the norm of F is the norm in the conjugate space $C(I)^*$ and the norm of g_F is the norm in $NBV(I)$, $V(g_F; a, b)$.

Proof: We shall assume that $I = [0, 1]$.

The space $C(I)$ is a closed subspace of the space $B(I)$ of bounded functions on I with the uniform norm. By the Hahn-Banach Theorem, 18.2, there is an extension $\hat{F}$ of F to $B(I)$ such that $\|\hat{F}\| = \|F\|$. For each x in $[0, 1]$ let u_x denote the function such that $u_x(t) = 0$ for $x \leqslant t \leqslant 1$ and $u_x(t) = 1$ otherwise. Define g on I by

$$g(x) = \hat{F}(u_x) \quad \text{for } x \in I.$$

If $\prod = \{t_0, t_1, \ldots, t_n\}$ is a partition of I then

$$\begin{aligned} V(g, \textstyle\prod) &= \sum_{j=1}^{n} |g(t_j) - g(t_{j-1})| \\ &= \sum_{j=1}^{n} (g(t_j) - g(t_{j-1}))sgn(g(t_j) - g(t_{j-1})) \\ &= \hat{F}\left\{\sum_{j=1}^{n} sgn(g(t_j) - g(t_{j-1}))(u_{t_j} - u_{t_{j-1}})\right\} \\ &\leqslant \|\hat{F}\| \left\|\sum_{j=1}^{n} sgn(g(t_j) - g(t_{j-1}))(u_{t_j} - u_{t_{j-1}})\right\| \\ &= \|F\|. \end{aligned}$$

It follows that g is in $BV(I)$. Since $g(0) = F(v_0) = F(0) = 0$, we conclude that $\|g\| \leqslant \|F\|$. If we define $g_F(t) = g(t+)$ for all t in I it follows that g_F has the same norm as g, and determines the same continuous linear functional on $C(I)$.

It remains to show that the functional determined by g is F. Given f in $C(I)$ define a sequence $\{f_n\}$ of functions in $B(I)$ by

$$f_n = \sum_{j=1}^{n} f\left(\frac{j}{n}\right)\{u_{j/n} - u_{(j-1)/n}\}.$$

Each f_n is a step function with value $f(j/n)$ on the interval $((j-1)/n, j/n]$. Since f is uniformly continuous on I it follows that the sequence $\{f_n\}$ converges to f in the uniform topology. For each n we have

$$\begin{aligned} \hat{F}(f_n) &= \sum_{j=1}^{n} f\left(\frac{j}{n}\right)\{\hat{F}(u_{j/n} - u_{(j-1)/n})\} \\ &= \sum_{j=1}^{n} f\left(\frac{j}{n}\right)\left\{g\left(\frac{j}{n}\right) - g\left(\frac{(j-1)}{n}\right)\right\}. \end{aligned}$$

The terms on the left converge to $F(f)$ since $\{f_n\}$ converges to f in $B(I)$ and F is continuous. The terms on the right are Riemann-Stieltjes sums that converge to

$\int_0^1 f\, dg = \int_0^1 f\, dg_F$. Therefore, we have

$$F(f) = \int_C^1 f\, dg_F$$

and by (c) of Theorem 25.3 it follows that $\|F\| \leqslant \|g_F\|$. ■

Exercises

A

1. What details in the proof of Theorem 25.6 must be modified to apply to an arbitrary interval $I = [a, b]$?
2. Determine the Riemann-Stieltjes integral $\int_a^b f\, dg$ for the following a, b, f, g:
   (a) $f = x^2$, $g = x$, $a = 0$, $b = 1$;
   (b) $f = x$, $g = [x]$ (greatest integer function), $a = 0$, $b = 10$;
   (c) $f = \cos x$, $g = x[x]$; $a = 0$, $b = 10\pi$.
3. (a) Prove that if g is in $C[a, b]$ then the function defined on $C[a, b]$ by $F_g(f) = \int_a^b f(t)g(t)\, dt$ is a continuous linear functional on $C[a, b]$.
   (b) Find h in $NBV[a, b]$ such that $F_g(f) = \int_a^b f\, dh$ for f in $C[a, b]$.

B

4. Suppose f is an increasing function on an interval $[a, b]$.
   (a) Show that for each $\varepsilon > 0$ there are at most finitely many points t at which $f(t+) - f(t-) > \varepsilon$.
   (b) Show that there are at most countably many points t at which f is discontinous.
   (c) Prove that every function f on $[a, b]$ of bounded variation has at most countably many points of discontinuity.
5. Suppose $g \in BV[a, b]$.
   (a) Show that if $\int_a^b f\, dg = 0$ for all f in $C[a, b]$ then g is constant.
   (b) Show that if $\int_a^b t^n\, dg = 0$ for all $n = 0, 1, \ldots$ then g is constant.
6. Construct an increasing function that is discontinuous at each rational number and continuous at each irrational number.
7. Prove Theorem 25.2.
8. Prove that the integral of Theorem 25.2 also holds if f is in $C(I)$ and g is in $BV(I)$ and then we have

$$f(b)g(b) - f(a)g(a) = \int_a^b f\, dg + \int_a^b g\, df$$

   where $\int_a^b g\, df$ is the number A that satisfies the conclusion of Theorem 25.2.
9. Prove Lemma 25.4.
10. (a) Do problem 3, assuming that g is integrable on $[a, b]$ instead of continuous.
    (b) Find an example of a continuous linear functional F on $C[a, b]$ for which there is no integrable function g such that $F(f) = \int_a^b fg\, dt$ for each f in $C[a, b]$.

26

Convergence in Measure

If $(S, \mathscr{A}, \mu)$ is a measure space, then the set $M(S, \mathscr{A}, \mu)$ of all real valued measurable functions on S, which we also denote by M, has been shown to be a linear space, an algebra, and a σ-lattice of functions. We shall consider f and g in M to be the same if $f(t) = g(t)$ almost everywhere so, strictly speaking, M consists of equivalence classes of functions rather than functions. The purpose of this section is to consider the relation between various kinds of convergence in M, namely, pointwise convergence, uniform convergence, and convergence in measure. The sequence $\{f_n\}$ of real valued functions on S is said to *converge almost everywhere* to a function f if there is a subset A of S having 0 measure such that $\lim_n f_n(x) = f(x)$ for each x in $S \backslash A$. The sequence $\{f_n\}$ is said to *converge uniformly* to f on a subset B of S if for each $\varepsilon > 0$ there is K such that $|f(s) - f_n(s)| < \varepsilon$ for each s in B when $n \geqslant K$. The concept of convergence in measure is weaker than both of these types of convergence. The sequence $\{f_n\}$ is said to *converge in measure to f* if for each $\varepsilon > 0$ there is K such that when $n \geqslant K$ the set of s, such that $|f_n(s) - f(s)| > \varepsilon$, has measure less than ε. Thus we shall have

$$\mu\{t\colon |f(t) - f_n(t)| > \varepsilon\} < \varepsilon.$$

If $\{f_n\}$ converges in measure to f, then for each pair $\delta, \varepsilon > 0$ there is K such that $\mu\{s\colon |f(s) - f_n(s)| > \delta\} < \varepsilon$ when $n \geqslant K$. To prove this it suffices to let η be the minimum of ε and δ and apply the definition. This latter condition then is equivalent to convergence in measure.

If $(S, \mathscr{A}, \mu)$ is a finite measure space, define δ on $M \times M$ for each $n = 1, 2, \ldots$ by the formula

$$\delta(f, g) = \int_S \frac{|f(t) - g(t)|}{1 + |f(t) - g(t)|}\, d\mu.$$

We shall show that δ is a metric on M, which makes M a complete metric space, and that convergence in measure is precisely convergence of a sequence with respect to the measure δ.

Theorem 26.1:

(a) The function δ is a metric on M, which makes M a complete metric space.

(b) A sequence (f_n) converges to f in (M, δ) if and only if (f_n) converges to f in measure.

Proof: We first show δ is a metric. It is clear that $\delta(f, g)$ is nonnegative and that $\delta(f, g) = \delta(g, f)$. If $\delta(f, g) = 0$ then

$$\int_S \frac{|f(t) - g(t)|}{1 + |f(t) - g(t)|}\, d\mu = 0$$

so that $f(t) - g(t) = 0$ for almost all t. That δ satisfies the triangle inequality follows from the inequality

$$\frac{a+b}{1+a+b} \leqslant \frac{a}{1+a} + \frac{b}{1+b},$$

which is true for all positive numbers a and b.

Before showing that M is complete with the metric δ we will prove part (b), that convergence in measure is equivalent to convergence with respect to the metric δ. Since S is a finite measure space, we may assume that $\mu(S)=1$. Suppose (f_n) converges in measure to f. Given $\varepsilon > 0$, let K be such that $\mu(D) < \varepsilon/2$ where $n \geqslant K$ and D is the set of all t such that $|f(t)-f_n(t)| > \varepsilon/2$. If $n > K$ we then have

$$\begin{aligned}\delta(f, f_n) &= \int_S \frac{|f-f_n|}{1+|f-f_n|}\,d\mu \\ &= \int_D \frac{|f-f_n|}{1+|f-f_n|}\,d\mu + \int_{D^\sim} \frac{|f-f_n|}{1+|f-f_n|}\,d\mu \\ &< \mu(D) + \mu(D^\sim)\frac{\varepsilon}{2} < \varepsilon.\end{aligned}$$

This proves that convergence in measure implies convergence with respect to the metric δ.

Next suppose (f_n) converges to f with respect to δ. Given $\varepsilon > 0$ let K be such that $\delta(f, f_n) < \varepsilon^2/(1+\varepsilon)$ when $n \geqslant K$. If $|f(t)-f_n(t)| > \varepsilon$, then since $x/(1+x)$ is an increasing function we shall have $|f(t)-f_n(t)|/(1+|f(t)-f_n(t)|) > \varepsilon/(1+\varepsilon)$. Thus the set D of all t such that $|f(t)-f_n(t)| > \varepsilon$ is contained in the set of all t such that $|f(t)-f_n(t)|/(1+|f(t)-f_n(t)|) > \varepsilon/(1+\varepsilon)$. Therefore, if n is greater than K, the set D cannot have measure more than ε. This proves that convergence in the metric implies convergence in measure.

Finally we show that the metric space (M, δ) is complete. Suppose (f_n) is a Cauchy sequence in M. Then we can use an argument resembling that in the previous paragraph to prove that the sequence (f_n) is *Cauchy in measure*, that is, for each $\varepsilon > 0$ there is K such that when m and n are greater than K, the set of all s such that $|f_m(s)-f_n(s)| > \varepsilon$ has measure $< \varepsilon$. Thus there is an increasing sequence $m(j)$ of integers $(m(j))$ such that for each j, the set D_j of all s for which $|f_p(s)-f_q(s)| > 2^{-j}$ with $p, q \geqslant m(j)$ has measure less than 2^{-j}. If $E_n = \bigcap_{j>n} D_j^\sim$ then $\mu(E_n^\sim) \leqslant \sum_{j=n}^{\infty} \mu(D_j) < 2^{-n+1}$, and the subsequence $(f_{m(j)}(s))$ converges uniformly on E_n because if s is in E_n, $K \geqslant n$ and both p and $q \geqslant K$ we have $|f_{m(p)}(s)-f_{m(q)}(s)| < 2^{-K}$. If $E = \bigcup_n E_n$ it follows that the complement of E has measure 0 and the sequence $(f_{m(j)})$ converges pointwise on E to some function f. By the Lebesgue Dominated Convergence Theorem we have

$$\lim_j \delta(f, f_{m(j)}) = \lim_j \int_S \frac{|f-f_{m(j)}|}{1+|f-f_{m(j)}|}\,d\mu = 0$$

so that a subsequence of (f_n) converges to f in M. But this implies that (f_n) converges to f. ■

The proofs of the following two corollaries are contained in the proof of Theorem 26.1. The implication in Corollary 26.2 that (a) $\Rightarrow$ (b) is called *Egoroff's Theorem.*

Corollary 26.2: Let $(S, \mathscr{A}, \mu)$ be a finite measure space. Consider the following three types of convergence of a sequence in $M(S, \mathscr{A}, \mu)$.

(a) (f_n) converges to f almost everywhere.

(b) For each $\varepsilon > 0$ there is a subset E of S with complement of measure $< \varepsilon$ such that (f_n) converges uniformly to f on E.

(c) (f_n) converges in measure to f.

Then (a) is equivalent to (b) and each imply (c).

Corollary 26.3: Let $(S, \mathscr{A}, \mu)$ be a finite measure space. If (f_n) is a sequence in M that converges in measure to f, then there is a subsequence $(f_{m(j)})$ of (f_n) that converges almost everywhere to f.

The following example shows that convergence in measure does not imply convergence almost everywhere.

Example 26.4:

Each x in $(0, 1)$ has a dyadic expansion

$$x = \frac{c_1}{2} + \frac{c_2}{4} + \frac{c_3}{8} + \cdots + \frac{c_n}{2^n} + \cdots$$

where c_n is 0 or 1. This expansion is unique except when $x = m/2^n$ for some pair of integers m, n. For each n, define $g_n(x)$ to be 0 if x is a dyadic rational and $g_n(x) = c_n$ otherwise. Consider the sequence (f_n)

$$g_1, \quad g_1 g_2, \quad (1 - g_1)g_2, \quad g_1 g_2 g_3, \quad g_1(1 - g_2)g_3, \quad (1 - g_1)(1 - g_2)g_3, \quad \ldots$$

where the nth block consists of 2^{n-1} products of g_n with all possible combinations of g_i or $(1 - g_i)$ for $i < n$. This sequence converges to 0 in measure because when there are more than n factors the set of s for which the product is equal to one has measure less than 2^{-n}. On the other hand this sequence does not converge to 0 almost everywhere because for each s in $(0, 1)$, $f_n(s) = 1$ for infinitely many n.

Exercises

A

1. Prove Corollary 26.2.
2. Prove Corollary 26.3.
3. Show that in Example 26.4 the subsequence $g_1, g_1 g_2, g_1 g_2 g_3, \ldots$ converges to 0 for all s in $(0, 1)$.
4. Prove that if f is in $M(S, \mathscr{A}, \mu)$ and f is bounded then for each $\varepsilon > 0$ there is a simple function g that assumes only rational values such that the set of all s with $|f(s) - g(s)| > \varepsilon$ has measure $< \varepsilon$. [*Hint*: Use the proof of Theorem 12.8 as a guide.]

5. Let $(S, \mathscr{A}, \mu)$ be a finite measure space. Prove that (f_n) converges to f in measure if and only if every subsequence of (f_n) contains a subsequence that converges almost everywhere to f.

6. Show that if (f_n) converges to f in measure and there is an integrable function g such that $|f_n| \leqslant g$ for each n, then

$$\lim_n \int_S |f_n - f| \, d\mu = 0.$$

7. If $(S, \mathscr{A}, \mu)$ is σ-finite but not finite, the conclusion of Egoroff's Theorem will not necessarily hold. Let $S = N = \{1, 2, \ldots\}$; let $\mathscr{A}$ be the collection of all subsets of N; let $\mu(\{n\}) = 1$ for each n. For each n, let f_n be the sequence such that $f_n(j) = 0$ for $j \leqslant n$ and $f_n(j) = 1$ for $j > n$. Show $\lim_n f_n(j) = 0$ for each j, but (f_n) does not converge uniformly to 0 on any subset of N having infinite measure.

B

8. If (f_n) converges in measure to f and (g_n) converges in measure to g show
   (a) $f_n + g_n$ converges in measure to $f + g$;
   (b) $f_n g_n$ converges in measure to fg;
   (c) $|f_n|$ converges in measure to $|f|$;
   (d) $\sup(f_n, g_n)$ converges in measure to $\sup(f, g)$;
   (e) $\inf(f_n, g_n)$ converges in measure to $\inf(f, g)$.

9. Prove that the Monotone Convergence Theorem (12.10), Fatou's Lemma (Corollary 12.11), and Lebesgue's Convergence Theorem (12.12) remain true when convergence almost everywhere is replaced by convergence in measure.

10. Let $S = [0, 1]$ and let m denote Lebesgue measure.
   (a) Prove that for each f in $M(S, \mathscr{A}, m)$ and $\varepsilon > 0$ there is a simple function g such that g takes on only rational values, for each nonzero value a assumed by g the set $g^{-1}(a)$ is a finite union of intervals with rational endpoints and $\delta(f, g) < \varepsilon$.
   (b) Prove that the space $M(S, \mathscr{A}, m)$ is separable.
   (c) Show that for each f and each $\varepsilon > 0$ there is a continuous function g such that $\delta(f, g) < \varepsilon$.
   (d) Show that for every Lebesgue measurable f function on $[0, 1]$ there is a sequence of continuous functions (f_n) that converge to f almost everywhere.

11. Show that if $(S, \mathscr{A}, \mu)$ is a σ-finite measure space with $S = \bigcup_n S_n$ with $\mu(S_n) < \infty$ for each n then $M(S)$ is a metric space with metric defined by $\delta(f, g) = \sum_n 2^{-n} \int_{S_n} |f - g|/(1 + |f - g|) \, d\mu$.

12. *(a)* Prove the metric δ is invariant under addition, that is, $\delta(f + h, g + h) = \delta(f, g)$ for all f, g, h in M.
   (b) Prove that addition and scalar multiplication are continuous on M. That is, if $f_n \to f$, and $g_n \to g$, then $f_n + g_n \to f + g$; and if $a_n \to a$ and $f_n \to f$, then $a_n f_n \to af$.
   (c) Prove that, δ is not a norm on M.

27

The Lebesgue Spaces

Let $(S, \mathscr{A}, \mu)$ be a σ-finite measure space, and suppose $p \geqslant 1$. A real valued measurable function f on S is said to be in $L^p(S, \mathscr{A}, \mu)$, or more briefly in L^p, if the function $|f|^p$ is integrable, that is, if

$$\int_S |f|^p \, d\mu < \infty.$$

If f and g are in L^p then so is $f+g$ because we have

$$|f+g|^p \leqslant 2^p\{|f|^p + |g|^p\},$$

and so is af for every real number a because $|af|^p = |a|^p |f|^p$. Thus, each L^p is a linear space. A norm-like function on L^p is defined by the formula

$$\|f\|_p = \left\{\int_A |f|^p \, d\mu\right\}^{1/p}.$$

The function $\|f\|_p$ is also written $\|f\|$ when p can be understood by its context. That $\|af\|_p = |a| \, \|f\|_p$ follows quickly from the definition. The triangle inequality for $\| \;\|_p$ is a result of the Minkowsky inequality that we shall prove below. The function $\| \;\|_p$ was called a "norm-like" function because $\|f\|_p = 0$ does not imply f is the zero function but only that $f = 0$ μ-almost everywhere. This difficulty vanishes if we agree to regard two members of L^p as the same if they are equal almost everywhere. More formally, L^p is defined as the collection of all equivalence classes of p-integrable functions under the relation of almost everywhere equality. With this understanding $\| \;\|_p$ will be a bona fide norm on L^p. The space $L^\infty(S, \mathscr{A}, \mu)$ is defined to be the set of all functions f such that $\|f\|_\infty < \infty$ where

$$\|f\|_\infty = \inf\{\sup\{|f(t)| \colon t \in A\} \colon A \in \mathscr{A}, \ \mu(S \backslash A) = 0\}.$$

The number $\|f\|_\infty$ is called the essential supremum of $|f|$ because it is the supremum of $|f|$ neglecting sets of measure 0. The space L^∞ is also a linear space and $\| \;\|_\infty$ is a norm on L^∞ provided we regard the space as consisting of classes of functions.

The following theorem is known as *Hölder's inequality*. See Theorem 8.11 on page 55 for a preliminary version of this result.

Theorem 27.1: Suppose p and q are nonnegative reals with $1/p + 1/q = 1$ and $(S, \mathscr{A}, \mu)$ is a σ-finite measure space. If $f \in L^p(S, \mathscr{A}, \mu)$ and $g \in L^q(S, \mathscr{A}, \mu)$, then $fg \in L^1(S, \mathscr{A}, \mu)$ and

$$\int |fg| \, d\mu \leqslant \|f\|_p \|g\|_q.$$

When both $\|f\|_p$ and $\|g\|_q$ are nonzero then equality holds if and only if there is $a > 0$ such that $|g|^q = a|f|^p$ almost everywhere.

Proof: The proof of this theorem is very much like that of Theorem 8.11. We have included the details for the benefit of those readers who have skipped that proof.

The first step is to prove the inequality

$$a^{1/p}b^{1/q} \leqslant \frac{a}{p} + \frac{b}{q} \tag{27–1}$$

for all nonnegative a, b. To do this, let $0 < \alpha < 1$, and consider the function

$$f(t) = t^{\alpha} - \alpha t + \alpha - 1.$$

Then $f'(t) = \alpha(t^{\alpha-1} - 1)$, so that $f(1) = f'(1) = 0$; $f'(t)$ is positive for $0 < t < 1$; and $f'(t)$ is negative for $t > 1$. This implies that $f(t) \leqslant 0$ for $t \geqslant 0$. For $b = 0$, inequality (27–1) is obvious. If $b > 0$ let $t = a/b$ and $\alpha = 1/p$. Then we have

$$f\left(\frac{a}{b}\right) = \left(\frac{a}{b}\right)^{1/p} - \left(\frac{1}{p}\right)\left(\frac{a}{b}\right) + \frac{1}{p} - 1 \leqslant 0.$$

If we multiply this through by b we obtain

$$a^{1/p}b^{(1-1/p)} \leqslant \frac{a}{p} + b\left(\frac{1-1}{p}\right),$$

which is clearly equivalent to (27–1). The function $f(t) = 0$ only when $t = 1$ so that equality holds in (27–1) only when $a = b$.

Now suppose $\|f\|_p = \|g\|_q = 1$. For each t in S we have

$$|f(t)g(t)| \leqslant \frac{|f(t)|^p}{p} + \frac{|g(t)|^q}{q} \tag{27–2}$$

by applying (27–1) with $a = |f(t)|^p$ and $b = |g(t)|^q$. If we integrate both sides of (27–2) over S we obtain

$$\begin{aligned}\int_S |fg| \, d\mu &\leqslant \int_S \frac{|f|^p}{p} \, d\mu + \int_S \frac{|g|^q}{q} \, d\mu \\ &= \frac{1}{p} + \frac{1}{q} = 1. \end{aligned} \tag{27–3}$$

If $\|f\|_p$ or $\|g\|_q$ are 0 then both sides of the Hölder inequality are 0 so we may assume that both are nonzero. Then we may apply the preceding paragraph to the functions $f/\|f\|_p$ and $g/\|g\|_p$, both of which have norm one to obtain

$$\int_A \left|\frac{f}{\|f\|_p}\right|\left|\frac{g}{\|g\|_q}\right| d\mu \leqslant 1$$

from which the Hölder inequality immediately follows by multiplying both sides by $\|f\|_p\|g\|_q$.

Equality can hold in (27–2) if and only if $|f(t)|^p = |g(t)|^q$ and equality can hold in (27–3) if and only if this equality holds for almost all t in S. Thus equality holds in the Hölder relation if and only if $\|g\|^q|f|^p = \|f\|^p|g|^q$ almost everywhere. ■

The first application of Hölder's inequality will be to derive the *Minkowski inequality*, which is the next result.

Theorem 27.2: If f or g is in $L^p(S, \mathscr{A}, \mu)$ where $1 < p < \infty$, then so is $f+g$, and we have $\|f+g\|_p \leqslant \|f\|_p + \|g\|_p$.

Proof: For each t in S we have

$$|f(t)+g(t)|^p \leqslant (|f(t)+g(t)|)^{p-1}|f(t)| + (|f(t)+g(t)|)^{p-1}|g(t)|$$

so that if we integrate both sides over S we obtain

$$\|f+g\|_p^p \leqslant \int_S |f+g|^{p-1}|f|\, d\mu + \int_S |f+g|^{p-1}|g|\, d\mu.$$

We use the Hölder inequality on the two integrals on the right-hand side to obtain

$$\int_S |f+g|^{p-1}|f|\, d\mu \leqslant \|f\|_p \| |f+g|^{p-1} \|_q$$

and

$$\int_S |f+g|^{p-1}|g|\, d\mu \leqslant \|g\|_p \| |f+g|^{p-1} \|_q.$$

By definition we have

$$\begin{aligned} \| |f+g|^{p-1} \|_q &= \left\{ \int_S |f+g|^{(p-1)q}\, d\mu \right\}^{1/q} \\ &= \|f+g\|_p^{p/q} \end{aligned}$$

because $(p-1)q = (p-1)(1-1/p)^{-1} = (p-1)p(p-1)^{-1} = p$. Combining these inequalities we find

$$\|f+g\|_p^p \leqslant \|f+g\|_p^{p/q}\{\|f\|_p + \|g\|_p\},$$

which gives the Minkowski inequality after we divide both sides by $\|f+g\|_p^{p/q} = \|f+g\|^{p-1}$. ■

The Minkowski inequality shows that each space L^p is a normed space with norm $\| \;\|_p$. The next result, known as the *Riesz-Fischer Theorem*, shows that L^p is a Banach space.

Theorem 27.3: For $p \geqslant 1$ the L^p spaces are complete.

Proof: We shall show that every absolutely convergent series in L^p converges and we can apply Theorem 17.8 to conclude that L^p is thus complete. Suppose $f_n \in L^p$ for each n and $\sum_n \|f_n\|_p = M < \infty$. For each k let $F_k = \sum_{n=1}^k |f_n|$. Then each F_k is measurable, and by Minkowski's inequality we have $\|F_k\|_p \leqslant \sum_{n=1}^k \|f_n\|_p \leqslant M$ so that $\int F_k^p\, d\mu \leqslant M^p$. The sequence $\{F_k\}$ must increase to an extended real function F for which $\int F^p\, d\mu \leqslant M^p$ by the Monotone Convergence

Theorem (12.10). This implies that F^p is integrable so that $F(t)$ is finite for almost all t.

For each t with $F(t) < \infty$, the series $\sum_n f_n(t)$ converges absolutely to a number $f(t)$. Since $f(t)$ is the limit almost everywhere of the partial sums $\sum_{n=1}^k f_n$ it follows that f is measurable. Since $|\sum_{n=1}^k f_n(t)|^p \leqslant F(t)^p$ for all t and k we conclude that f is in L^p and $\|f\|_p \leqslant \sum_{n=1}^\infty \|f_n\|_p$ by the Lebesgue Convergence Theorem (12.12).

If we apply the above reasoning to $f - \sum_{n=1}^k f_n$ for each k we find that $\|f - \sum_{n=1}^k f_n\|_p \leq \sum_{n=k+1}^\infty \|f_n\|$. This means that the series $\sum_n f_n$ converges to f. ■

Our next task is to derive a representation theorem for the dual space of $L^p(S, \mathscr{A}, \mu)$. It will be shown that the dual of L^1 is L^∞ and that the dual of L^p for $1 < p$ is L^q where $1/p + 1/q = 1$. We shall prove our results under the assumption that S is a finite measure space, and let you extend the results to the σ-finite case. Throughout the rest of this section we shall assume that p is a number greater than 1 and q is a number greater than 1 with $1/p + 1/q = 1$.

Lemma 27.4: Suppose g is an integrable function on the finite measure space $(S, \mathscr{A}, \mu)$ for which there is $M > 0$ such that

$$\left|\int_S g\phi \, d\mu\right| \leqslant M \|\phi\|_p$$

for all simple functions ϕ on S. Then g is in L^q and $\|g\|_q \leqslant M$.

Proof: The inequality in the hypothesis implies the stronger inequality $\int_S |g\phi| \, d\mu \leqslant M \|\phi\|_p$ for all simple functions ϕ on S. For, let $B = \{s\colon g(s) \geqslant 0\}$ and let $C = \{s\colon g(s) < 0\}$. If ϕ is any simple function $\phi = \sum a_i \chi_{A_i}$ define $\phi_1(s) = |\phi(s)|$ for $s \in B$ and $\phi_1(s) = -|\phi(s)|$ for $s \in C$. Then ϕ_1 is also a simple function and $\|\phi_1\|_q = \|\phi\|_q$ while $\int_S g\phi_1 \, d\mu = \int_S |g\phi| \, d\mu$.

If ψ is a simple function with $0 \leqslant \psi \leqslant |g|^q$ then $\psi^{1/q}$ is a simple function with $\psi^{1/q} \leqslant |g|$ so that for every simple function ϕ we shall have

$$\int_S |\psi^{1/q}\phi| \, d\mu \leqslant \int_S |g\phi| \, d\mu \leqslant M \|\phi\|_p.$$

If we take ϕ to be $\psi^{1/p}$ this inequality becomes

$$\int_S \psi \, d\mu \leqslant M \|\psi^{1/p}\|_p = M \left\{\int_S \psi \, d\mu\right\}^{1/p},$$

which gives us

$$\left\{\int_S \psi \, d\mu\right\}^{1/q} \leqslant M$$

or $\int_S \psi \, d\mu \leqslant M^q$. If h is any bounded measurable function with $0 \leqslant h \leqslant |g|^q$ the definition of the integral $\int_S h \, d\mu$ implies that $\int_S h \, d\mu \leqslant M^q$. Therefore, we have $\int_S |g|^q \leqslant M^q$, which is the desired conclusion. ■

The following theorem is called the *Riesz Representation Theorem.*

Theorem 27.5: Suppose $(S, \mathscr{A}, \mu)$ is a finite measure space. Each member g of $L^q(S, \mathscr{A}, \mu)$ determines a continuous linear functional F_g on $L^p(S, \mathscr{A}, \mu)$ by the formula

$$F_g(f) = \int_S fg \, d\mu. \tag{27–4}$$

For each continuous linear functional F on $L^p(S, \mathscr{A}, \mu)$ there is g in L^q such that $F = F_g$ as determined by formula (27–4). Moreover, the correspondence between g and F_g is an isometry since $\|g\|_q$ is equal to the norm $\|F_g\|$ as a member of $(L^p)^*$.

Proof: If $g \in L^q$ then for each f in L^p the quantity $F_g(f)$ defined by (27–4) is finite by the Hölder inequality and we have

$$|F_g(f)| \leqslant \|g\|_q \|f\|_p. \tag{27–5}$$

The linearity of $F_g(f)$ in f results from the linearity of the integral. From (27–5) we conclude that F_g is bounded and $\|F_g\| \leqslant \|g\|_q$.

Suppose now F is a continuous linear functional on L^p. Define a function on $\mathscr{A}$ by the formula

$$v(A) = F(\chi_A) \quad \text{for } A \in \mathscr{A}$$

where χ_A denotes the characteristic function of A. That χ_A is in L^p follows since $(S, \mathscr{A}, \mu)$ is a finite measure space. We next verify that v is countably additive and it is thus a signed measure. Suppose $\{A_n\}$ is a sequence of disjoint subsets in $\mathscr{A}$ with $A = \bigcup_n A_n$. If χ_k denotes the characteristic function of $\bigcup_{n=1}^k A_n$ then we shall have

$$\lim_k \|\chi_A - \chi_k\|_p = \lim_k \mu\left\{A \setminus \bigcup_{n=1}^{k} A_n\right\}^{1/p} = 0$$

so that (χ_k) converges in L^p to χ_A. Since F is continuous it follows that $F(\chi_A) = \lim_k F(\chi_k) = \sum_{n=1}^{\infty} F(\chi_{A_n})$. If $\mu(A) = 0$ then $\|\chi_A\|_p = 0$ so that $F(\chi_A) = v(A) = 0$. This implies that v is absolutely continuous with respect to μ. Therefore, by the Radon-Nikodym Theorem (15.2) there is a measurable function g on S such that for each $A \in \mathscr{A}$ we have

$$F(\chi_A) = v(A) = \int_A g \, d\mu.$$

By the linearity of F and of the integral it follows that

$$F(\phi) = \int_S \phi g \, d\mu$$

for every simple function ϕ. Since F is a bounded linear functional on L^p we have the relation

$$\left|\int_S \phi_g \, d\mu\right| = |F(\phi)| \leqslant \|F\| \, \|\phi\|_p$$

for every simple function ϕ. By Lemma 27.4 it follows that g is in L^q and $\|g\|_q \leqslant \|F\|$.

We now know that the continuous linear functionals F and F_g coincide on the simple functions. It remains to prove that the collection of simple functions is dense in L^p. Let f in L^p and $\varepsilon > 0$ be given. Since $\int_S |f|^p \, d\mu$ is finite and equal to the sum

$$\sum_{n=1}^{\infty} \int_{S_n} |f|^p \, d\mu \quad \text{where } S_n = \{s \colon n-1 \leqslant |f(s)| < n\}$$

it follows that there is N such that $\sum_{n>N} \int_{S_n} |f|^p \, d\mu < \varepsilon^p$. Let B denote the set $\bigcup_{n>N} S_n$ and let A denote $B^{\sim}$. Then if s is in A we have $-N \leqslant f(s) < N$. Let m be such that $2N/m < \varepsilon$, and for $j = -m, -(m-1), \ldots, 0, \ldots, m-1$, let A_j be the set of all s such that $Nj/m \leqslant f(s) < N(j+1)/m$. Let ϕ be the simple function

$$\phi = \sum_{j=-m}^{m-1} \left(\frac{j}{m}\right) \chi_{A_j}.$$

Then we shall have

$$\begin{aligned}\int_S |f-\phi|^p \, d\mu &= \int_A |f-\phi|^p \, d\mu + \int_B |f|^p \, d\mu \\ &\leqslant \sum_{j=-m}^{m-1} \varepsilon^p \mu(A_j) + \varepsilon^p \\ &\leqslant (1+\mu(S))\varepsilon^p.\end{aligned}$$

Since $\varepsilon > 0$ is arbitrary it follows that we can approximate members of L^p as closely as desired by simple functions. ∎

It is also true that the dual space of L^1 is L^∞ as the following version of the Riesz Representation Theorem asserts. It is not the case, however, that the dual of L^∞ is L^1.

Theorem 27.6: Suppose $(S, \mathscr{A}, \mu)$ is a finite measure space. Each member g of $L^\infty(S, \mathscr{A}, \mu)$ determines a continuous linear functional F_g on $L^1(S, \mathscr{A}, \mu)$ by the formula

$$F_g(f) = \int_S fg \, d\mu. \tag{27–6}$$

For each continuous linear functional F on $L^1(S, \mathscr{A}, \mu)$ there is g in L^∞ such that $F = F_g$ as determined by formula (27–6). Moreover, the correspondence between g and F_g is an isometry since $\|g\|_\infty$ is equal to the norm $\|F_g\|$ as a member of $(L^\infty)^*$.

The details of the proof of Theorem 27.6 resemble those of Theorem 27.5 and they are left to you, the reader.

Exercises

A

1. Verify that $|f+g|^p \leqslant 2^p\{|f|^p+|g|^p\}$. [*Hint*: First show that for all $0<t\leqslant 1$ we have $(1+t)^p<2^p$. Next show that if a and b are both positive we have $(a+b)^p \leqslant 2^p(a^p+b^p)$ by applying the first inequality to $(1+b/a)^p$ if $a=\max(a, b)$.]
2. Show that the collection of f in L^p with $\|f\|_p=0$ forms a subspace of L^p.
3. Define f on $[0, 1]$ by $f(r)=p$ when r is a rational of the form p/q where p, q have greatest common divisor equal to 1 and $f(t)=1$ if t is irrational. Show $\sup\{|f(t)|: 0\leqslant t\leqslant 1\}=\infty$ but the essential supremum $\|f\|_\infty$ of f on $[0, 1]$ is 1.
4. Show $\| \quad \|_\infty$ is a norm on L^∞.
5. Show that if $f\in L^p$ and $g\in L^\infty$ then $fg\in L^p$ and $\|fg\|_p\leqslant\|f\|_p\|g\|_\infty$.
6. Prove that the space L^2 is a Hilbert space with inner product given by

$$\langle f, g\rangle=\int_S f\bar{g}\, d\mu.$$

7. Use the Riesz Representation Theorem to determine representations for the dual spaces of the following: $L^{3/2}$, $L^{8/7}$, L^2, L^3, L^5, $L^{13/3}$.
8. If $f\in L^1$ show there is g in L^∞ such that $\|g\|_\infty=1$ and

$$\int_S fg\, d\mu=\|f\|_1.$$

B

9. Define L^p for $0<p<1$ analogously to L^p. Show that L^p also forms a linear space of functions.
10. Suppose μ is a finite measure.
    (a) Show that if $1\leqslant p\leqslant r$ then $L^r(S, \mathscr{A}, \mu)\subset L^p(S, \mathscr{A}, \mu)$.
    (b) Show that $L^\infty(S, \mathscr{A}, \mu)$ is contained in each L^p.
    (c) Let $\mathscr{M}$ denote the σ-algebra of Lebesgue measurable subsets of $I=[0, 1]$ and let m denote Lebesgue measure on I. For $1<r$ define $f_r(t)=t^{-1/r}$ for $t>0$, $f_r(0)=0$. If $r>p$ then $f_r\in L^p[0, 1]$ while if $r<p$ then f_r is not in $L^p[0, 1]$.
    (d) Show that if $f\in L^\infty(S, \mathscr{A}, \mu)$, then $\|f\|_\infty=\lim_{p\to\infty}\|f\|_p$.
    (e) Find f, which is in $L^p(I)$, for each p but not in $L^\infty(I)$.
11. The space l^p consists of all sequences (t_n) such that $\sum_v |t_n|^p<\infty$.
    (a) Explain how l^p is a special case of an L^p space.
    (b) Show that if $p<q$ then $l^p\subset l^q$.
    (c) Show that if $t_n=(1/n)^{1/r}$ for $r>1$ then $(t_n)\in l^p$ for $p>r$ but not in l^p for $p\leqslant r$. The space l^∞ consists of all bounded sequences; this space is also denoted by m.
    (d) Find a sequence in l^∞ but not in $\bigcup_p l^p$.
    (e) Show that if (t_n) is in l^p for some p then $\|(t_n)\|_\infty=\lim_p\|(t_n)\|_p$.

12. Show that no inclusion relations exist among the spaces $L^p(R, \mathcal{M}, m)$ where m denotes Lebesgue measure on the real line $\mathbb{R}$.
13. *(a)* Prove that $L^2[0, 2\pi]$ is a Hilbert space, and the sequence $\{1/\sqrt{2\pi}, (1/\sqrt{\pi}) \cos nx, (1/\sqrt{\pi}) \sin nx\}$ is a *CON* sequence in $L^2[0, 2\pi]$.
    (b) Prove that if f is in $L^2[0, 2\pi]$ then $\lim_n \int_0^{2\pi} f(t) \cos nt \, dt = 0$.
14. Furnish a proof for Theorem 27.6.
15. Show that if $g \in L^\infty$, then for each $\varepsilon > 0$ there is $f \in L^1$ with $\|f\|_1 = 1$ such that $\int_S fg \, d\mu \geqslant \|g\|_\infty - \varepsilon$. [*Hint*: Use the Riesz Representation Theorem.]
16. Show that L^∞ is dense in L^p for all $p \geqslant 1$. [*Hint*: This means that for each f in L^p and each $\varepsilon > 0$ there is g in L^∞ such that $\|f - g\|_p < \varepsilon$.]
17. Show that l^∞ and $L^\infty[0, 1]$ are not separable spaces.

C

18. Let $I = [0, 1]$. For each $n = 1, 2, \ldots$ let $I_{n1} = [0, 1/n]$, $I_{nj} = (j - 1/n, j/n]$ for $j = 2, \ldots, n$; let χ_{nj} denote the characteristic function of I_{nj}. For f in $L(I, \mathcal{M}, m)$, $1 \leqslant p < \infty$ define $P_n f$ by the formula

$$P_n f = \sum_{j=1}^{n} \left\{ \int_{I_{nj}} f \, dt \right\} \chi_{nj}.$$

(a) Prove that each P_n is a linear operator from L^p into L^p such that $\|P_n\| = 1$.
(b) Prove that for each f in L^p we have $\lim_n P_n f = f$.
(c) Prove that a subset B of L^p is compact if and only if B is closed and $\{P_n\}$ converges to the identity uniformly on B, that is, for each $\varepsilon > 0$ there is M such that $\|f - P_n f\| < \varepsilon$ for all f in B when $n > M$.
(d) For each $k = 1, 2, \ldots$ and each $j = 1, 2, \ldots, 2^k$, let $h_{2^k + j - 1} = \chi_{2^k j}$. Show that the sequence $\{h_n : n = 1, 2, \ldots\}$ forms a Schauder basis of L^p.

28

Spaces of Sequences

Among the Banach spaces of sequences that we have so far encountered are the following

m or l^∞: The space of all bounded sequences (t_n) with the norm

$$\|(t_n)\| = \sup_n |t_n|.$$

l^p: For $1 \leqslant p < \infty$ with the norm $\|(t_n)\| = \{\sum_n |t_n|^p\}^{1/p}$.

c_0: The space of all sequences that converge to 0 with the norm

$$\|(t_n)\| = \sup_n |t_n|.$$

c: The space of all convergent sequences with the norm

$$\|(t_n)\| = \sup_n |t_n|.$$

bv: The space of all sequences (t_n) such that $\sum_n |t_{n+1} - t_n| < \infty$ with the norm $\|(t_n)\| = |t_1| + \sum_n |t_{n+1} - t_n| < \infty$.

cs: The space of all sequences (t_n) such that $\sum_n t_n$ converges with the norm $\|(t_n)\| = \sup_k |\sum_{n=1}^k t_n|$.

bs: The space of all sequences (t_n) such that $\sup_k |\sum_{n=1}^k t_n| < \infty$ with the norm $\|(t_n)\| = \sup_k |\sum_{n=1}^k t_n| < \infty$.

bv_0: The space of all sequences (t_n) in *bv* that converge to 0 with the same norm as *bv*.

A *BK-space* is a Banach space of sequences S such that for each k the coordinate functional E_k defined by $E_k(t_n) = t_k$ is continuous on S. All of the sequence spaces listed above are *BK*-spaces as you can easily see by examining the norm. For example, *bs* is a *BK*-space because

$$|E_k(t_n)| = |t_k| \leqslant \left|\sum_{n=1}^{k} t_n\right| + \left|\sum_{n=1}^{k-1} t_n\right| \leqslant 2\|(t_n)\|_{bs}.$$

An infinite matrix (a_{ij}) is said to *determine a linear operator from a sequence space S into a sequence space T* if for each (s_j) in S the infinite series $\sum_j a_{ij}s_j$ converges and the sequence $(\sum_j a_{ij}s_j)_i = (\sum_j a_{1j}s_j, \sum_j a_{2j}s_j, \ldots)$ is in T. If the matrix is denoted by A and the sequence (s_j) is denoted by s then we write the result of the operation described above as As. The following result is a stunning application of the Closed Graph Theorem in the setting of *BK*-spaces.

Theorem 28.1: Every matrix mapping A from a *BK*-space S to a *BK*-space T is continuous.

Proof: By the Closed Graph Theorem (20.4) it suffices to prove that the graph of A is closed, that is, if $(s^{(n)})$ is a sequence (of sequences) in S that converges to s and if $(As^{(n)})$ converges in T to t then it follows that $As = t$. Since two sequences are equal if and only if they are equal at each coordinate it suffices to show that

$$E_k(As) = \sum_j a_{kj}s_j = E_k(t) = t_k$$

for each k. Assume the index k is fixed. For each integer $m = 1, 2, \ldots,$ the linear functional defined by $f_m(r) = \sum_{j=1}^m a_{kj}r_j$ is continuous on S because it is a linear combination of the functionals E_j that are known to be continuous. In fact, we have $f_m = \sum_{j=1}^m a_{kj}E_j$. Since $\lim_m f_m(r)$ converges for each r in S by the hypothesis that A is defined on all of S, it follows from the Uniform Boundedness Principle (19.1) that $\sup_m \|f_m\| < \infty$; and it follows from the Banach-Steinhaus Theorem (19.3) that the linear functional $f(r) = \lim_m f_m(r) = \sum_j a_{kj}r_j$ is continuous on S. Since $(s^{(n)})$ converges to s in S it follows that

$$\sum_j a_{kj}s_j = \lim_n f(s^{(n)}) = \lim_n \sum_j a_{kj}s_j^{(n)}$$

but T is also a *BK*-space so $\lim_n \sum_j a_{kj}s_j^{(n)} = \lim_n E_k(As^{(n)}) = E_k(t)$. ∎

Since the inclusion operator is determined by the matrix I which has 1's on the diagonal and 0's elsewhere, it follows that if S and T are *BK*-spaces with $S \subset T$

then the inclusion from S into T is continuous. This means that if $s_n \to s$ in S then $s_n \to s$ in T as well. An implication of this is that the BK-topology on a BK-space is unique because if $\mathscr{T}_1$ and $\mathscr{T}_2$ are two topologies on S that make S a BK-space, then the identity matrix I is continuous from $(S, \mathscr{T}_1)$ onto $(S, \mathscr{T}_2)$ and from $(S, \mathscr{T}_2)$ onto $(S, \mathscr{T}_1)$. Therefore, I is a topological isomorphism.

Using Theorem 28.1 we see that cs is isomorphic to c via the operator $A(t_n) = (t_1, t_1 + t_2, t_1 + t_2 + t_3, \ldots)$, which is represented by the matrix

$$\begin{matrix} 1 & 0 & 0 & 0 & \cdots \\ 1 & 1 & 0 & 0 & \cdots \\ 1 & 1 & 1 & 0 & \cdots \\ \cdots & \cdots & \cdots & \cdots & \end{matrix}\ .$$

The inverse B of A from c onto cs is given by the formula

$$B(s_n) = (s_1, s_2 - s_1, s_3 - s_2, \ldots).$$

The matrix form of B is

$$\begin{matrix} 1 & 0 & 0 & 0 & 0 & \cdots \\ -1 & 1 & 0 & 0 & 0 & \cdots \\ 0 & -1 & 1 & 0 & 0 & \cdots \\ \cdots & \cdots & \cdots & \cdots & \cdots & \end{matrix}\ .$$

The matrix A is also an isomorphism from l^1 onto bv with inverse B.

A BK-space S is said to have AD if the space ϕ of finitely nonzero sequences is dense in S. It is said to have AK if the set $\{e_n: n = 1, 2, \ldots\}$ is an s-basis for S where e_n is the sequence having 1 in the nth place and 0's elsewhere. Thus S has AK if

$$(s_n) = \sum_n s_n e_n$$

for each s in S. The β-dual, S^β, of a sequence space S is the set of all sequences (t_n) such that $\sum_n s_n t_n$ converges for all s in S. It is easy to see that S^β is always a sequence space.

Theorem 28.2: Suppose S is a BK-space having AD. Then S has AK if and only if (*) for each continuous linear functional f on S there is a sequence t^f in S^β such that

$$f(s) = \sum_n s_n t_n^f \quad \text{for each } s \text{ in } S.$$

Proof: Suppose $\{e_n\}$ is a basis for the BK-space S and f is a continuous linear functional on S. Then for each s in S we have

$$f(s) = f\left(\sum_n s_n e_n\right) = \sum_n s_n f(e_n)$$

so we may take $t_n^f = f(e_n)$ for all n. This proves the necessity of (*).

In order to prove the sufficiency of (*) define P_k on S by

$$P_k s = \sum_{n=1}^{k} s_n e_n = \sum_{n=1}^{k} E_n(s) e_n.$$

Then each P_k is continuous because each E_n is continuous. If (*) is valid then $t_n^f = f(e_n)$ for each n so that $f(P_k s)$ converges to $f(s)$ for all s in S. Therefore, the sequence $\{P_k s\colon k = 1, 2, \ldots\}$ is weakly bounded for each s in S so by Theorem 21.6 on page 165, it is strongly bounded. For each n it is clear that $P_k e_n \to e_n$, and since S has AD the sequence $\{e_n\}$ is fundamental in S. By the Banach-Steinhaus Theorem (21.3) it follows that $P_k s \to s$ for all s in S, but $\lim_k P_k s = \sum_n s_n e_n$, which means $\{e_n\}$ forms a basis for S. ■

Theorem 28.2 is a general representation theorem. It implies that l^1 is the dual of c_0, l^q is the dual of l^p when $p > 1$, and $1/p + 1/q = 1$, bv is the dual of cs, and bs is the dual of bv_0, and l^∞ is the dual of l^1.

The dual of l^∞ is not a space of sequences as we shall now show. A real valued function μ on an algebra $\mathscr{A}$ of sets is called a *finitely additive signed measure or fas-measure* if for every finite collection $\{A_1, A_2, \ldots, A_n\}$ of disjoint subsets of $\mathscr{A}$ we have $\mu(\bigcup_{j=1}^n A_j) = \sum_{j=1}^n \mu(A_j)$. The *variation* $\|\mu\|$ of a fas-measure μ is the supremum of all numbers $\sum_{j=1}^n |\mu(A_j)|$ as $\{A_1, \ldots, A_n\}$ ranges over all finite collections of disjoint subsets in $\mathscr{A}$. The fas-measure μ is called *finite* if $\|\mu\| < \infty$. We denote by m_0 the set of all sequences that have finitely many values. For example, the sequence $(1, -1, 0, 1, -1, 0, 1, -1, 0, \ldots)$ is in m_0 because it has only the values 1, -1, and 0. Another way to envision m_0 is as the linear span of all sequences of 1's and 0's. It is not hard to see that m_0 is a subspace of l^∞.

Lemma 28.3: The space m_0 is dense in l^∞.

Proof: Suppose (s_n) is a sequence in l^∞. Since (s_n) is bounded there is $M > 0$ such that $-M < s_n < M$ for all n. Given $\varepsilon > 0$, let K be an integer such that $M/K < \varepsilon$. For each K there is a smallest $j \geqslant -K$ such that $jM/n \leqslant s_n < (j+1)M/n$; denote jM/n by t_n. Since (t_n) assumes only the values $\{jM/n\colon j = 0, \pm 1, \pm 2, \ldots, \pm n\}$ it follows that (t_n) is in m_0. For each n we have $0 \leqslant s_n - t_n \leqslant (j+1)M/n - jM/n = M/n < \varepsilon$. Therefore, we conclude that $\|(s_n) - (t_n)\| < \varepsilon$. ■

Let $\mathscr{A}$ denote the σ-algebra of all subsets of the set N of integers. If μ is a fas-measure on $\mathscr{A}$ then μ determines a linear functional on m_0 by the formula

$$f_\mu(s_n) = \sum_{j=1}^{k} a_j \mu\{n\colon s_n = a_j\}$$

where $a_1, a_2, \ldots, a_k$ denote the totality of values assumed by (s_n). If the variation $\|\mu\|$ of μ is finite, then the functional f_μ is bounded on m_0 with the relative topology of m because we have

$$\begin{aligned} |f_\mu(s_n)| &\leqslant \sup_j |a_j| \sum_{j=1}^{k} |\mu\{n\colon s_n = a_j\}| \\ &\leqslant \|\mu\| \, \|(s_n)\|. \end{aligned}$$

Since f_μ is continuous and linear on a dense subset of l^∞ it has a unique extension to l^∞ by Theorem 21.2 on page 163. We denote this unique extension by f_μ. An alternative way to write $f_\mu(s_n)$ is in the integral form $\int_N (s_n)\, d\mu$.

Theorem 28.4: If μ is a fas-measure on N having bounded variation then the formula

$$f(s_n) = \int_N (s_n)\, d\mu \tag{28–1}$$

determines a continuous linear functional on l^∞. Moreover, if f is any continuous linear functional on l^∞ there is a fas-measure μ on N such that f is given by (28–1), and we shall have $\|f\| = \|\mu\|$ where the first norm is the norm of f as a member of $(l^\infty)^*$ and the second norm is the variation of μ.

Proof: We have already noted that every fas-measure μ of bounded variation determines a member of $(l^\infty)^*$ and that $\|f_\mu\| \leq \|\mu\|$. Suppose, then, that f is a continuous linear functional on l^∞. The restriction of f to m_0 is continuous on m_0 and has the same norm. Define μ on the collection $\mathscr{A}$ of subsets of N by the formula $\mu(A) = f(\chi_A)$ where χ_A denotes the characteristic function of A. By the linearity of f it follows that for each (s_n) in m_0 we have

$$f(s_n) = \sum_{j=1}^{k} a_j f(\chi_{A_j}) = \sum_{j=1}^{k} a_j \mu(A_j)$$

where $A_j = \{n\colon s_n = a_j\}$. This shows that $f = f_\mu$ on m_0. Since the extensions of continuous functional to l^∞ are unique we have $f = f_\mu$ on all of l^∞. That $\|\mu\| \leq \|f\|$ follows because if $\{A_1, A_2, \ldots, A_n\}$ are disjoint subsets of N, the sequence (s_n) for which $s_n = sgn(\mu(A_j))$ if $n \in A_j$ and $s_n = 0$ otherwise has norm one. Hence we shall have

$$\sum_{j=1}^{n} |\mu(A_j)| = f(s_n) \leq \|f\|\,\|(s_n)\| \leq \|f\|.$$

Taking the supremum over all possible $\{A_1, A_2, \ldots, A_n\}$, we conclude that $\|\mu\| \leq \|f\|$. ■

If f is a continuous linear functional on l^∞, let $t_i = f(e_i)$ for each i. Since $\sum_{j=1}^{n} |f(e_j)| = f(\sum_{j=1}^{n} sgn(f(e_j))e_j) \leq \|f\|$ it follows that $(t_i) \in l^1$. If we define f_0 by $f_0(s_j) = f(s_j) - \sum_j t_j s_j$, then $f_0(e_j) = 0$ for each j. This means that each f in $(l^\infty)^*$ has a unique representation $f = f_0 + f_1$ where $f_0(s) = 0$ for each s in c_0, which is the closed linear span of $\{e_j\}$ in l^∞, and f_1 is determined by a member of l^1. The functional f_1 is the uniquely determined countably additive part of f. Consequently, we have $(l^\infty)^*$, the direct sum of two closed subspaces Z_1, which is isometric to l^1 and Z_0, that consists of functionals determined by fas-measures that are 0 on finite sets. There is no way to construct fas-measures that are 0 on finite sets; their existence can be shown only by appealing to the Axiom of Choice.

Exercises

A

1. *(a)* Show that the function on c defined by $L(s_n) = \lim_n s_n$ is a continuous linear functional on c, and $\|L\| \leq 1$.
   (b) If e denotes the sequence with $e_j = 1$ for each j, show that each s in c has a unique representation $s = t + L(s)e$ where $t \in c_0$.

(c) Show that every continuous linear functional f on c can be represented in the form

$$f(s) = L(s)u_0 + \sum_n t_n u_n$$

where $t \in c_0$ is the sequence determined in (b), $u_0 \in \mathbb{R}$ and $u \in l^1$.

2. For S a BK-space and $t = (t_n)$ a sequence let tS consist of all sequences $(t_n s_n)$ as (s_n) ranges over S.
(a) Prove that tS is a linear space.
(b) Prove that if $t_n \neq 0$ for each n then tS is a BK-space isomorphic to S with the norm $\|(t_n s_n)\| = \|(s_n)\|$.
(c) Prove that if $t_n \neq 0$ for each n and S has AK then so does tS.

3. Let $w_n = (-1)^{n+1}$ for $n = 1, 2, \ldots$.
(a) Show that if $s \in l^\infty$ then the sequence $(s_1 - s_2, s_2 - s_3, s_3 - s_4, \ldots)$ is in bs while the sequence $(s_1 + s_2, s_2 + s_3, s_3 + s_4, \ldots)$ is in wbs where wbs is defined as in Exercise 2.
(b) Prove that $l^\infty = bs + wbs$ where the sum of two sequence spaces $S + T$ is defined as the set of all sums $s + t$ where $s \in S$ and $t \in T$.
(c) Prove that $c_0 = cs + wcs$.

4. *(a)* Show that $bv \subset c$ and that the function $L(s) = \lim_n s_n$ is continuous on bv.
(b) Show that bv_0 is the closed linear span of $\{e_n\}$ in bv.

5. Prove that a matrix $A = (a_{ij})$ determines a linear operator from l^1 into l^∞ if and only if $\sup_{ij} |a_{ij}| < \infty$.

6. *(a)* Show directly without using Theorem 28.1 that if S and T are BK-spaces with $S \subset T$ then the inclusion from S into T is continuous.
(b) Find a sequence $\{t^{(n)}\}$ in l^1 such that $t^{(n)} \to 0$ in l^∞ but not in l^1.
(c) Same as (b) for l^2 and l^1.

7. *(a)* Use the Hahn-Banach Theorem to show there exists a linear functional f on l^∞ such that $|f(s)| \leqslant \limsup_n |s_n|$ for each s in l^∞.
(b) Show that such a functional f has $\|f\| \leqslant 1$ and $f(e_n) = 0$ for each n.

8. Suppose (t_n) is in S^β where S is a BK-space. Prove that the linear functional f defined by $f(s_n) = \sum_n s_n t_n$ is bounded on S.

B

9. *(a)* Let $\sigma_1 = s_1$, $\sigma_n = s_1 + s_2 + \cdots + s_n$. Show that for each n, $\sum_{j=1}^n s_j t_j = \sum_{j=1}^n \sigma_j (t_j - t_{j+1})$.
(b) Use (a) to show that (i) if $S \in bs$ and $t \in bv$ then $st = (s_n t_n) \in bs$; (ii) if $s \in cs$ and $t \in bv$ then $st \in cs$; (iii) if $s \in bs$ and $t \in bv_0$ then $st \in cs$; (iv) $bs = bv_0^\beta$; (v) $bv = cs^\beta$.

10. *(a)* Prove that if S and T are BK-spaces with norms $\| \ \|_S$ and $\| \ \|_T$ respectively then $S \cap T$ is a BK-space with the norm $\|s\| = \sup\{\|s\|_S, \|s\|_T\}$.
(b) Show that if S and T have AK then so does $S \cap T$.
(c) Prove that for each $p > 1$, $cs \cap l^p$ is a BK-space with AK.

11. For each partition $\prod = \{A_1, A_2, \ldots, A_n\}$ of N let $b(j)$ be the smallest subscript in A_j. Define a function P_Π from l^∞ into m_0 by the formula $P_\Pi s = t$ where $t_i = s_b(j)$ if $i \in A_j$.

(a) Show that each P_Π is a linear operator on l^∞ with $\|P_\Pi\| = 1$.
(b) Show that $P_\Pi^2 = P_\Pi$ for each $\prod$ so that P_Π is a projection.
(c) Show that for each s in l^∞, $\lim_\Pi P_\Pi s = s$ in the sense that for each $\varepsilon > 0$ there is a partition $\prod_0$ of $\mathbb{N}$ such that $\|s - P_\Pi s\| < \varepsilon$ when $\prod$ refines $\prod_0$.

12. Suppose S is a BK-space with AK; define $P_n s = \sum_{j=1}^n s_j e_j$. Show that a subset A of S is compact if and only if
(a) A is closed and
(b) $P_n s$ converges uniformly to s on A; that is, for each $\varepsilon > 0$ there is K such that when $n > K$ we have $\|s - P_n s\| < \varepsilon$ for each s in A.
(c) If $s \in l^p$ for $1 \leqslant p < \infty$ then the set of all t in l^p with $|t_j| \leqslant |s_j|$ for each j is compact in l^p.
(d) The statement analogous to (c) is false for l^∞.

13. *(a)* Show that a matrix $A = (a_{ij})$ maps l^1 into a BK-space S if and only if the columns of A, namely $\{(a_{11}, a_{21}, a_{31}, \ldots), (a_{12}, a_{22}, a_{32}, \ldots), (a_{13}, a_{23}, a_{33}, \ldots), \ldots\}$ form a bounded set in S.
(b) Prove the matrix A maps l^1 into l^p for $1 \leqslant p < \infty$ if and only if

$$\sup_j \sum_i |a_{ij}|^p < \infty.$$

14. *(a)* Prove that a matrix $A = (a_{ij})$ maps a BK-space S into l^∞ if and only if (i) each row $(a_{i1}, a_{i2}, \ldots)$ is in S^β; (ii) the sequence of continuous linear functionals $\{f_i\}$ defined by $f_i(s_j) = \sum_j a_{ij} s_j$ is bounded in S^*.
(b) Prove that a matrix A maps l^p into l^∞ if and only if

$$\sup_i \sum_j |a_{ij}|^q < \infty$$

where $1/p + 1/q = 1$.

15. If S and T are BK-spaces let $M(S, T)$ denote the set of all sequences (u_n) such that $(u_n s_n) \in T$ for each (s_n) in S.
(a) Prove that with the norm $\|(u_n)\| = \sup\{\|(u_n s_n)\|_T : \|(s_n)\|_S \leqslant 1\}$, $M(S, T)$ is a Banach space.
(b) Prove that if $e_n \in S$ for each n then $M(S, T)$ is a BK-space.
(c) Prove that for any BK-space S, $S^\beta = M(S, cs)$.
(d) Prove that if $p > r$, $M(l^p, l^r) = l^s$ where $1/s = 1/r - 1/p$.
(e) Prove that if $p \leqslant r$, $M(l^p, l^r) = l^\infty$.

16. *(a)* Prove that if S is a separable Banach space then there is a bounded sequence $\{f_n\}$ in S^* such that if $f_n(s) = 0$ for each n, we must have $s = 0$.
(b) Prove that if S is a separable Banach space then there is a one-to-one continuous linear mapping from S onto a closed subspace of l^∞.

C

17. *(a)* Prove that the spaces l^p, $1 \leqslant p < \infty$, c_0, and c are symmetric in the sense that if (t_n) is in one of these spaces S and π is any permutation of the integers then $(t_{\pi(n)})$ is also in S and $\|(t_{\pi(n)})\| = \|(t_n)\|$.
(b) Show that the spaces bs, cs, bv, and bv_0 are not symmetric.

(c) Show that if (u_n) is any sequence of nonnegative real numbers, let $\lambda^p(u_n)$ denote the set of all sequences (t_n) such that $(t_{\pi(n)}u_n)$ is in l^p for every permutation π of the integers.

(d) Show that if (u_n) is an unbounded sequence then $\lambda^p(u_n) = \phi$, the space of sequences that are 0 except for a finite number of n.

(e) Show that if (u_n) is in ϕ then $\lambda^p(u_n)$ is the space ω of all sequences.

(f) Show that if (u_n) is bounded but not in c_0 then $\lambda^p(u_n) = l^p$.

(g) Show that if (u_n) is in l^p but not in ϕ then $\lambda^p(u_n) = l^\infty$.

(h) Show that if (u_n) is bounded but not in l^p then $\lambda^p(u_n)$ is a symmetric BK-space with the norm $\|t\| = \{\|t_{\pi(n)}u_n)\|_p$: π is a permutation of the integers$\}$ where $\| \ \|_p$ is the norm on l^p.

18. Let S be a separable Banach space and let D be a countable dense subset of the unit sphere of S, that is, $U = \{s: s \in S, \|s\| = 1\}$.

(a) Show that if $s \in U$ there are $s_1, s_2, \ldots$ in D and (b_n) in l^1 with $s = \sum_j b_j s_j$. [*Hint*: Let s_1 in D be such that $\|s - s_1\| = a_1 < \frac{1}{2}$. Let s_2 in D be such that $\|(s - s_1)/a_1 - s_2\| = a_2 < \frac{1}{2}$. Then we have $\|s - s_1 - a_1 s_2\| = a_1 a_2 < \frac{1}{4}$. Suppose we have chosen $s_1, s_2, \ldots, s_n$ and $a_1, a_2, \ldots, a_n$ such that $\|s - s_1 - a_1 s_2 - \ldots a_1 a_2 \ldots a_{n-1} s_n\| = a_1 a_2 \ldots a_n < 1/2^n$. Let s_{n+1} in D be such that

$$\left\| \frac{(s - s_1 - a_1 s_2 - \cdots - a_1 a_2 \ldots a_{n-1} s_n)}{(a_1 a_2 \ldots a_n)} - s_{n+1} \right\| = a_{n+1} < \frac{1}{2}.$$

For each $n > 1$ let $b_n = a_1 a_2 \ldots a_{n-1}$, and let $b_1 = 1$.]

(b) Prove that if S is a separable Banach space then there is a continuous linear mapping T from l^1 onto S. Thus every separable Banach space is isomorphic to a quotient of l^1.

29

Spaces of Differentiable Functions

In this section we shall introduce several classes of spaces of differentiable functions defined on a compact interval $I = [a, b]$ or on the compact disk in the complex plane $D = \{z: |z| \leqslant 1\}$. These spaces and their relatives are of interest in the study of differential equations and analytic functions.

The space $C^{(n)}(I)$ consists of all real functions f on $I = [a, b]$ such that the nth derivative $f^{(n)}(t)$ of f exists at each t in (a, b) and has a continuous extension to all of I. If g is a continuous function on $[a, b]$ let $\|g\|_\infty$ denote the $C[a, b]$ norm of g, that is,

$$\|g\|_\infty = \sup\{|g(t)|: a \leqslant t \leqslant b\}.$$

We shall occasionally write $f^{(0)}$ to denote f.

Theorem 29.1: With norm given by the formula

$$\|f\| = \|f\|_\infty + \|f'\|_\infty + \|f''\|_\infty + \cdots + \|f^{(n)}\|_\infty,$$

$C^{(n)}(I)$ is a Banach space that is isomorphic to the space $C(I) \times \mathbb{R}^n$.

Proof: That $\| \ \|$ is a norm on $C^{(n)}(I)$ follows from the linearity of the derivative and the fact that $\| \ \|_\infty$ is a norm. If $f \in C^{(n)}$ then $f' \in C^{(n-1)}$. Thus we can define T from $C^{(n)}$ into $C^{(n-1)} \times \mathbb{R}$ by the formula $Tf = (f', f(a))$. It is clear that T is linear. If $C^{(n-1)} \times \mathbb{R}$ is given the norm $\|(g, b)\| = \|g\|_{n-1} + |g(a)|$, then T is continuous because

$$\begin{aligned} \|f'\|_{n-1} + |f(a)| &= |f(a)| + \|f'\|_\infty + \cdots + \|f^{(n)}\|_\infty \\ &\leqslant \|f\|_\infty + \|f'|_\infty + \cdots + \|f^{(n)}\|_\infty \\ &= \|f\|. \end{aligned}$$

The inverse T^{-1} of T is given by the formula

$$\{T^{-1}(g, c)\}(t) = c + \int_a^t g(s)\, ds.$$

Since we have the inequality

$$\left\| c + \int_a^t g(s)\, ds \right\|_\infty \leqslant |c| + (b-a)\|g\|_\infty$$

it follows that

$$\|T^{-1}(g, c)\| \leqslant (1 + b - a)(|c| + \|g\|)$$

so that T^{-1} is also continuous. Therefore, $C^{(n)}$ is isomorphic as a normed space to $C^{(n-1)} \times \mathbb{R}$. By continuing this process through n steps we conclude that $C^{(n)}$ is isomorphic to $C(I) \times \mathbb{R}^n$. Since $C(I)$ is complete, $\mathbb{R}^n$ is complete and the product of complete normed spaces is complete, it follows that $C^{(n)}$ is a Banach space. ■

Related to the spaces $C^{(n)}(I)$ are the *Sobolev spaces* $W_p^{(n)}(I)$ that consist of all functions f on $I = [a, b]$ such that $f^{(n)}(t)$ exists for almost all t in $[a, b]$ and is in $L^p(I)$. For g a function in $L^p(I)$ let $\|g\|_p$ denote its L^p – norm, that is,

$$\|g\|_p = \left\{ \int_a^b |g(t)|^p\, dt \right\}^{1/p}.$$

The proof of the following theorem resembles that of Theorem 29.1, and we leave it to you, the reader.

Theorem 29.2: For $1 \leqslant p \leqslant \infty$, each space $W_p^{(n)}$ is a Banach space with the norm

$$\|f\| = \|f\|_p + \|f'\|_p + \cdots + \|f^{(n)}\|_p$$

and is isomorphic to the space $L^p(I) \times \mathbb{R}^n$.

If $r > p \geqslant 1$ then we have

$$C(I) \subset L^{\infty}(I) \subset L^{r}(I) \subset L^{p}(I)$$

from which it follows that for each $n = 1, 2, \dots$ we have

$$C^{(n)}(I) \subset W_{\infty}^{(n)}(I) \subset W_{r}^{(n)}(I) \subset W_{p}^{(n)}(I).$$

It is also true that the norm on a smaller space determines a stronger topology on that space than the norm on a larger space so that all of the inclusions are continuous. The following theorem gives another such inclusion result.

Theorem 29.3: If $n > 1$ then $W_{p}^{(n)} \subset C^{(n-1)}$ for all p and the inclusion is continuous.

Proof: If $f \in W_{p}^{(n)}$ then $f^{(n)}$ is in L^{p} so $f^{(n)}$ is in L^{1}. Thus we have the equation

$$f^{(n-1)}(t) = f^{(n-1)}(a) + \int_{a}^{t} f^{(n)}(s)\, ds$$

so $f^{(n-1)}$ is absolutely continuous. This verifies that $f \in C^{(n-1)}$.

For each $1 \leqslant k \leqslant n$, we have

$$|f^{(k-1)}(a) - f^{(k-1)}(t)| \leqslant \int_{a}^{t} |f^{(k)}(s)|\, ds \leqslant \|f^{(k)}\|_{1}.$$

From this it follows that

$$\begin{aligned} |f^{(k-1)}(a)| &= (b-a)^{-1} \int_{a}^{b} |f^{(k-1)}(a)|\, ds \\ &\leqslant (b-a)^{-1} \left\{ \int_{a}^{b} |f^{(k-1)}(a) - f^{(k-1)}(s)|\, ds + \int_{a}^{b} |f^{(k-1)}(s)|\, ds \right\} \\ &\leqslant (b-a)^{-1} \left\{ \int_{a}^{b} \|f^{(k)}\|_{1}\, ds + \|f^{(k-1)}\|_{1} \right\} \\ &= \|f^{(k)}\|_{1} + (b-a)^{-1} \|f^{(k-1)}\|_{1}. \end{aligned}$$

Thus for each t in $[a, b]$ we have

$$\begin{aligned} |f^{(k-1)}(t)| &= \left| f^{(k-1)}(a) + \int_{a}^{t} f^{(k)}(s)\, ds \right| \\ &\leqslant |f^{(k-1)}(a)| + \|f^{(k)}\|_{1} \\ &\leqslant (b-a)^{-1} \|f^{(k-1)}\|_{1} + 2\|f^{(k)}\|_{1} \end{aligned}$$

so that

$$\|f^{(k-1)}\|_{\infty} \leqslant (b-a)^{-1} \|f^{(k-1)}\|_{1} + 2\|f^{(k)}\|_{1}.$$

If we sum from $k = 1$ to n we obtain

$$\begin{aligned} &\|f\|_{\infty} + \|f'\|_{\infty} + \cdots + \|f^{(n-1)}\|_{\infty} \\ &\qquad \leqslant (b-a)^{-1} \|f\|_{1} + (b-a+2)\|f'\|_{1} + \cdots + 2\|f^{(n)}\|_{1} \\ &\qquad \leqslant \{(b-a)^{-1} + 2\}\{\|f\|_{1} + \|f'\|_{1} + \cdots + \|f^{(n)}\|_{1}\}. \end{aligned}$$

This proves the inclusion from $W_1^{(n)}(I)$ into $C^{(n-1)}(I)$ is continuous. We have already noted that the inclusion from $W_p^{(n)}$ into $W_1^{(n)}$ is continuous for $p>1$. ■

Corollary 29.4: For $n \geqslant 1$ the functionals defined on $W_p^{(n)}(I)$ by $F_t^{(j)}(f)=f^{(j)}(t)$ are continuous for all $j<n$.

Theorem 29.5: For each $n \geqslant 1$ the space $W_2^{(n)}$ is a Hilbert space with the inner product

$$\langle f, g\rangle = \sum_{j=0}^{n} \int_a^b f^{(j)} g^{(j)}\, dt.$$

Proof: That $\langle , \rangle$ is an inner product follows from the inner product properties of the integral and the sum. The inner product determines the norm

$$\|f\|_* = \left\{ \sum_{j=0}^{n} \|f^{(j)}\|_2^2 \right\}^{1/2}.$$

Since we obviously have

$$\|f\|_* \leqslant \sum_{j=0}^{n} \|f^{(j)}\|_2, \quad \text{and}$$

$$\sum_{j=0}^{n} \|f^{(j)}\|_2 \leqslant \sqrt{n}\|f\|_*$$

by the Hölder inequality, it follows that the norm $\|f\|_*$ is equivalent to the original norm defined for $W_2^{(n)}$. ■

Suppose $\{A_j(t, s): j=1, 2, \ldots, n\}$ are measurable functions on $[a, b] \times [a, b]$ such that for each f in $W_p^{(n)}$ the function g defined by

$$\begin{aligned} g(t) &= \sum_{j=0}^{n} \int_a^b A_j(t, s) f^{(j)}(s)\, ds \\ &= \int_a^b \sum_{j=0}^{n} A_j(t, s) f^{(j)}(s)\, ds \end{aligned} \tag{29–1}$$

is in $W_r^{(m)}$. The operator $Tf=g$ is linear because of the linearity of the integral and the sum. We shall prove that such operators must also be continuous. First note, however, that the Integration by Parts Formula shows that if $j<n$ we have

$$\begin{aligned} \int_a^b A_j(t, s) f^{(j)}(s)\, ds &= \left\{ \left(\int_a^s A_j(t, r)\, dr \right) f^{(j)}(s) \right\}_{s=a}^{b} \\ &\quad - \int_a^b \left\{ \left(\int_a^s A_j(t, r)\, dr \right) f^{(j+1)}(s) \right\} ds \\ &= \left\{ \int_a^b A_j(t, r)\, dr \right\} f^{(j)}(b) \\ &\quad - \int_a^b \left\{ \int_a^s A_j(t, r)\, dr \right\} f^{(j+1)}(s)\, ds. \end{aligned}$$

We can continue to reduce formula (29–1) until only $f^{(n)}$ appears within an integral. At that point we obtain the equivalent formula

$$g(t) = \sum_{j=0}^{n-1} f^{(j)}(b)u_j(t) + \int_a^b B(t, s)f^{(n)}(s)\, ds \qquad \textbf{(29–2)}$$

where each u_j is in $W_r^{(m)}$. By Corollary 29.4, $f^{(j)}(b)$ is continuous on $W_p^{(n)}$ for each $j < n$. Therefore, we conclude that $\sum_{j=0}^{n-1} f^{(j)}(b)u_j(t)$ depends continuously on f. It remains to show if $\int_a^b B(t, s)f^{(n)}(s)\, ds$ is in $W_q^{(r)}$ for all f in $W_p^{(n)}$ then the integral depends continuously on f. By the way we defined the norm on $W_p^{(n)}$ the correspondence from f to $f^{(n)}$ is continuous from $W_p^{(n)}$ into L^p. We define the mapping S from L^p into $W_q^{(r)}$ by the formula

$$Sh(t) = \int_a^b B(t, s)h(s)\, ds.$$

Since for each h in L^p there are several f in $W_p^{(n)}$ for which $f^{(n)} = h$. It follows that Sh is in $W_r^{(m)}$ for each h in L^p. We shall employ the Closed Graph Theorem to prove that S is continuous. To that end, suppose $h_n \to h$ in $W_p^{(n)}$ and $Sh_n \to v$ in $W_r^{(m)}$. For each t the integral $\int_a^b B(t, s)w(s)\, ds$ exists for each w in L^p. This implies that $B(t, s)$ is in L^q $(1/p + 1/q = 1)$ for each t and that the functional $F_t(w) = \int_a^b B(t, s)w(s)\, ds$ is continuous on L^p. From this we conclude that Sh_n converges almost everywhere to Sh. By Fatou's Lemma (12.11) we conclude

$$\begin{aligned}\int_a^b |Sh - v|\, ds &\leq \lim_n \int_a^b |Sh_n - v|\, ds \\ &= \lim_n \|Sh_n - v\|_1 \\ &\leq \lim_n \|Sh_n - v\| \quad (\text{norm of } W_r^{(m)}) \\ &= 0.\end{aligned}$$

This proves that $Sh = v$ so that S is a closed, hence continuous, operator from L^p into $W_r^{(m)}$. By composition of continuous operators it follows that $\int_a^b B(t, s)f(s)\, ds$ depends continuously upon f. We present these conclusions in the following theorem.

Theorem 29.6: Suppose $g = Tf$ is given by (29–1) or equivalently by (29–2). If Tf is in $W_r^{(m)}$ for each f in $W_p^{(n)}$ then the operator S is continuous from $W_p^{(n)}$ into $W_r^{(m)}$.

Denote by D the unit disc in the complex plane, that is, the set of all complex z with $|z| \leq 1$; by D° the interior of D, which is the set of all z with $|z| < 1$ and by ∂D the boundary of D, the set of all z with $|z| = 1$. The set of all complex functions f on D, which are analytic on D° and continuous on D, is denoted by $A(D)$. With the norm

$$\|f\| = \sup\{|f(z)| : z \in D\}$$

$A(D)$ is a Banach space; for suppose f is the uniform limit of a sequence $\{f_n\}$ in

$A(D)$. It first follows that f is continuous on D by the completeness of $C(D)$. For every closed contour Γ in D° we have $\int_\Gamma f_n(z)\,dz = 0$ by the Cauchy Integral Theorem. Thus $\int_\Gamma f(z)\,dz = 0$ so by Morera's Theorem, f is analytic at each z in D°.

Exercises

A

1. *(a)* Prove that if $r > p \geqslant 1$ then $C(I) \subset L^\infty(I) \subset L^r(I) \subset L^p(I)$.
   (b) Show that if f is in $C(I)$ then the norm of f as a member of $C(I)$ is the same as the norm of f as a member of $L^\infty(I)$.
   (c) Show that if f is in $L^\infty(I)$ then $\|f\|_r \leqslant (b-a)^{1/r}\|f\|_\infty$ where $I = [a, b]$.
   (d) Show that if f is in $L^r(I)$ and $p < r$ then $\|f\|_p \leqslant (b-a)^d \|f\|_r$ where $d = (r-p)/rp$.
2. Use Exercise 1 to show that $C^{(n)}(I) \subset W_\infty^{(n)}(I) \subset W_r^{(n)}(I) \subset W_p^{(n)}(I)$ and that the inclusions are continuous.
3. *(a)* Prove that the function defined on $C^{(n)}[a, b]$ by the formula

$$F(f) = \sum_{j=1}^{n-1} c_j f^{(j)}(a) + \int_a^b f(t)\,dg(t),$$

   where g is in $BVN[a, b]$ is a continuous linear functional on $C^{(n)}[a, b]$, and every continuous linear functional on $C^{(n)}[a, b]$ can be described in this way.
   (b) Suppose $1 < p < \infty$ and let $1/p + 1/q = 1$. Show that if g is in $W_q^{(n)}$ the function F defined on $W_p^{(n)}$ by the formula

$$F(f) = \sum_{j=1}^{n-1} f^{(j)}(a) g^{(j)}(a) + \int_a^b f^{(n)} g^{(n)}\,dm$$

   is a continuous linear functional on $W_p^{(n)}$ and every continuous linear functional on $W_p^{(n)}$ can be so obtained.
   (c) Determine a representation for the dual space of $W_1^{(n)}$.
4. *(a)* Prove the norm $\|f\|_*$ on $C^{(n)}[a, b]$ given by the formula

$$\|f\|_* = |f(a)| + |f'(a)| + \cdots + \|f^{(n)}\|$$

   is equivalent to the norm described in Theorem 29.1.
   (b) Show that the norm on $W_p^{(n)}$ given by $|f(a)| + |f'(a)| + \cdots + |f^{(n-1)}(a)| + \|f^{(n)}\|_p$ is equivalent to the norm described in Theorem 29.2.
   (c) Determine other equivalent norms for $C^{(n)}$ and $W_p^{(n)}$ in terms of the derivatives at other points in $[a, b]$.
5. Prove that if $\|\ \|_*$ is any norm on $C^{(n)}(I)$ or $W_p^{(n)}(I)$ that makes the space a Banach space and such that the inclusion from the space into $L^1(I)$ is continuous, then $\|\ \|_*$ is equivalent to the original norm on the space.
6. For f in $L^1(I)$ define Sf by $Sf(t) = \int_a^t f(s)\,ds$.
   (a) Prove that S is an isomorphism from $L^1(I)$ onto the closed subspace of $W_1^{(1)}$, which consists of all g in $W_1^{(1)}$ with $g(a) = 0$.
   (b) Show that S is an isomorphism from $L^p(I)$ onto a closed subspace of $W_p^{(1)}$ consisting of all g with $g(a) = 0$.

(c) Prove that S^n is an isomorphism from $L^p(I)$ onto the closed subspace $Z_p^{(n)}$ of $W_p^{(n)}$, which consists of all g with $g(a) = g'(a) = \ldots = g^{(n-1)}(a) = 0$.

(d) Suppose $\{f_j\}$ is an s-basis for $L^p(I)$. Show that the sequence $\{1, x-a, (x-a)^2, \ldots, (x-a)^{n-1}\} \cup \{S^n f_j\}$ is an s-basis for $W_p^{(n)}(I)$.

7. Define spaces analogous to $W_p^{(n)}(I)$ and $C^{(n)}(I)$ for a rectangle in $\mathbb{R}^2$ and prove these spaces are Banach spaces. Define these spaces over an arbitrary set S in $\mathbb{R}^n$ such that S is the closure of a nonempty open set.

B

8. (a) Show that $A(D)$ is a closed subspace of $C(D)$.

(b) If $f \in A(D)$ define the functions f_r on D for $0 < r < 1$ by $f_r(z) = f(rz)$. Show that for each f in $A(D)$ we have $\lim_{r \uparrow 1} f = f$.

(c) Show that if $\sum_{n=0}^{\infty} a_n z^n$ is the Taylor series for f then $f_r = \sum_{n=0}^{\infty} a_n r^n z^n$ is the Taylor series for f_r, and this series converges absolutely to f_r in $A(D)$.

(d) Show that a continuous linear functional F on $C(D)$ is 0 on $A(D)$ if and only if $F(z^n) = 0$ for $n = 0, 1, 2, \ldots$.

(e) Use the Hahn-Banach Theorem to show that $A(D)$ is the closed linear span of $\{z^n : n = 0, 1, 2, \ldots\}$ in $C(D)$.

9. (a) Suppose C is a closed convex subset of $W_2^{(n)}$ and f is a function in $W_2^{(n)}$. Show that there is a unique g in C such that $\|f - g\|$ is a minimum.

(b) Suppose F is a closed subspace of $W_2^{(n)}(I)$ and $t_1, t_2, t_3, \ldots, t_n$ are points of I such that (*) for each j there is t_j in F such that $g_j(t_j) \neq 0$ and $g_j(t_i) = 0$ for $i \neq j$. Show that for each f in $W_2^{(n)}$ there is a unique g in F such that $g(t_i) = f(t_i)$ for $i = 1, 2, \ldots, n$; and $\|f - g\|$ is a minimum.

(c) A real valued function f on I is called *convex* if for each s, t in I, and $0 < a < 1$ we have $f(as + (1-a)t) \leqslant af(s) + (1-a)f(t)$. Show that if F is a subspace of $W_2^{(n)}$ then the set C of convex functions in F is a closed convex subset of F.

(d) Prove that if f is a function in $W_2^{(n)}$ that is not convex, then there is a convex function g in $W_2^{(n)}$ such that $\|f - g\|$ is a minimum.

10. Suppose f is in $C^{(n)}(I)$ and $\{g_j\}$ is a sequence of polynomials that converges in $C(I)$ to f. Let f_j be the polynomial that has nth derivative g_j and such that $f_j^{(i)}(a) = f^{(i)}(a)$ for $i < n$, where $I = [a, b]$. Prove that $\{f_j\}$ converges to f in $C^{(n)}(I)$. Thus the space of polynomials is dense in $C^{(n)}(I)$.

11. Prove that the space of polynomials on I is dense in $W_p^{(n)}(I)$ for all p and all I.

12. The sequence $\{1, \cos nx, \sin nx : n = 1, 2, \ldots\}$ is a complete orthogonal sequence in $W_2^{(n)}$. [*Hint*: Show that if $\langle f, h \rangle = 0$ for each member h of this sequence, then $\langle f, g \rangle = 0$ for each polynomial g, and use Exercise 10.]

6

Applications

Divergent series are a creation of the devil and one should base upon them no demonstration whatsoever. N. H. Abel
The series is divergent; therefore, we may be able to do something with it. O. Heaviside

This chapter contains a sampler of the numerous applications of abstract analysis to other branches of mathematics. At first glance it may disappoint someone looking for "real world" applications, but with further study and experience the reader will learn that much of the mathematics in this chapter is as useful in some specific problem as arithmetic is in preparing income tax returns. In fact, acquiring knowledge is like learning how to swim: it is pleasant to learn for its own sake; you may only need to use it once or twice in your life; but when you need it, nothing else will replace it.

In Section 30 we define the Gateaux differential of a real valued function on a normed space, and we use this concept to derive the Euler-Lagrange equations, which are the foundation of the calculus of variations. This requires little more functional analysis than the concept of normed space.

The next two sections, 31 and 32, concern the existence of solutions to differential equations. In Section 31 we prove the Picard Theorem on existence and uniqueness, which is a nice illustration of Banach's Contraction Mapping Theorem. In Section 32 we prove Peano's Existence Theorem, which is based on the Arzela-Ascoli Theorem.

Section 33 describes the application of functional analysis to summability theory. We prove the Silverman-Toeplitz Theorem, which describes when a summability method preserves the limit of convergent sequences. Then we describe some of the main classes of summability methods such as the Cesaro and Hausdorff means.

Section 34 shows how the operational calculus of O. Heaviside can be placed upon a firm theoretical footing using the integral operator instead of the Laplace transform. The results are theoretical and computational.

The development of a theory of compact operators was an early triumph of functional analysis. We describe the essentials of this theory in Section 35. Among the main results are the theorem of Schauder, which states that an operator is compact if and only if its dual operator is as well; and the spectral theorem for compact operators, which asserts that the spectrum of a compact operator is a countable set with no limit point other than 0.

Section 36 is a brief introduction to ergodic theory. One striking feature of this subject is the clever combination of measure theory and functional analysis to draw the conclusion of ergodicity. In this section we present two types of ergodic theorem: an individual ergodic theorem and a mean ergodic theorem.

30

The Euler-Lagrange Equation

Suppose F is a function from a subset of a normed space E into the scalar field, x is in the interior of the domain of F, and v is any point of E. If the limit

$$dF(x; v) = \lim_{t\to 0} \frac{F(x+tv) - F(x)}{t}$$

exists it is called the *Gateaux differential of F at x in the direction v*. The differential may exist for some v and not exist for others. If the differential exists at x for all v in E then we say F is *Gateaux differentiable* at x. We assume x is in the interior of the domain of F so that $F(x+tv)$ will be defined for sufficiently small t. The reader may show that $dF(x; v)$ is the derivative of the real valued function ϕ at 0 where $\phi(t) = F(x+tv)$.

Example 30.1:

Let G be an integrable real valued function on $[0, 1] \times [0, 1]$ and define F from $C[0, 1]$ into $\mathbb{R}$ by

$$F(f) = \int_0^1 G(f(s), s)\, ds.$$

Then for g in $C[0, 1]$ we have

$$\begin{aligned} dF(f; g) &= \lim_{t\to 0} (1/t)\left\{\int_0^1 G(f(s)+tg(s), s)\, ds - \int_0^1 G(f(s), s)\, ds\right\} \\ &= \lim_{t\to 0} \int_0^1 \left\{\frac{G(f(s)+tg(s), s) - G(f(s), s)}{t}\, ds\right\} \\ &= \int_0^1 G_1(f(s), s)g(s)\, ds, \end{aligned}$$

provided that G has a partial derivative with respect to its first variable.

A real valued function f on a normed space E is said to *have a local maximum at x in E* if there is a neighborhood U of x such that $f(u) \leqslant f(x)$ for all $u \neq x$ in U. The concept of local minimum has an analogous definition.

Theorem 30.2: Suppose the real valued function F is Gateaux differentiable on the normed space E. If F has a local maximum or local minimum at $x \in F$, then $dF(x; v) = 0$ for all v in E.

Proof: If $dF(x; v) > 0$, then there is $\varepsilon > 0$ such that

$$\frac{\{F(x+tv) - F(x)\}}{t} > 0 \quad \text{for } |t| < \varepsilon.$$

If $t > 0$ then $F(x+tv) > F(x)$ for $|t| < \varepsilon$ so F cannot have a local maximum at x; if $t < 0$ then $F(x+tv) < F(x)$ for $|t| < \varepsilon$ so F cannot have a local minimum at x. A corresponding argument shows that F cannot have a local maximum or minimum at x if $dF(x; v)$ is negative. ■

The classical problem of the calculus of variations is stated as follows: Suppose G is a real valued function on $\mathbb{R}^3$ that has continuous second partial derivatives with respect to the first two of its arguments; that is, if $G = G(x, y, z)$, we mean $\partial^2 G/\partial x^2$, $\partial^2 G/\partial x \partial y$, and $\partial^2 G/\partial y^2$. Let $C_0^{(1)}[0, 1]$ consist of all f in $C^{(1)}[0, 1]$ with $f(1) = f(0) = 0$. Determine that function f in $C_0^{(1)}[0, 1]$ that minimizes the functional

$$J(f) = \int_0^1 G(f(s), f'(s), s)\, ds.$$

The equation in the following theorem is called the *Euler-Lagrange equation.*

Theorem 30.3: A necessary condition for f in $C_0^{(1)}[0, 1]$ to minimize the functional J is that

(E–L) $(d/dt)G_2(f, f', t) - G_1(f, f', t) = 0$

Proof: By Theorem 30.2 we know that if J has a minimum at f then $dJ(f; g) = 0$ for all g in $C_0^{(1)}$. The Gateaux differential is calculated as follows

$$\begin{aligned} dJ(f; g) &= \left[\frac{d}{dt} J(f+tg)\right]_{t=0} \\ &= \left[\frac{d}{dt}\int_0^1 G(f+tg, f'+tg', s)\, ds\right]_{t=0} \\ &= \int_0^1 \{G_1(f, f', s)g(s) + G_2(f, f', s)g'(s)\}\, ds. \end{aligned}$$

If $(d/dt)G_2$ is continuous then we can integrate the second term above by parts to obtain

$$\begin{aligned} \int_0^1 G_2(f, f', s)g'(s)\, ds &= [G_2 g]_0^1 - \int_0^1 \frac{d}{dt} G_2(f, f', s)g(s)\, ds \\ &= -\int_0^1 \frac{d}{dt} G_2(f, f', s)g(s)\, ds. \end{aligned}$$

Setting $dJ(f; g)$ to zero we obtain

$$\int_0^1 \left\{G_1(f, f', s) - \frac{d}{dt} G_2(f, f', s)\right\} g(s)\, ds = 0$$

for all g in $C_0^{(1)}$. Denote the quantity in the braces by $H(t)$. If there were an interval (a, b) on which $H(t) > 0$ let g_{ab} *be* defined as follows

$$g_{ab}(t) = 0 \quad \text{for } 0 \leqslant t < a \text{ and } b \leqslant t < 1$$

$$g_{ab}(t) = (t-a)^2 (t-b)^2 \quad \text{on } [a, b].$$

Then g_{ab} is in $C_0^{(1)}[0, 1]$ and $\int_0^1 H(s)g(s)\, ds > 0$, which is a contradiction. Therefore, $H(s) \leqslant 0$ for all s in $[0, 1]$; similarly we can show $H(s) \geqslant 0$ for all s in $[0, 1]$ so that H is identically 0 on $[0, 1]$. ■

If we want to solve the problem: find f in $C^{(1)}[0, 1]$ such that $f(0) = 0$, $f(1) = b$ and f minimizes $J(f)$ we can make the substitution $g = f - bt$. Then g is in $C_0^{(1)}[0, 1]$ and minimizes

$$\int_0^1 G(g + bt, g' + b, t)\, dt$$

so that Theorem 30.3 requires that

$$\frac{d}{dt} G_2(g + bt, g' + b, t) - G_1(g + bt, g' + b, t) = \frac{d}{dt} G_2(f, f', t) - G_1(f, f', t) = 0,$$

which shows the Euler-Lagrange equation is still valid but now with the boundary conditions $f(0) = 0$, $f(1) = b$. Analogously we can show that the Euler-Lagrange equation is necessary for minimizing $J(f)$ with the conditions $f(0) = a$, $f(1) = b$.

Example 30.4:

One of the most famous problems in the calculus of variations is the following: A mass slides on a wire connecting points P and Q in a vertical plane. For what shape of wire is the time of descent a minimum?

Take P as the origin $x = y = 0$. During the motion energy is conserved so we have $\frac{1}{2}mv^2 = mgy$ where m is the mass, v is the velocity, and g is the acceleration due to gravity. Hence, v is $\sqrt{2gy}$, and the problem is to find the curve $y = y(x)$ that minimizes

$$J(y) = \int_P^Q dt = \int_P^Q \frac{ds}{v} = \int_0^a \left\{\frac{[1 + (y')^2]}{[2gy]}\right\}^{1/2} dx$$

with $y(0) = 0$ and $y(a) = b$. Since

$$G(y, y') = \left\{\frac{[1 + (y')^2]}{[2gy]}\right\}^{1/2}$$

is not explicitly a function of x then if y satisfies the Euler-Lagrange equation we shall have

$$0 = y'((G_2)' - G_1) = y''G_2 + y'(G_2)' - y'G_1 - y''G_2$$
$$= \frac{d}{dx}\{y'G_2 - G\},$$

which implies that $y'G_2 - G$ is constant. By direct calculation we have $G_2 = y'/(2gyG)$ so the equation becomes

$$(y')^2/(2gyG) - G = C$$
$$(y')^2 - 2gy(G)^2 = C(2gyG)$$
$$(y')^2 - (1 + (y')^2) = C\{2gy[(1 + (y')^2]\}^{1/2}$$

that we can transform into

$$y(1 + (y')^2) = B^2$$

where $B^2 = -1/(2gC^2)$. (Note that g is negative.) We solve this equation for y' to obtain

$$y' = \left\{\frac{B^2 - y}{y}\right\}^{1/2}$$

and then separate variables to find

$$dx = \left\{\frac{y}{B^2 - y}\right\}^{1/2} dy.$$

If we make the substitution $y = B^2 \sin^2 u$, we can integrate the equation above to obtain

$$x = B^2 \frac{2u - \sin 2u}{2}$$

where B^2 is determined so that $y(a) = b$. The pair of equations

$$x = B^2 \frac{2u - \sin 2u}{2}$$
$$y = B^2 \sin^2 u$$

are parametric equations for the curve known as the *Cycloid*.

The Euler-Lagrange equation can be extended to a necessary condition for f to minimize the functional

$$J(f) = \int_0^1 G(f_1(t), \ldots, f_n(t), f'_1(t), \ldots, f'_n(t), t)\, dt$$

subject to $f(0) = a$, $f(1) = b$. Namely we must have

$$\frac{d}{dt} G_{n+j} - G_j = 0 \quad \text{for } j = 1, 2, \ldots, n$$

as can be proved by a straightforward induction.

Exercises

A

1. Show that $dF(x; v)$ is the derivative at 0 of the function ϕ defined by $\phi(t) = F(x + tv)$.
2. Find the Gateaux differentials of the following functionals F defined on $C[0, 1]$.
   (a) $F(f) = \int_0^1 (f(t))^2 \, dt$.
   (b) $F(f) = \{\int_0^1 f(t) \, dt\}^2$.
   (c) $F(f) = f(0) + f(\frac{1}{2})^2 + 2f^3(1)$.
3. Find the function f that maximizes or minimizes $J(f)$ for
   (a) $J(f) = \int_0^1 f(s)f'(s) \, ds$, $f(0) = 0$, $f(1) = 1$;
   (b) $J(f) = \int_a^b \{1 + (f(s)')^2\}/f(s) \, ds$, $f(a) = f(b) = 1$.
4. Find the Euler-Lagrange equations for the functional

$$J(f) = \int_{\mathscr{G}} f_1^2 + f_2^2 + fF(x, y) \, dx \, dy$$

   where $f = f(x, y)$ is a function of two variables.

B

5. If $f = (f_1, f_2, \ldots, f_n)$ minimizes the functional

$$J(f) = \int_0^1 G(f_1, \ldots, f_n, f'_1, \ldots, f'_n, t) \, dt$$

   show that $(d/dt) \, G_{n+j} - G_j = 0$ for $j = 1, 2, \ldots, n$.
6. A science fiction writer suggested that one could travel from one location upon the earth to another by digging straight through the earth between the two locations. A vehicle dropping into the hole would fall downward for half of the distance and then reemerge using the acquired kinetic energy.
   (a) Calculate the time for such a trip, neglecting friction and show the trip would be faster if the path between the two locations were a cycloid.
   (b) Show that the cycloid would make the trip between two points in the shortest time.
7. Suppose an electric supplier has two methods of generation: a low-cost method to produce a constant load at cost $L(x)$ and a high-cost method that can be quickly brought on line to produce electricity at cost $H(x)$. Assume L and H are increasing and L is differentiable. Show that if the demand for electricity at time t, $0 \leqslant t \leqslant 1$, is $F(t)$ then the minimum cost is realized if the base (constant) load x satisfies

$$L'(x) = \int_0^1 H'(F(s) - x) \, ds.$$

8. Find an Euler-Lagrange equation that is necessary to minimize the functional

$$J(f) = \int_0^1 G(f, f', f'', t) \, dt \quad \text{for } f \text{ in } C^{(2)}[0, 1]$$

   by calculating the Gateaux differential of J and setting it equal to 0.

9. (a) For a given function f such that $f(0)=a_1$, $f(1)=b_1$, show that the function g such that $g(0)=a_2$, $g(1)=b_2$, and for which the functional

$$J(f, g)=\int_0^1 \{f'(t)^2-g'(t)^2\}^{1/2}\, dt$$

is a minimum is given by

$$g(t)=\left\{\frac{b_2-b_1}{a_2-a_1}\right\}\{f(t)-a_1\}+b_1.$$

(b) In general $J(f, g)$ is minimized without regard for terminal conditions when $g=Af+B$.

(c) Find the minimum of the functional

$$H(f)=\int_0^1 \{f'(t)^2+f''(t)^2\}^{1/2}$$

such that $f(0)=1$, $f(1)=e$, and $f'(0)=1$. What will $f'(1)$ have to be?

(d) Find the minimum of $H(f)$ when $f(0)=0$ and $f(1)=1$.

31

Existence Theorems I Picard's Theorem

A striking application of the Contraction Mapping Theorem, Theorem 9.5 on page 62 is the proof of *Picard's Theorem* on the existence of a unique local solution to the differential equation

(DE) $dy/dx=f(x, y)$, $y(x_0)=y_0$,

which we give below as Theorem 31.1. That y is a local solution of the above equation means that there is $\delta>0$ such that $y'(x)=f(x, y(x))$ for all x in a nbd $(x_0-\delta, x_0+\delta)$ of x_0.

Theorem 31.1: Suppose f is a continuous real valued function on an open subset D of $\mathbb{R}^2$ for which there is $M>0$ such that

(L) $|f(x, y_1)-f(x, y_2)|\leqslant M|y_1-y_2|$

for all (x, y_1), (x, y_2) in D. Then for every (x_0, y_0) in D there is $\delta>0$ and a unique differentiable function $y(x)$ defined on the interval $(x_0-\delta, x_0+\delta)$ such that $y(x_0)=y_0$ and

$$y'(x)=f(x, y(x)) \quad \text{for all } x \text{ in } (x_0-\delta, x_0+\delta).$$

Proof: Since D is open there is $\delta_1>0$ with

$$D_1=\{(x, y)\colon \|(x, y)-(x_0, y_0)\|\leqslant\delta_1\}\subset D.$$

Now D_1 is compact so there is $B > 1$ such that $|f(x, y)| \leqslant B$ for (x, y) in D_1. Let $\delta > 0$ be such that

(a) the set D_1 contains the rectangle $\{(x, y): |x - x_0| \leqslant \delta, |y - y_0| \leqslant \delta\}$ and
(b) $\delta < 1/M$.

Let S be the set of all continuous functions ϕ on the interval $I = [x_0 - \delta, x_0 + \delta]$ such that $\phi(x_0) = y_0$ and $|\phi(x) - y_0| \leqslant B\delta$ for all x in I. It is not hard to see that S is a closed subset of $C(I)$ and hence a complete metric space.

Define a function T on S by the formula

$$(T\phi)(x) = y_0 + \int_{x_0}^{x} f(t, \phi(t))\, dt \qquad \phi \in S,\ x \in I.$$

To verify that T maps S into S first note that $(T\phi)(x_0) = y_0$, that $T\phi$ is continuous (in fact, differentiable), and for $x \in I$

$$\begin{aligned} |(T\phi)(x) - y_0| &= \left| \int_{x_0}^{x} f(t, \phi(t))\, dt \right| \\ &\leqslant \int_{x_0}^{x} B\, dt \leqslant B\delta. \end{aligned}$$

Next we show that T is a contraction. If ϕ_1 and ϕ_2 are in S then for all x in I we have

$$\begin{aligned} |(T\phi_1)(x) - (T\phi_2)(x)| &= \left| \int_{x_0}^{x} f(t, \phi_1(t)) - f(t, \phi_2(t))\, dt \right| \\ &\leqslant \left| \int_{x_0}^{x} f(t, \phi_1(t)) - f(t, \phi_2(t)) \right| dt \\ &\leqslant M\delta \|\phi_1 - \phi_2\| \end{aligned}$$

where $\|\phi_1 - \phi_2\| = \sup\{|\phi_1(t) - \phi_2(t)|: t \in I\}$. Taking the supremum over all t in I we conclude that

$$\|T\phi_1 - T\phi_2\| \leqslant M\delta \|\phi_1 - \phi_2\|,$$

which implies that T is a contraction. Therefore, by Theorem 9.5 there is a unique y in S such that $Ty = y$. This means that

$$y(x) = y_0 + \int_{x_0}^{x} f(t, y(t))\, dt$$

for each x in $(x_0 - \delta, x_0 + \delta)$so by the Fundamental Theorem of Calculus we have

$$y'(x) = f(x, y(x)) \quad \text{for } |x_0 - x| < \delta. \ \blacksquare$$

Condition (L) above is called a *Lipschitz condition.* Thus (L) means that on the open set D, f satisfies a Lipschitz condition in y. If y is a local solution for (DE) on the interval $(x_0 - \delta, x_0 + \delta)$, then the set of points $\{(x, y(x)): |x_0 - x| < \delta\}$ is a curve contained in the domain D. The question is, how far can the solution curve be extended? The next theorem shows that if D is connected, the curve can be extended between boundaries of D.

Theorem 31.2: Suppose f is continuous on a bounded open connected set D and satisfies the Lipschitz condition (L) on D. Then for every (x_0, y_0) in D there is a unique solution of the differential equation $dy/dx = f(x, y)$ such that $y_0 = y(x_0)$ and such that the curve $y = y(x)$ runs between the boundaries of D.

Proof: By Theorem 31.1 we know that there are real numbers $s < x_0 < t$ such that the differential equation has a unique solution y on (s, t) with $y(x_0) = y_0$. Let u be the infimum of all such s, and v be the supremum of all such t. If $u < s < x_0$, let s_1 be some number between u and s. Since there is a unique solution y on the interval (s_1, t_1) for some $t_1 > x_0$ it follows that this solution is defined for s. Therefore, there is a unique solution y on the interval (u, v) with $y(x_0) = y_0$.

It remains to show that the closures of the two sets of points

$$\{(x, y(x)): u < x \leqslant x_0\} \quad \text{and} \quad \{(x, y(x)): x_0 \leqslant x < v\}$$

do not lie in D. This will show that both ends of the curve $\{(x, y(x)): u < x < v\}$ approach the boundary of D. We discuss only the first set. Suppose $\{s_n\}$ converges to u and $y(s_n)$ converges to w. If (u, w) were in D we could find $\varepsilon > 0$ such that the set

$$\{(x, y): |u - x| < \varepsilon, |w - y| < \varepsilon\}$$

is contained in D. Suppose k is such that

$$u < s_k < u + \frac{\varepsilon}{4} \quad \text{and} \quad |w - y(s_k)| < \frac{\varepsilon}{4}.$$

By an argument that imitates the proof of Theorem 31.1, we can show that there is a unique solution z of the given equation on the interval $(s_k - \varepsilon/4, s_k + \varepsilon/4)$ such that $z(s_k) = y(s_k)$. But this means that z coincides with y on the interval (u, v) and extends this solution to $(s_k - \varepsilon/4, s_k + \varepsilon/4) \cup (u, v)$ contradicting the definition of u as the infimum of all such s. Therefore, it cannot be the case that (u, w) is in D so that the closure of the first set is not in D. ■

Exercises

A

1. A real function f on a subset S of R is said to satisfy a *Lipschitz condition* (*with constant* M) on S if there is $M > 0$ such that $|f(x) - f(y)| \leqslant M|x - y|$ for all x, y in S.
   (a) Show that if f satisfies a Lipschitz condition on S then f is continuous on S.
   (b) Show that if f has a continuous derivative on an open set containing $[a, b]$ then f satisfies a Lipschitz condition on $[a, b]$.
   (c) Prove that $x \sin(1/x)$ is continuous but does not satisfy a Lipschitz condition on $[0, 1]$.

2. *(a)* Show that if f_1 satisfies a Lipschitz condition on $[a, b]$ with constant M_1 and f_2 satisfies a Lipschitz condition on $[a, b]$ with constant M_2 then $f_1 + f_2$ satisfies a Lipschitz condition on $[a, b]$ with constant $M_1 + M_2$.
   (b) Prove that the set of all functions f that satisfies a Lipschitz condition on $[a, b]$ forms a linear space of functions.
3. Let $\mathscr{Y}$ denote the collection of all curves $\{(x, y(x)): u < x < v\}$ where y satisfies (DE) and the curve runs between the boundaries of the open connected set D in the sense of Theorem 31.2. Show that every point of D is contained in precisely one member of $\mathscr{Y}$.

B

4. Show that the S of all functions f on $[a, b]$ that satisfies a Lipschitz condition with $M \leqslant K$ and with $\|f\| \leqslant K$ forms a compact subset of $C[a, b]$. [*Hint*: Use the Arzela-Ascoli Theorem (24.6).]
5. Prove that the set of all functions f on $[a, b]$ that satisfies a Lipschitz condition forms a Banach space with the norm

$$\|f\|_L = \sup\{|f(x) - f(y)|/|x - y|: x \neq y;\ x, y \in [a, b]\} + \sup\{|f(x)|: x \in [a, b]\}.$$

6. Suppose K is a continuous real valued function on a square $\{(x, y): a \leqslant x \leqslant b, a \leqslant y \leqslant b\}$, M is the maximum of $|K(x, y)|$ on this square, and g is continuous on $[a, b]$. If $|c| < 1/\{M(b-a)\}$ show there is a unique solution to the integral equation

$$f(x) = c \int_a^b K(x, y) f(y)\, dy + g(x), \quad x \in [a, b]$$

by showing that the function T from $C[a, b]$ into itself defined by

$$(T\phi)(x) = c \int_a^b K(x, y) f(y)\, dy + g(x)$$

is a contraction.
7. Prove that if $0 < r < 1$ and g is continuous on $[0, r]$ then there is a unique f in $C[0, r]$ that satisfies the functional equation

$$f(x) = g(x) + \cos(xf(x)).$$

[*Hint*: First use the Mean Value Theorem to show that

$$|\cos u - \cos v| \leqslant |u - v| \text{ for all } u, v.]$$

32

Existence Theorems II The Peano Existence Theorem

Whereas Picard's existence theorem asserts that the differential equation

(DE) $dy/dx = f(x, y),\ y(x_0) = y_0$

has a unique solution on a domain D where $f(x, y)$ is continuous and satisfies a Lipschitz condition in y, the *Peano Existence Theorem* requires only that f be continuous on D. However, the Peano Theorem does not assert the uniqueness but only the existence of a solution. For purposes of applications, uniqueness is at least as important as existence. We shall prove the Peano Existence Theorem as an application of the Arzela-Ascoli Theorem (24.6).

Theorem 32.1: Suppose f is a continuous real valued function on an open subset D of $\mathbb{R}^2$. For every (x_0, y_0) in D there is $\varepsilon > 0$ and a differentiable function $y(x)$ on the interval $(x_0 - \varepsilon, x_0 + \varepsilon)$ such that $y(x_0) = y_0$ and

$$y'(x) = f(x, y(x)) \quad \text{for all } x \text{ in } (x_0 - \varepsilon,\ x_0 + \varepsilon).$$

Proof: Since D is open there is $c > 0$ such that the square

$$S = \{(x, y)\colon |x_0 - x| \leqslant c \quad \text{and} \quad |y_0 - y| \leqslant c\}$$

is contained in D. Since S is compact and f is continuous on S there is $M > 1$ such that $|f(x, y)| < M$ for all (x, y) in S. Let $\varepsilon = c/M$, and denote $(x_0 - \varepsilon, x_0 + \varepsilon)$ by I. The next step is to construct a sequence $\{\psi_n\}$ of approximate solutions of (DE) whereby:

(1) Each ψ_n is a piecewise linear and continuous function on I.

This means that there is a partition $x_0 - \varepsilon = u_0 < u_1 < \ldots < u_m = x_0 + \varepsilon$ of I such that ψ_n is linear on $[u_{i-1}, u_i]$.

(2) $|\psi_n'(x) - f(x, \psi(x))| < 1/n$ if $x \in I$ is not one of the points in the partition.

(3) $\psi_n(x_0) = y_0$.

We show how to construct ψ_n on the interval $[x_0, x_0 + \varepsilon]$; a symmetric construction works on $[x_0 - \varepsilon, x_0]$. Since f is uniformly continuous on S there is $\delta > 0$ such that

$$|f(s, t) - f(u, v)| < \frac{1}{n} \quad \text{when } |s - u| \quad \text{and} \quad |t - v| < \delta.$$

Let $x_0 = u_0 < u_1 \ldots < u_k = x_0 + \varepsilon$ be such that $|u_i - u_{i-1}| < \delta/M$ for $i = 1, 2, \ldots, k$. Let $\psi_n(x_0) = y_0$ and for x in $[u_0, u_1]$ let $\psi_n(x)$ be linear with slope $f(x_0, y_0)$, that is,

$$\psi(x) = f(x_0, y_0)(x - x_0) + y_0.$$

Since ψ_n is to be continuous this determines $\psi_n(u_1)$. On $[u_1, u_2]$ define ψ_n to be linear with slope $f(u_1, \psi_n(u_1))$. Continue this process until you reach u_k. It is clear that ψ_n satisfies (1) and (3). That it satisfies (2) follows because if x is in $[u_{i-1}, u_i]$ then we have

$$|u_{i-1} - x| < \frac{\delta}{M} < \delta \quad \text{and}$$

$$|\psi_n(x) - \psi_n(u_{i-1})| = |f(u_{i-1}, \psi_n(u_{i-1}))(x - u_{i-1})| < M\left(\frac{\delta}{M}\right) < \delta$$

so that we may conclude

$$|\psi_n'(x) - f(x, \psi_n(x)| = |f(u_{i-1}, \psi_n(u_{i-1})) - f(x, \psi_n(x))| < \frac{1}{n}.$$

The sequence (ψ_n) is equicontinuous because

(U) $|\psi_n(s) - \psi_n(t)| \leq M|s-t|$

for all s and t in I and all n. Moreover, (ψ_n) is bounded because of (U) and the fact that $\psi_n(x_0) = y_0$ for each n. Thus, the closure of $\{\psi_n : n = 1, 2, \ldots\}$ is compact by Theorem 24.6 on page 186, the Arzela-Ascoli Theorem. It follows that (ψ_n) contains a subsequence (ψ_{n_k}) that converges to some y in $C(I)$. The reader can verify that for each n and each x in I, we have

$$\psi_n(x) = y_0 + \int_{x_0}^{x} \psi_n'(t)\, dt$$

from which we obtain the identity

(*) $\psi_n(x) = y_0 + \int_{x_0}^{x} \{f(t, \psi_n(t)) + [\psi_n'(t) - f(t, \psi_n(t))]\}\, dt.$

At the partition points ψ_n' can be defined arbitrarily; choose the value so the term contained in the brackets vanishes. Since f is uniformly continuous on S, $f(t, \psi_n(t))$ converges uniformly to $f(t, y(t))$. By property (2) of the definition of ψ_n it follows that $[\psi_{n_k}'(t) - f(t, \psi_{n_k}(t))]$ converges uniformly to 0. Therefore, in the limit equation (*) becomes

$$y(x) = y_0 + \int_{x_0}^{x} f(t, y(t))\, dt.$$

If we integrate the right-hand side of this identity we see y satisfies (DE) on I. ■

Exercises

A

1. Show that $y_1 = \sqrt{x}$ and $y_2 = -\sqrt{x}$ both satisfy the differential equation $y' = 1/(2y)$ on $(0, \infty)$ and $y_1(0) = y_2(0) = 0$. Where does the hypothesis of Picard's Theorem not hold in this case?
2. Let $f(x) = \exp(-x^{-2})$, $x \neq 0$, $f(0) = 0$. For each real number c show that cf is a solution to the differential equation $y' = 2x^{-3}y$ and that $cf(0) = 0$.

B

3. Suppose that for each $n = 1, 2, \ldots$, f_n is a continuous real function on an open subset D of R^2 with $(x_0, y_0) \in D$; suppose y_n is a continuous function on $I = [x_0 - \varepsilon, x_0 + \varepsilon]$ that satisfies the differential equation $dy/dx = f_n(x, y)$ on $(x_0 - \varepsilon, x_0 + \varepsilon)$ and $y_n(x_0) = y_0$.
   (a) Show that if (f_n) converges uniformly to f on D then (y_n) converges uniformly on I to a function y that satisfies the equation $dy/dx = f_n(x, y)$ on $(x_0 - \varepsilon, x_0 + \varepsilon)$ and $y(x_0) = y_0$.

(b) Use (a), the Weierstrass Polynomial Approximation Theorem and Picard's Theorem to derive the Peano Existence Theorem.

33

Topics in Summability Theory

The main task of summability theory is to assign a limit to a divergent sequence, or sum to a divergent series in such a way that the sequence or series can be manipulated as though it were convergent. The motivation for this came from those divergent series that arose from differential equations, and Fourier expansions which seemed to have a "natural" sum. For example, the mathematician Euler assigned the sum $\frac{1}{2}$ to the divergent series $\sum_{n=1}^{\infty}(-1)^{n-1}$ because:

(a) If $x = \sum_{n=1}^{\infty}(-1)^{n-1}$ then $1 - x = 1 - \sum_{n=1}^{\infty}(-1)^{n-1} = x$;
(b) For $|t| < 1$, $F(t) = 1/(1+t) = \sum_{n=1}^{\infty}(-1)^{n-1}t^n$ and $F(1) = \frac{1}{2}$.

The most common way to sum a divergent sequence or series is to use an infinite matrix. If $A = (a_{ij})$ is an infinite matrix and $s = (s_j)$ is a sequence we define the A *limit of* s to be b and write $\lim_A s = b$ if

(S1) $\sum_{j=1}^{\infty} a_{ij}s_j$ converges for each i, and
(S2) $\lim_i \{\sum_{j=1}^{\infty} a_{ij}s_j\} = b$.

One of the main questions that arises in summability theory is: when does a matrix assign the usual limit to each convergent sequence? The answer to this question is given by the *Silverman-Toeplitz Theorem.* The original "hard analysis" proof of this theorem is very difficult, and the short functional analytic proof is an early triumph of functional analysis. S. Banach presented this proof in his 1932 classic, *Théorie des Opérations Linéaires.*

A matrix $A = (a_{ij})$ is said to be *regular* if

(C) $\lim_A s = \lim_i s_i$

for each $s = (s_i)$ in the space c of convergent sequences.

Theorem 33.1: The matrix $A = (a_{ij})$ is regular if and only if the following three conditions hold

(ST1) $\sup_i \sum_{j=1}^{\infty} |a_{ij}| < \infty$;
(ST2) $\lim_i a_{ij} = 0$ for all j;
(ST3) $\lim_i \sum_{j=1}^{\infty} a_{ij} = 1$.

Proof: Necessity. For $n = 1, 2, \ldots$ let e_n denote the sequence with 1 in the nth coordinate and 0's elsewhere and let e_0 denote the sequence with 1 in each coordinate. Condition (ST2) is necessary since $\lim_A e_n = 0$ for each n; and condition (ST3) is required since $\lim_A e_0 = 1$. Thus these two conditions are clearly necessary.

In order to verify the necessity of (ST1) we first note that for each i, the functional f_i defined on c by the equation

$$f_i(s) = \sum_j a_{ij} s_j$$

is continuous. This is a special case of Theorem 28.1 (page 205) with S the space c and T the scalar field. This fact is also established in the proof of that theorem. That $\lim_i f_i(s)$ exists for each s in c implies that $\sup_i |f_i(s)| < \infty$ for each s in c. From the Uniform Boundedness Principle, Theorem 21.1, it follows that $\sup_i \|f_i\| < \infty$. For each i and each n the sequence $(t_j^{i,n})$ for which $t_j^{i,n} = sgn(a_{i,j})$ for $j \leqslant n$ and $t_j^{i,n} = 0$ for $j > n$ is in c and has norm at most one. Therefore, we have $\|f_i\| \geqslant \sum_j |a_{ij}|$ for each i so that (ST1) is valid.

Sufficiency. Define the sequence of functionals (f_i) as above. For each s in c we have

$$|f_i(s)| \leqslant \sum_j |a_{ij} s_j| \leqslant \sup_j |s_j| \sum_j |a_{ij}|$$

so that (ST1) implies that (f_i) is uniformly bounded on c, that is, that $\sup_i \|f_i\| < \infty$. Conditions (ST2) and (ST3) imply that

$$\lim_i f_i(x) = \lim x$$

for $x = e_i$, $i = 0, 1, 2, \ldots$. Thus the theorem will follow from the Banach-Steinhaus Theorem (21.3 on page 163) if we show $\{e_i : i = 0, 1, \ldots\}$ is a fundamental subset of c. We have already seen that $\{e_i : i = 1, 2, \ldots\}$ is a basis for c_0 the subspace of c consisting of all sequences that converge to 0. That $\{e_i : i = 0, 1, \ldots\}$ is fundamental now follows from the fact that c is the direct sum of c_0 and the linear span $[e_0]$ of e_0, which we leave to you to prove. ■

Clearly, a matrix A is regular if and only if it maps c into c and preserves the limit functional. The next theorem known as the *Kojima-Schur Theorem* states conditions under which A maps c into c. The proof mimics that of Theorem 33.1 so we leave it to you, the reader.

Theorem 33.2: A matrix $A = (a_{ij})$ maps c into c if and only if it satisfies the following three conditions: (ST1),

(KS2) $\lim_i a_{ij}$ exists for each j;

(KS3) $\lim_i \sum_j a_{ij}$ exists.

If A and B map c into c then by (KS2) each column of B is convergent, hence bounded; and by (ST1) each row of A is absolutely summable. Therefore, the series $\sum_j a_{ij} b_{jk}$ converges absolutely for each i and each k. This implies that the matrix product AB is defined. The next theorem provides further information about this matrix product.

Theorem 33.3: If A and B map c into c then so does AB; if A and B are regular then so is AB.

Proof: We have just noted that the product matrix AB exists. Next we show that for s in c we have

$$(AB)s = A(Bs).$$

This associativity is not obvious because it has not been established that AB represents the composition of A and B on s as linear operators. However, the nth coordinate of $A(Bs)$ is

$$\sum_i a_{ni} \sum_j b_{ij}s_j = \sum_i \sum_j a_{ni}b_{ij}s_j.$$

The second sum converges absolutely because

$$\sum_i \sum_j |a_{ni}b_{ij}s_j| \leqslant \left\{\sum_i |a_{ni}|\right\}\left\{\sup_i \sum_j |b_{ij}|\right\} \sup_j |s_j| < \infty$$

because of (ST1). Therefore, the nth coordinate of $A(Bs)$ is

$$\sum_j \sum_i a_{ni}b_{ij}s_j,$$

which is the same as the nth coordinate of $(AB)s$. This verifies the first assertion. To check the second assertion, suppose A and B are regular and s has limit b. Then Bs has limit b so $A(Bs) = (AB)s$ has limit b. ■

There are numerous examples of regular matrices that have been the object of intense research. These examples have been organized into several classes. We shall describe a few of the more important matrices and classes.

Perhaps the most obvious matrix is that determined by simple averaging. This process transforms the sequence $s = (s_1, s_2 \ldots)$ into the sequence

$$(s_1, \tfrac{1}{2}s_1 + \tfrac{1}{2}s_2, \tfrac{1}{3}s_1 + \tfrac{1}{3}s_2 + \tfrac{1}{3}s_3, \ldots).$$

Thus it is given by the matrix

$$\begin{bmatrix} 1 & 0 & 0 & 0 & . & . \\ \frac{1}{2} & \frac{1}{2} & 0 & 0 & . & . \\ \frac{1}{3} & \frac{1}{3} & \frac{1}{3} & 0 & . & . \\ \frac{1}{4} & \frac{1}{4} & \frac{1}{4} & \frac{1}{4} & 0 & . \\ . & . & . & . & . & . \end{bmatrix}$$

It is obvious that this matrix satisfies (ST1), (ST2), and (ST3) and therefore, is regular. The summability method defined by this matrix is called the *first Cesaro mean*, and it is denoted by $(C, 1)$. We shall also denote the matrix for $(C, 1)$ summability by $(C, 1)$.

From Theorem 33.3 it follows that the powers $(C, 1)^k$ of $(C, 1)$ are regular for all $k = 1, 2, \ldots$. These powers give the result of taking averages of averages. The higher order Cesaro means, however, are defined in a more sophisticated way than this. Let E denote the matrix (d_{ij}) where $d_{ij} = 1$ if $j \leqslant i$ and $d_{ij} = 0$ otherwise; let D denote the diagonal matrix with entries $1, \frac{1}{2}, \frac{1}{3}, \ldots$ along the diagonal. Then we have $(C, 1) = DE$. Let D_n denote the diagonal matrix whose kth entry is the reciprocal of the sum of the kth column of the matrix E^n. The *nth Cesaro mean* is defined to be the summability method determined by the matrix $(C, n) = D_n E^n$. It is evident that (C, n) satisfies (ST1) and (ST3) but it is not immediately clear that (C, n) satisfies (ST2). We leave to you, the reader to determine that it does.

The first column of E^n consists of the coefficients of the power series obtained by raising the power series $1 + t + t^2 + t^3 + \cdots$ to the nth power. The kth column of this matrix is the same as the first column offset by $k - 1$, zeros. This fact can be used to extend the definition of Cesaro means to noninteger values. For $\alpha > 0$ let E^α denote the matrix whose first column consists of the coefficients of the Maclaurin expansion of $(1 - t)^\alpha$ and whose kth column consists of $k - 1$ terms equal to 0's followed by a repetition of the first column. Let D_α be the diagonal matrix whose kth entry is the reciprocal of the sum of the kth row of E^α. Define (C, α) to be $D_\alpha E^\alpha$. It can be shown that (C, α) is a regular matrix. This matrix is said to determine the α *Cesaro mean*, which is denoted by (C, α).

Let $H = h_{ij}$ denote the lower triangular matrix whose kth row consists of the coefficients in the expansion of $(a - b)^{k-1}$ followed by zeros, that is,

$$h_{ij} = (-1)^{j-1}\binom{i-1}{j-1} \quad \text{for } j \leqslant i \quad \text{and} \quad h_{ij} = 0 \quad \text{for } j > i.$$

The series $\sum_{j=1}^{\infty} h_{ij}h_{jk}$ consists entirely of 0's when $i < k$; when $i = k$, the term $h_{ij}h_{ii} = 1$ and the other terms are 0. When $i > k$ the series is

$$\begin{aligned}\sum_{j=k}^{i} (-1)^{j+k-2}\binom{i-1}{j-1}\binom{j-1}{k-1} &= (-1)^{k-1}\sum_{j=k}^{i} (-1)^{j-1}\binom{i-k}{j-k}\binom{i-1}{k-1} \\ &= \binom{i-1}{k-1}\sum_{j=1}^{i-k+1} (-1)^{j}\binom{i-k}{j-1} \\ &= \binom{i-1}{k-1}(1-1)^{i-k} = 0.\end{aligned}$$

This means that H is its own inverse, $H^2 = I$. A *Hausdorff mean* is a summability method determined by the matrix HDH where D is a diagonal matrix. If the diagonal of D is $r = (r_i)$ then HDH is written H_r. Since the j, k entry of DH is $r_j h_{jk}$, the i, k entry $H_r(i, k)$ of H_r for $k \leqslant r$ is given by the formula

$$\begin{aligned}H_r(i, k) = \sum_j h_{ij} r_j h_{jk} &= \sum_{j=k}^{i} (-1)^{k+j-2} r_j \binom{i-1}{j-1}\binom{j-1}{k-1} \\ &= \binom{i-1}{k-1}\left\{\sum_{j=k}^{i} (-1)^{j-k}\binom{i-k}{j-k} r_j\right\}. \qquad (*)\end{aligned}$$

The expression within the braces above is denoted by $\Delta^{i-k} r_k$. The following lemma is the inverse formula for (*).

Lemma 33.4: For each m and $i > m$ we have

$$r_m = \sum_{k=1}^{i} \frac{(k-1)\ \ldots\ (k-m)}{(i-1)\ \ldots\ (i-m)} H_r(i, k).$$

Proof: Each term in the sum is 0 if $k \leqslant m$ so it may be written

$$\sum_{k=m}^{i} \left\{ \binom{k-1}{m-1} \Big/ \binom{i-1}{m-1} \right\} H_r(i, k)$$

$$= \sum_{k=m}^{i} \left\{ \binom{k-1}{m-1} \Big/ \binom{i-1}{m-1} \right\} \binom{i-1}{k-1} \sum_{j=k}^{i} (-1)^{j+k} \binom{i-k}{j-k} r_j$$

$$= \sum_{k=m}^{i} \sum_{j=k}^{i} (-1)^{j+k} \binom{i-m}{k-m} \binom{i-k}{j-k} r_j$$

$$= \sum_{j=m}^{i} \binom{i-m}{j-m} \left\{ \sum_{k=m}^{j} (-1)^{j+k} \binom{j-m}{k-m} \right\} r_j$$

$$= \sum_{j=m}^{i} \binom{i-m}{j-m} \left\{ \sum_{k=0}^{j-m} (-1)^{k} \binom{j-m}{k} \right\} r_j$$

$$= \sum_{j=m}^{i} \binom{i-m}{j-m} \{1-1\}^{j-m} r_j = r_m. \ \blacksquare$$

Lemma 33.5: For every m we have

$$\lim \left| \left\{ \frac{k}{i} \right\}^m - \binom{k}{m} \Big/ \binom{i}{m} \right| = 0$$

uniformly in k.

Proof: Let $\varepsilon > 0$ be given. By Sterling's Formula there is $M > 0$ such that

$$(1-\varepsilon) k^m e^{-m} < \binom{k}{m} < k^n e^{-m} \quad \text{for } k > M,$$

where e is the base of the natural logarithms. Let N be such that

$$(**) \quad \left| \left\{ \frac{k}{i} \right\}^m - \binom{k}{m} \Big/ \binom{i}{m} \right| < \varepsilon \quad \text{for } i > N,\ k \geqslant M$$

Then for $M < k \leqslant i$ we have

$$(1-\varepsilon)(k^m e^{-m})(i^m e^{-m}) \binom{k}{m} \Big/ \binom{i}{m} < \frac{1}{(1-\varepsilon)} \frac{k^m e^{-m}}{i^m e^{-m}}$$

$$< (1+\varepsilon) \frac{k^m}{i^m}$$

so that (**) holds for $k > M$ as well. $\blacksquare$

For every sequence $s = (s_1, s_2, \ldots)$ let Δs be the difference sequence $(s_2 - s_1, s_3 - s_2, \ldots)$; we can iterate the operator Δ to take second differences $\Delta^2 s = \Delta(\Delta s)$, third differences $\Delta^3 s = \Delta(\Delta^2 s)$, etc. A sequence s is called *completely monotone* if for every k the kth difference $\Delta^k s$ is nonnegative. It is called a *Hausdorff moment sequence* if there is a function F in $NBV[0, 1]$ such that

$$s_n = \int_0^1 t^{n-1}\, dF(t), \quad \text{for } n = 1, 2, \ldots.$$

The following theorem is called the *Hausdorff Moment Theorem.*

Theorem 33.6: The following three statements are equivalent for a sequence $s = (s_i)$.

(a) s is a moment sequence;
(b) s is the difference of two completely monotone sequences;
(c) H_s maps l^∞ into l^∞.

Proof:
(a) ⇒ (b). If $s_n = \int_0^1 t^n\, dF$, let $F = F^+ - F^-$ where F^+ and F^- are nonnegative. Then we have

$$s_n = \int_0^1 t^n\, dF^+ - \int_0^1 t^n\, dF^- \quad \text{for all } n.$$

Thus it suffices to show that if G is a nonnegative function in NBV, the sequence $(\int_0^1 t^n\, dG)$ is totally monotone. For such a G we have

$$\begin{aligned}\Delta^k\left(\int_0^1 t^{j-1}\, dG\right) &= \sum_{i=0}^{k} (-1)^i \binom{k}{i} \int_0^1 t^{j+i-1}\, dG \\ &= \int_0^1 \left\{\sum_{i=0}^{k} (-1)^i \binom{k}{i} t^i\right\} t^{j-1}\, dG \\ &= \int_0^1 (1-t)^k t^{j-1}\, dG \geqslant 0.\end{aligned}$$

(b) ⇒ (c). It suffices to assume that s is completely monotone. In this case for each i we have

$$\sum_k |H_s(i, k)| = \sum_k H_s(i, k) = s(1)$$

by Lemma 33.4. This condition shows that H_s maps l^∞ into l^∞. See Exercise 14 of Section 28 on page 210.

(c) ⇒ (a). If H_s maps l^∞ into l^∞ then by Exercise 14 of Section 28 there is $M > 0$ such that

(BDD) $\sum_k |H_s(i, k)| \leqslant M$

for each i. If continuous linear functionals Φ_i are defined on $C[0, 1]$ by the formula

$$\Phi_i(f) = \sum_{k=1}^{i} H_s(i, k) f\left(\frac{k-1}{i-1}\right)$$

then the sequence (Φ_i) is uniformly bounded on $C[0, 1]$ by (BDD). By Lemmas

33.4 and 33.5 we have

$$\lim_i |s_n - \Phi_i(t^{n-1})| = \lim_i \left| \sum_{k=1}^{i} \left\{ \frac{(k-1)\dots(k-m)}{(i-1)\dots(i-m)} - \frac{(k-1)}{(i-1)} \right\} H_s(i, k) \right|$$

$$= 0.$$

Therefore, we have $\lim_i \Phi_i(t^{n-1}) = s_n$ for each n.

By the Weierstrass Polynomial Approximation Theorem (24.1 on page 181) the set $\{t^n: n = 0, 1, \dots\}$ is fundamental in $C[0, 1]$, so by the Banach-Steinhaus Theorem (21.3), $\lim_i \Phi_i(f)$ converges for all f in $C[0, 1]$ to a continuous linear functional $\Phi(f)$ on $C[0, 1]$ for which $\Phi(t^{n-1}) = s_n$. The conclusion now follows from the Riesz Representation Theorem for the dual of $C[0, 1]$, Theorem 25.5 on page 191. ■

Theorem 33.7: The Hausdorff matrix H_s is regular if and only if

(H1) s is a moment sequence;

(H2) $s_1 = 1$;

(H3) $\lim_i H_s(i, 1) = 0$.

Proof: Condition (H1) is equivalent to (BDD), which is equivalent to (ST1). By Lemma 31.4 the sum of each row of H_s is s_1 so condition (H2) is equivalent to condition (ST3).

If s is a moment sequence with $s_n = \int_0^1 t^{n-1}\, dG$ for each n, then by (*) we have

$$H_s(i, k) = \binom{i-1}{k-1} \left\{ \sum_{j=k}^{i} (-1)^{j-k} \binom{i-k}{j-k} \int_0^1 t^{j-k}\, dG \right\}$$

$$= \binom{i-1}{k-1} \int_0^1 t^{k-1}(1-t)^{i-k}\, dG.$$

Thus, for $k > 1$ we have

$$\lim_{i\to\infty} H_s(i, k) = \int_0^1 t^{k-1} \lim_{i\to\infty} \binom{i-1}{k-1} (1-t)^{i-k}\, dG = 0$$

since $i(i-1)\dots(i-k)(1-t)^{i-k} \to 0$ as $i \to \infty$. This means that (H1) and (H3) together imply (ST2). On the other hand, (H3) is a special case of (ST2). ■

The Cesaro means are a subcollection of the Hausdorff means. The reader may show that (C, s) for $s > -1$ is the Hausdorff matrix H_r where

$$r_n = \binom{n+s-1}{n}^{-1} = s \int_0^1 t^{n-1}(1-t)^{s-1}\, dt.$$

Two other important means that are subcases of the Hausdorff means are:

(a) The *Hölder means*, $H^{(n)}$, $n = 1, 2, \dots$. These were originally defined as the nth power of $(C, 1)$, that is, $H^{(n)} = (C, 1)^n$. It can be shown that $H^{(n)}$ is the Hausdorff matrix H_s where

(Höl) $s_k = k^{-n} = ((n-1)!)^{-1} \int_0^1 t^{n-1}(\log 1/t)^{n-1}\, dt$.

(b) The *Euler means*, $E^{(\alpha)}$, $0 < \alpha < 1$, are defined as H_s where
(Eul) $s_k = \alpha^{n-1} = \int_0^1 t^{n-1}\, dF_\alpha$.

Here $F_\alpha(t) = 1$ if $t = \alpha$ and 0 if $t \neq \alpha$.

Exercises

A

1. Show that if $a_{ij} \geqslant 0$ for each i, j, then conditions (ST2) and (ST3) together imply (ST1).
2. Show that a matrix $A = (a_{ij})$ maps l^∞ into l^∞ if and only if (ST1) holds.
3. Show that if A and B map l^∞ into l^∞ then AB is defined and maps l^∞ into l^∞.
4. Calculate $(H, 2) = (C, 1)^2$ and $(C, 2)$ and show they are different.
5. *(a)* If H_r and H_s are two regular Hausdorff matrices show that $H_r H_s = H_{rs}$ and H_{rs} is also regular.
   (b) Show that if r and s are moment sequences so is rs.
6. The summability field c_A of a matrix A is the set of all sequences s such that $\lim_A s$ exists.
   (a) Prove that c_A is a linear space;
   (b) Prove that $(1, -1, 1, -1, \ldots) \in c_{(C,1)}$ and $\lim_{(C,1)}(-1)^{n-1} = 0$.
7. Show that the summability field of a Hausdorff matrix H_r is precisely c if and only if both r and $r^{-1} = (r_1^{-1}, r_2^{-1}, \ldots)$ are moment sequences.
8. Show that the $(n+1)$st term of Hs is the first term of $\Delta^n s$.
9. Show that the first column of E^n consists of the coefficients of the power series obtained by raising the power series $(1 + t + t^2 \ldots)$ to the nth power.
10. Calculate $(C, 2)$ and $(C, 3)$ by means of the definition $(C, n) = D_n E^n$.
11. Give the proof of Theorem 33.2.
12. Verify that $(C, 1)$, $(C, 2)$, and $(C, 3)$ are Hausdorff means H_r where $r_n = \binom{n+s-1}{n}^{-1}$ for $s = 1, 2, 3$.
13. Verify that $(H, 1)$, $(H, 2)$, and $(H, 3)$ are Hausdorff means H_s where s is given by $s_k = k^{-n}$ for $n = 1, 2, 3$.

B

14. Let $r = r_1, r_2, \ldots$ be a sequence of positive numbers, and let $R_n = \sum_{j=1}^n r_j$. The Nörlund matrix N_r determined by r is

$$\begin{bmatrix} 1 & 0 & 0 & 0 & \cdots \\ r_2/R_2 & r_1/R_2 & 0 & 0 & \cdots \\ r_3/R_3 & r_2/R_3 & r_3/R_3 & 0 & \cdots \\ r_4/R_4 & r_3/R_4 & r_2/R_4 & r_1/R_4 & \cdots \\ \cdot & \cdot & \cdot & \cdot & \cdots \end{bmatrix}$$

Thus if $N = (c_{ij})$, we have $c_{ij} = r_{i-j+1}/r_i$ if $j \leqslant i$ and 0 otherwise.
(a) Show that for each n, (C, n) is the Nörlund mean determined by the first column of E^n.

(b) Prove that the Nörlund mean determined by r is regular if and only if $\lim_i R_i/R_{i+1} = 1$.

15. Show that (C, α) is regular for all $\alpha > 0$.

16. Use the definition of (C, α) as $D_\alpha E$ to show that the entry in the jth column of the ith row is equal to $\binom{j+\alpha-1}{j}/\{\sum_{k=0}^{i} \binom{k+\alpha-1}{k}\}$ where $\binom{j+\alpha-1}{j} = \alpha(1+\alpha)\dots(j+\alpha-1)/j!$.

17. Suppose $A = (a_{ij})$ is a lower triangular matrix with no zeros on the diagonal, that is, $a_{ii} \neq 0$ for all i, $a_{ij} = 0$ for $j > i$. Prove c_A is a BK-space with the norm $\|s\| = \sup_i |\sum_{j=1}^{i} a_{ij} s_j|$.

18. Show that $(1, -1, 1, -1, \dots)$ is not a moment sequence.

19. Show that the space of all moment sequences is a BK-space isometric to $NBV[0, 1]$.

20. Prove that the following statements are equivalent for an infinite matrix $A = (a_{ij})$:
 (a) A maps l^1 into l^1;
 (b) $\sup_j \sum_i |a_{ij}| < \infty$;
 (c) A^T the transpose of A maps l^∞ into l^∞ ($A^T = (b_{ij})$ means $b_{ij} = b_{ji}$ for each i, j).

21. (a) Let $(k(n))$ be a strictly increasing sequence of integers. Let $A = (a_{ij})$ be the matrix such that $a_{ik(i)} = 1$ and $a_{ij} = 0$ otherwise. Show A is regular.
 (b) Prove that if t is a sequence with a convergent subsequence then there is a regular matrix A such that $\lim_A t$ exists.
 (c) Show that if t is a sequence with no convergent subsequence then there is a subsequence (t_{n_k}) of t such that $\lim_k t_{n_k}/t_{n_{k+1}} = 0$.
 (d) Prove that if $\lim_k t_k/t_{k+1} = 0$ then the matrix defined by

$$A = \begin{bmatrix} 1 & -t_1/t_2 & 0 & 0 & . & . \\ 0 & 1 & -t_2/t_3 & 0 & . & . \\ 0 & 0 & 1 & -t_3/t_4 & . & . \\ . & . & . & . & . & . \end{bmatrix}$$

 is regular and $\lim_A t = 0$.
 (e) Verify that for every divergent sequence s and every number c there is a regular matrix A such that $\lim_A s = c$.

34

Quasi-Nilpotence

In the first half of the twentieth century mathematical analysts made a strong effort to construct a rigorous theoretical basis for the operational calculus that the electrical engineer Oliver Heaviside (1850–1945) invented to solve linear differential equations. When Heaviside encountered a linear differential equation of the form

$$\text{(LDE)}\quad a_n x^{(n)} + a_{n-1} x^{(n-1)} + \cdots + a_1 x' + a_0 x = f(t)$$

he would consider it as an equation in x involving the differential operator D,

$$(a_n D^n + a_{n-1} D^{n-1} + \cdots + a_1 D + a_0)x = f(t).$$

He would then treat the linear operator D precisely as one would treat a matrix operator and invert the form $(a_n D^n + \cdots + a_0)$ to solve for x. Realizing that there were deficiencies in his method, he did not trust solutions found in this way until he replaced them in the given equations. He engaged in frequent arguments with contemporary mathematicians regarding his method.

Several modern mathematical theories justify Heaviside's operational calculus. The most sophisticated of these is the theory of distributions created by Laurant Schwartz. There is also the theory of Laplace transforms, which appears in elementary texts on differential equations.

The simplest justification for the operational calculus is to make the integral substitution $u = y^{(n)}$ or, equivalently, $y = S^n u$ where S denotes an integral operator

$$Su = \int_a^t u(s)\, ds.$$

This substitution transforms the differential equation (LDE) into the integral equation

$$\textit{(LIE)}\quad a_n u + a_{n-1} Su + \cdots + a_1 S^{n-1} u + a_0 S^n u = S^n f.$$

Note that the roots of the resulting polynomial are the reciprocals of the roots of the polynomial resulting from (LDE). The theoretical problems for treating equation (LIE) evaporate since the integral operator, unlike the differential operator, is continuous on almost any space on which it is defined. Indeed, it has the strong property of quasi-nilpotence. A linear operator T from a normed space E into itself is called *quasi-nilpotent* if the series $\sum_{n=0}^{\infty} a^n T^n$ converges for each scalar a. Also the presence of $S^n f$ on the right-hand side of the equation "upgrades" f so that it need not surprise us if f is a Heaviside function

$$f(x) = 0 \quad \text{if } x < 0, \qquad f(x) = 1 \quad \text{if } x \geqslant 0$$

or a Dirac function, that is, the measure μ such that $\mu(A) = 1$ if $0 \in A$ and $\mu(A) = 0$ otherwise.

Let $X[a, b]$, or X if there is no ambiguity, denote a Banach space of functions from the interval $[a, b]$ into the scalar field that satisfies the following hypotheses.

(A) X contains all polynomials in t.

(B) Each x in X is integrable and whenever x is in X so is $S_a x$ defined by $S_a x(t) = \int_a^t x(s)\, ds$.

(C) There is defined on X a family $\{N_t\colon a \leqslant t \leqslant b\}$ of seminorms for which

- *(C1)* If $|x_1(s)| \leqslant |x_2(s)|$ for $s \leqslant t$ then $N_t(x_1) \leqslant N_t(x_2)$;
- *(C2)* $N_t(x) \leqslant N_s(x)$ when $t \leqslant s$ for all x in X;
- *(C3)* $\|x\|_X = N_b(x)$ for all x in X.

Although hypotheses (A)–(C) may seem technical, there are many examples of Banach function spaces that satisfy them. Here are a few examples:

$C[a, b]$ with $N_t(x) = \sup\{|x(s)|\colon a \leqslant s \leqslant t\}$, $a \leqslant t \leqslant b$.

$L^p[a, b]$ with $N_t(x) = \{\int_a^t |x(s)|^p\, ds\}^{1/p}$.
$L^\infty[a, b]$ with $N_t(x) = ess\ \sup\{|x(s)|: a \leqslant s \leqslant t\}$.
$W_p^{(n)}[a, b]$ with the norms

$$N_t(x) = \sum_{k=0}^{n-1} \sup\{|x^{(k)}(s)|: a \leqslant s \leqslant t\} + \left\{\int_a^t |x^{(n)}(s)|^p\, ds\right\}^{1/p}$$

An appropriate modification is made in the final term when $p = \infty$.

A linear operator T from $X[a, b]$ into itself is called *p-tempered* for $1 \leqslant p < \infty$ if there is $K > 0$ such that for all t in $[a, b]$ and all x in $X[a, b]$ we have

(PT) $N_t(Tx) \leqslant K\{\int_a^t N_s(x)^p\, ds\}^{1/p}$.

A 1-tempered operator will simply be called *tempered.*

We leave to you to derive the following elementary facts about p-tempered operators.

(a) If T is p-tempered, then it is q-tempered for $q > p$.
(b) If T and U are p-tempered and c is a constant, then cT and $T + U$ are p-tempered.

Theorem 34.1: If T is a p-tempered operator from $X[a, b]$ into itself, then for all $n = 1, 2, \ldots$ all x in X and all t in $[a, b]$ the following equation holds

(QN1) $N_t(T^n x) \leqslant K^n N_t(x)\{(t-a)^n/n!\}^{1/p}$

where K is the constant in (PT).

Proof: For $n = 1$ the inequality (QN1) follows directly from (PT). If (QN1) is true for n it is also true for $n + 1$ since

$$N_t(T^{n+1}x) \leqslant K\left\{\int_c^t N_s(T^n x)^p\, ds\right\}^{1/p}$$

by (PT) and by (QN1) applied to n we have

$$\begin{aligned} K\left\{\int_a^t N_s(T^n x)^p\, ds\right\}^{1/p} &\leqslant K\left\{\int_a^t (K^n N_s(x))^p \frac{(s-a)^n}{n!}\, ds\right\}^{1/p} \\ &\leqslant K^{n+1}\left(\frac{N_t(x)}{n!}\right)^{1/p} \left\{\int_a^t (s-a)^n\, ds\right\}^{1/p} \\ &= K^{n+1} N_t(x)\left\{\frac{(s-a)^{n+1}}{(n+1)!}\right\}^{1/p}. \end{aligned}$$

Therefore, (QN1) is true for all n by induction. ∎

Corollary 34.2: If T is a p-tempered operator from X into X for all λ, the series $\sum_{n=0}^{\infty} \lambda^n T^n$ converges absolutely to the continuous inverse of the operator $(I - \lambda T)$.

Proof: The first assertion follows since the series

$$\sum_n K^n \lambda^n \left(\frac{\lambda^n}{n!}\right)^{1/p}$$

converges absolutely for all λ by the ratio test. The second assertion is true because $\sum_{n=0}^{\infty} \lambda^n T^n$ converges to $(I-\lambda T)^{-1}$. ■

In other words, a p-tempered operator T is quasi-nilpotent. The prototypical p-tempered operator defined on the spaces X listed in the examples is the operator S_a defined by

$$(S_a x)(t) = \int_a^t x(s)\, ds.$$

It is obvious that S_a is tempered for $X = L^\infty$ or C. For $X = L^p$, S_a is p-tempered since

$$\begin{aligned} N_t(S_a x) &= \left\{\int_a^t |S_a x(s)|^p\, ds\right\}^{1/p} \\ &= \left\{\int_a^t \left|\int_a^s x(r)\, dr\right|^p ds\right\}^{1/p} \\ &\leqslant \left\{\int_a^t \left\{\int_a^s |x(r)|^p\, dr\right\}(s-a)^{p/p'}\, ds\right\}^{1/p} \end{aligned}$$

by the Hölder inequality, where $1/p + 1/p' = 1$. Thus we have

$$N_t(S_a x) \leqslant (b-a)^{1/p'} \left\{\int_a^t N_s(x)^p\, ds\right\}^{1/p}.$$

If $X = W_q^p[a, b]$, S_a is tempered because for each x in W_q^p we have

$$\begin{aligned} N_t(S_a x) &= \sum_{k=0}^{q-1} \sup\{|S_a x^{(k)}(s)|: a \leqslant s \leqslant t\} + \left\{\int_a^t |S_a x^{(q)}(s)|^p\, ds\right\}^{1/p} \\ &\leqslant \left\{\sum_{k=0}^{q-1} \int_a^t |x^{(k)}(s)|\, ds + \left(\int_a^t |x^{(q-1)}(s)|^p\, ds\right)^{1/p}\right\} \\ &\leqslant \left\{(b-a)^{1/p'} \sum_{k=0}^{q-1} \left(\int_a^t |x^{(k)}(s)|^p\, ds\right)^{1/p}\right. \\ &\qquad \left. + \left(\int_a^t |x^{(q-1)}(s)|^p\, ds\right)^{1/p}\right\} \qquad (1/p + 1/p' = 1) \\ &\leqslant (1 + (b-a)^{1/p'}) \left\{\sum_{k=0}^{q-1} \left(\int_a^t |x^{(k)}(s)|^p\, ds\right)^{1/p}\right\} \\ &\leqslant (1 + b - a)^{1/p'} \left\{\int_a^t \left(\sum_{k=0}^{q-1} \sup\{|x^{(k)}(r)|: a \leqslant r \leqslant s\}\right)^p ds\right\}^{1/p} \\ &\leqslant (1 + b - a)^{1/p'} \left(\int_a^t N_s(x)^p\, ds\right)^{1/p} \end{aligned}$$

where $1/p + 1/p' = 1$.

We now show how the idea of p-tempered operators can be used to derive existence results for a linear initial value problem. Consider the equations

(IV1) $y^{(n)} + r_{n-1}(t)y^{(n-1)} + \cdots + r_1(t)y' + r_0(t) = g.$

(IV2) $y(a) = c_0,\ y'(a) = c_1,\ \ldots,\ y^{(n-1)}(a) = c_{n-1}.$

Here g is a known member of $X[a, b]$; $r_0, \ldots, r_{n-1}$ are continuous scalar valued mappings on $[a, b]$; $c_0, c_1, \ldots, c_{n-1}$ are scalars and y is an unknown member of X. Let f be the polynomial defined by

$$f(t) = c_0 + c_1(t-a) + \cdots + c_{n-1}(t-a)^{n-1}/(n-1)!.$$

For every u in X the function $S^n u + f$ satisfies (IV2). If we substitute $y = S^n u + f$ into (IV1) we obtain

$$u + r_{n-1}Su + r_{n-2}S^2 u + \cdots + r_1 S^{n-1}u + r_0 S^n u$$
$$= g - (r_0 f + r_1 f' + \cdots + r_{n-1} f^{(n-1)}).$$

If S is a p-tempered operator from X into X and if $\sup\{|r_k(t)|: a \leqslant t \leqslant b\} = M_k < \infty$ for each k, then the operator T_k defined by

$$(T_k x)(t) = [r_k(t)s^{n-k}x](t)$$

is a tempered operator from X into X because

$$N_t(T_k x) \leqslant M_k N_t(S^{n-k}x) \leqslant M_k \left\{\int_a^t N_s(S^{n-k-1}x)^p\, ds\right\}^{1/p},$$

for each x in X. If we define T from X into X by

$$T = -T_{n-1} - T_{n-2} - \cdots - T_0$$

equations (IV1) and (IV2) become

$$(I - T)u = g - (r_0 f - \cdots + r_{n-1} f^{(n-1)}).$$

If the function $r_0(t)f(t) + \cdots + r_{n-1}(t)f^{(n-1)}(t)$ happens to be in X we obtain

$$u = (I - T)^{-1}\{g - (r_0 f + r_1 f' + \cdots + r_{n-1} f^{(n-1)})\}$$

so that the solution to (IV1), (IV2) is

(SOL) $y = S^n(I - T)^{-1}\{g - (r_0 f + r_1 f' + \cdots + r_{n-1} f^{(n-1)})\} + f.$

This establishes the following result.

Theorem 34.3: Suppose X is a Banach space of scalar valued functions on $[a, b]$ that satisfies hypotheses (A)–(C) and upon which the linear operator S is p-tempered. If for $j = 0, 1, \ldots, \{|r_j(t)|: a \leqslant t \leqslant b\}$ is bounded, and g is in X, then the initial value problem (IV1), (IV2) has a unique solution in X.

Suppose the coefficients are constant in the initial value problem (IV1), (IV2), that is, $r_j(t) = a_j$ for $j = 0, \ldots, n-1$. Then the solution (SOL) assumes the form

(SLN) $y = (I + a_{n-1}S + \cdots + a_0 S^n)^{-1} S^n\{g - (a_0 f + a_1 f' + \cdots + a_{n-1} f^{(n-1)})\} + f.$

If we can factor the polynomial $1 + a_{n-1}s + \cdots + a_0 s^n$ then we can solve (SLN) either by successively finding inverses of the factors or by using a fractional decomposition of $s^n(1 + a_{n-1}s + \cdots + a_0 s^n)^{-1}$. This provides an alternative to the Laplace transform method for solving linear differential equations with constant coefficients. The polynomial encountered is related to the polynomial in the Laplace transform method by the substitution $s \to 1/s$.

In order to apply the method outlined above it is necessary to solve the first-order equation

(ORD1) $(I - \lambda S)y = G$

where $G = S^n(g - \{a_0 f + a_1 f' + \cdots + a_{n-1} f^{(n-1)}\})$. This is easy to accomplish by differentiating both sides of (ORD1) and solving a linear differential equation with initial condition $y(a) = G(a)$. The solution is

(SOLN1) $y = e^{\lambda t} \int_a^t G' e^{-\lambda s}\, ds + G(a) e^{\lambda(t-a)}$.

We can take the derivative of G because G is an nth order integral having the form $S^n F$ for some F. This means that in the case of the initial value problem (IV1), (IV2) with scalar coefficients the operator S^n "upgrades" the right-hand side of (IV1). We have established the following result.

Theorem 34.4: In order that the initial value problem (IV1), (IV2) with scalar coefficients has a solution in the space X it suffices that $S^n g$ be a member of X.

For example, when $X = C[a, b]$ we will have a solution to (IV1), (IV2) which is in X for $n = 1$ when g is a Heaviside function, for $n = 2$ when g is a Dirac function, and for $n = 3$ when g is the derivative of a Dirac function.

We now illustrate the method of solution described above with two examples.

Example 34.5:

$y'' + 4y = \sin t$, $y(0) = 0$, $y'(0) = 1$: The substitution is $y = S^2 u + t$. Then we have

$$u + 4S^2 u + 4t = \sin t$$

$$u = (I + 4S^2)^{-1}(\sin t - 4t)$$

$$y = S^2(I + 4S^2)^{-1}(\sin t - 4t) + t$$

$$= \frac{I}{4} - \frac{(I + 2iS)}{8} - \frac{(I - 2iS)^{-1}}{8}(\sin t - 4t) + t.$$

By (SOLN1) we have

$$(I - \lambda S)^{-1} \sin t = \frac{\lambda \cos t}{1 + \lambda^2} + \frac{\sin t}{1 + \lambda^2} - \frac{\lambda e^{-\lambda t}}{1 + \lambda^2}$$

so that

$$\{(I - 2iS)^{-1} + (I + 2iS)^{-1}\} \sin t = -\tfrac{2}{3} \sin t + \tfrac{4}{3} \sin 2t \quad \text{and}$$

$$\{(I - 2iS)^{-1} + (I + 2iS)^{-1}\} t = \sin 2t.$$

Therefore, we have

$$y = \tfrac{1}{4}\sin t - t - \tfrac{1}{8}(-\tfrac{2}{3}\sin t + \tfrac{4}{3}\sin 2t) + \tfrac{1}{2}\sin 2t + t$$
$$= \tfrac{1}{3}\sin t + \tfrac{1}{3}\sin 2t.$$

Example 34.6:

We consider the following initial value problem

$$x' + z' - z = 0$$
$$-x' + 2z' + x + y = \tanh t$$
$$2x'' + y'' + z'' + z = -2\sinh t \cosh^{-3} t$$
$$x(0) = x'(0) = y(0) = y'(0) = z(0) = z'(0) = 0.$$

Let $x = S^2X$, $y = S^2Y$, $z = S^2Z$. Then the system of equations becomes

$$\begin{aligned} SX \qquad\qquad &+ (S - S^2)Z = 0 \\ (S^2 - S)X + S^2Y \qquad &- 2SZ = \tanh t \\ 2X \quad + Y &+ (S^2 + I)Z = -2\sinh t \cosh^{-3} t \\ &= (\tanh t)''. \end{aligned}$$

By differentiating the first two equations we obtain

$$\begin{aligned} X \qquad\qquad &+ (I - S)Z = 0 \\ (S - I)X + SY \qquad &- 2Z = (\tanh t)' \\ 2X \quad + Y &+ (S^2 + I)Z = (\tanh t)''. \end{aligned}$$

Let

$$H = \begin{vmatrix} I & 0 & I - S \\ S - I & S & -2I \\ 2I & I & S^2 + I \end{vmatrix} = (S^2 + I)(S + I).$$

Then by Cramer's Rule we find

$$X = H^{-1}\begin{vmatrix} 0 & 0 & I - S \\ (\tanh t)' & S & -2I \\ (\tanh t)'' & I & S^2 + I \end{vmatrix}$$
$$= H^{-1}((I - S)(\tanh t)' - (I - S)S(\tanh t)'')$$
$$= (-S(S^2 - I)^{-1} + (S + I)^{-1})1 = \sin t + e^{-t}$$
$$x = S^2X = \sin t + e^t - 1$$
$$y = S^2Y$$
$$= S^2H^{-1}\begin{vmatrix} I & 0 & I - S \\ S - I & (\tanh t)' & -2I \\ 2I & (\tanh t)'' & S^2 + I \end{vmatrix} = H^{-1}\begin{vmatrix} I & 0 & I - S \\ S - I & S(\tanh t) & -2I \\ 2I & \tanh t - t & S^2 + I \end{vmatrix}$$

(by application of S^2 to the second column)

$$= (S^2+I)^{-1}(S+I)^{-1}\{(S^2+I)S \tanh t + (-I+2S-S^2)(\tanh t - t)$$
$$-2(I-S)S \tanh t + 2(\tanh t - t)\}$$
$$= (S^2+I)^{-1}(S+I)^{-1}\{(S^3+S^2+S+I) \tanh t + (-I-2S+S^2)(t)\}$$
$$= \tanh t + 1 - 2 \sin t - e^{-t}$$

$$z = S^2 Z = S^2 H^{-1} \begin{vmatrix} I & 0 & 0 \\ S-I & S & (\tanh t)' \\ 2I & I & \tanh t - t \end{vmatrix} = \begin{vmatrix} I & 0 & 0 \\ S-I & S & S \tanh t \\ 2I & I & \tanh t - t \end{vmatrix}$$

$$= (S^2+I)^{-1}(S+I)^{-1}\{S(\tanh t - t) - S \tanh t\}$$
$$= -\tfrac{1}{2}\{(S-I(S^2+I)^{-1} + (S+I)^{-1}\}(1) = \tfrac{1}{2}\{-\sin t + \cos t - e^{-t}\}.$$

Exercises

A

1. Transform the following linear differential equations into the form (LIE) and determine their solution.
   (*a*) $y'' + 2y' + y = x$; $y(0) = 2$, $y'(0) = -2$
   (*b*) $y'' + 5y' + 6y = x + 1$; $y(0) = 0$, $y'(0) = 0$
   (*c*) $y''' + y'' + y' + y = \cos x$; $y(0) = 1$, $y'(0) = 1$, $y''(0) = 0$
2. A linear operator T on a linear space E is called *nilpotent* if there is an integer k such that $T^k x = 0$ for all x in E. Show that a nilpotent operator on a normed space is quasi-nilpotent, but that the operator S is not nilpotent on any of the spaces given as examples.
3. (*a*) Prove that no nonzero diagonal linear operator from l^p into l^p can be quasi-nilpotent.
   (*b*) Show that the mapping from l^p into l^p defined by
   $$Tx = (x_2/2!, x_3/3!, x_4/4!, \ldots)$$
   is quasi-nilpotent but not nilpotent.
4. Let f be the piecewise linear function whose graph joins the points (0, 0), (1, 0), (2, 1), (3, 1), (4, 0), (5, 0). Find $S^2 f$ and determine the solution to the equation $y'' - 5y' + 4y = f$; $y(0) = 1$, $y'(0) = 0$ on $[0, 5]$.
5. Show $C^{(n)}[a, b]$ satisfies (A)–(C) with N_t the same family of seminorms used on $W_\infty^{(n)}[a, b]$.
6. Prove assertions (a) and (b), which precede Theorem 34.1.
7. Verify that if $\sum_{n=0}^{\infty} \lambda^n T^n$ converges in the uniform operator topology, then it must converge to $(I - \lambda T)^{-1}$.
8. Define T from $C[a, b]$ into itself by $(Tf)(t) = \int_a^t sf(s)\, ds$. Show that T is tempered.
9. If ν is a measure defined on the Borel subsets of $[a, b]$ define $S\nu$ to be the function given by $S\nu(t) = \nu([a, t])$.

(a) Let δ be the Dirac function, that is, the measure with $\delta(A) = 1$ if $0 \in A$ and $\delta(A) = 0$ otherwise.

(b) Show that $S\delta$ is the Heaviside function, that is, $S\delta(t) = 0$ if $t < 0$, $S\delta(t) = 1$ if $t > 0$. Find $S^n\delta$ on $[-1, 1]$ for $n = 2, 3, 4, \ldots$.

(c) Solve $d^2y/dx^2 - y = \delta$, $y(-1) = 1$, $y'(-1) = 0$ on $[-1, 1]$.

10. Verify that the solution of $(I + \lambda^2 S^2)y = G$ is

$$y = \cos \lambda t \int_a^t G'(s) \cos \lambda s \, ds + \sin \lambda t \int_a^t G'(s) \sin \lambda s \, ds + G(a) \cos \lambda(t - a).$$

11. Use the result in Exercise 11 to solve $y'' + 4y = \delta$, $y(0) = y'(0) = 0$ where δ is the Dirac function.

12. Suppose g is in $C[0, 1]$ but does not have a continuous derivative. Show the equation $y^{(n)} + a_{n-1}y^{(n-1)} + \cdots + a_1 y' + a_0 y = g$, $y^{(j)}(0) = a_j, j = 0, 1, \ldots, n-1$; has a solution that is in $C^{(n)}[0, 1]$ but not in $C^{(n+1)}[0, 1]$.

13. Let A be the matrix $\{\begin{smallmatrix} 0 & 1 \\ 0 & 0 \end{smallmatrix}\}$. Show A and A^T, the transpose of A, are nilpotent but $A + A^T$ is not nilpotent.

B

14. Prove that a quasi-nilpotent linear operator T from a finite dimensional space E into itself is nilpotent.

15. Prove that the operator T from a normed space into itself is quasi-nilpotent if and only if $\sup_n \|\lambda^n T^n\| < \infty$ for all scalars λ, where $\| \ \|$ denotes the operator norm.

16. For T, a p-tempered operator, on the space $X[a, b]$ define $\|T\|_{PT}$ to be the infimum of all K that satisfies (PT).

(a) Show $\|T\|_{PT}$ can be used for K in (PT).

(b) Show that the collection $PT(X)$ of all p-tempered operators from X into X is a Banach space with the norm $\|T\|_{PT}$.

17. Solve $x' + 4y' + x = H(t-1)$, $x' - 2y' + y = 0$, $x(0) = y(0) = z(0) = 0$ on the interval $[0, 2]$ where H denotes the Heaviside function.

18. If T is a quasi-nilpotent linear operator on a normed space E then T does not have a continuous inverse. [*Hint*: Let $x \neq 0$ be arbitrary. For each n let $y_n = (I - nT)^{-1}x$. Show $y_n/n - Ty_n \to 0$. Thus there is M such that $\|Ty_n\|/\|y_n\| < 2/n$ for $n > M$.]

35

Compact Operators

A linear operator T from a Banach space E into a Banach space F is *compact* if the image $T(B_E)$ of the unit ball of E has a compact closure in F. We leave to you the proof of the following characterizations of compact operators.

Theorem 35.1: Let T be a linear operator from a Banach space E into a Banach space F. The following statements are equivalent:

(a) T is compact;

(b) If S is a bounded subset of E then $T(S)$ has compact closure in F;

(c) If $\{x_n\}$ is a bounded sequence in E then $\{Tx_n\}$ contains a convergent subsequence.

Example 35.2:

Let ϕ be a continuous scalar valued function on the square $[0, 1] \times [0, 1]$. Define T from $C[0, 1]$ into itself by

$$(Tf)(t) = \int_0^1 \phi(s, t) f(s)\, ds.$$

Given $\varepsilon > 0$ let δ be such that

$$|\phi(s_1, t_1) - \phi(s_2, t_2)| < \varepsilon \quad \text{when } \|(s_1, t_1) - (s_2, t_2)\| < \delta.$$

Then if $\|f\| \leqslant 1$ and $|t_1 - t_2| < \delta$ we have

$$|Tf(t_1) - Tf(t_2)| \leqslant \int_0^1 |\phi(s, t_1) - \phi(s, t_2)\| f(s)|\, ds$$
$$\leqslant \int_0^1 \varepsilon \|f\|\, ds = \varepsilon.$$

Therefore, we see that the set $\{Tf\colon \|f\| \leqslant 1\}$ is equicontinuous in $C[0, 1]$; so that by the Arzela-Ascoli Theorem (24.6) the set has a compact closure. Consequently the operator T is compact.

If T is a continuous linear operator from E into F then a continuous linear operator T^* called the *adjoint of* T is defined from F^*, the dual space of F, into E^* by the formula $T^*f = f \circ T$ where "$\circ$" denotes composition. In other words, the value of T^*f on x in E is $f(Tx)$. It is easy to see that T^* is continuous from F^* into E^*, and not hard to show that $\|T^*\| = \|T\|$. The following theorem due to Schauder relates the compactness of T to that of T^*.

Theorem 35.3: A continuous linear operator from a Banach space E into a Banach space F is compact if and only if T^* is compact.

Proof: If T is compact then $T(B)$ has compact closure, where B denotes the unit ball of E. If B^* denotes the unit ball of F^* then B^* is equicontinuous on $\overline{T(B)}$ because $|f(x) - f(y)| \leqslant \|x - y\|$ for x, y in $\overline{T(B)}$. By the Arzela-Ascoli Theorem it follows that B^* is a compact subset of $C(\overline{T(B)})$. If $\{g_n\}$ is a sequence in B^* we can find a subsequence $\{g_{n_k}\}$ that converges uniformly on $\overline{T(B)}$. This implies that $\{g_{n_k} \circ T\}$ converges uniformly on B, from which it follows that the closure of $T^*(B^*)$ is sequentially compact, hence compact.

If T^* is compact then $T^{**} = (T^*)^*$ is compact from E^{**} into F^{**}. Therefore, the closure of the image of B^{**} the unit ball of E^{**} is compact in F^{**}. Since $J(T(B)) \subset T^{**}(B^{**})$ it follows that $J(T(B))$ has a compact closure, where J is the

James mapping from E into E^{**}. But J is an isometry so that $T(B)$ has a compact closure in E. ■

The *spectrum* of a linear operator T from a normed space E into itself is the set of all scalars λ such that the operator $(\lambda I - T)$ does not have a continuous inverse from E into E. Our next goal is to develop a theory originated by F. Riesz that illustrates the striking similarity of the spectral properties of a compact operator with those of a matrix operator on a finite dimensional space.

Theorem 35.4: Suppose T is a compact operator from a Banach space E into itself and suppose λ is a nonzero scalar in the spectrum of T.

(a) The operators $\lambda I - T$ and $\lambda I - T^*$ both have closed ranges.

(b) Either there is x in E such that $Tx = \lambda x$ or there is f in E^* such that $T^*f = \lambda f$.

Proof:

(a) We shall prove that $\lambda I - T$ has a closed range. The proof that $\lambda I - T^*$ has a closed range proceeds similarly. Let $\{x_n\}$ be a sequence in $(\lambda I - T)E$ with limit x. We shall show that x is in $(\lambda I - T)E$. Since each x_n is in the range of $(\lambda I - T)$ there is y_n in E with $x_n = \lambda y_n - Ty_n$.

The first task is to show $\{y_n\}$ contains a bounded subsequence. Otherwise we shall have $\|y_n\| \to \infty$. If we set $z_n = y_n/\|y_n\|$ then we have $(\lambda I - T)z_n \to 0$ and $\|z_n\| = 1$ for all n. Since T is compact there is a subsequence $\{z_{n_h}\}$ of $\{z_n\}$ such that $\{Tz_{n_h}\}$ converges. Since $z_{n_h} - \lambda^{-1}Tz_{n_h} \to 0$ it follows that $\{z_{n_h}\}$ converges. If we set $z = \lim z_{n_h}$, then $\|z\| = 1$ and $(\lambda I - T)z = 0$ contradicting the fact that $(\lambda I - T)$ is one-to-one.

Therefore, $\{y_n\}$ contains a bounded subsequence $\{y_{n_k}\}$ so $\{Ty_{n_k}\}$ contains a convergent subsequence $\{Ty_{n_h}\}$ because T is compact. Since $\{x_{n_h}\}$ converges, $\{\lambda y_{n_h}\}$ converges as well. Because $\lambda \neq 0$ it follows that $\{y_{n_h}\}$ converges say to y. Then we have

$$(\lambda I - T)y = \lim_{n_h}(\lambda I - T)y_{n_h} = \lim_n x_n = x$$

so that x is in the range of $(\lambda I - T)$.

(b) If there is no x in E such that $Tx = \lambda x$ then $(\lambda I - T)$ is one-to-one. By (a) $(\lambda I - T)$ has a closed range F and it has continuous inverse on F by the Closed Graph Theorem (22.4 on page 169). We have assumed that λ is in the spectrum of T so it cannot be the case that F is all of E. Therefore, by Corollary 20.4 on page 157 there is a nonzero continuous linear functional f on E, which is 0 on F. This means that $f((\lambda I - T)x) = 0$ for all x in E or $(\lambda I - T^*)f = 0$. ■

Lemma 35.5: If F is a closed subspace of a normed space E, which is not all of E, then for each r with $0 < r < 1$ there is x_r with $\|x_r\| = 1$ and

$$\inf\{\|y - x_r\| : y \in F\} > r.$$

Proof: Take any point u in $E \backslash F$ and set

$$d = \inf\{\|u - y\| : y \in F\}.$$

Then $d > 0$ because F is closed. Choose $y_0 \in F$ such that

$$d \leqslant \|y_0 - u\| < d(2 - r)$$

and set

$$x_r = \frac{(u - y_0)}{\|u - y_0\|}.$$

It is clear that x_r has norm one. If y is any member of F we have

$$\|y - x_r\| = \left\| y - \frac{u - y_0}{\|u - y_0\|} \right\|$$

$$= \|u - y_0\|^{-1} \|y\|u - y_0\| + y_0 - u\|.$$

Since $y\|u - y_0\| + y_0$ is in F we have

$$\|y - x_r\| \geqslant \frac{d}{\|u - y_0\|} > \frac{d}{d(2 - r)} > r. \blacksquare$$

Theorem 35.6:

(a) The spectrum of a compact operator T on a normed space E is countable and has no limit point except possibly 0.

(b) If $\lambda \neq 0$ is in the spectrum of T then the space F_λ of all x in E such that $Tx = \lambda x$ has finite dimension.

Proof: First suppose $\{\lambda_n\}$ is a sequence of distinct numbers in the spectrum of T. We show $\lim_n \lambda_n = 0$. Otherwise, there is $\varepsilon > 0$ and a subsequence $\{\lambda_{n_k}\}$ such that $|\lambda_{n_k}| > \varepsilon$ for each k. By Theorem 35.4 we must have for each k a vector x_k in E with $Tx_k = \lambda_{n_k} x_k$ or a functional f_k in E^* with $T^* f_k = \lambda_{n_k} f_k$. Therefore, one of these equations must hold for infinitely many k. We continue the proof under the assumption that it is the first case; the second case can be pursued similarly by the reader. In the first case we have a sequence $\{y_n\}$ in E and a sequence $\{\mu_n\}$ of scalars such that $|\mu_n| > \varepsilon$ for each n and $Ty_n = \mu_n y_n$ for all n. The vectors y_n are li because if we could find some smallest n with y_n in $[y_1, \ldots, y_{n-1}]$ we would have

$$y_n = a_1 y_1 + \cdots + a_{p y_p}$$

for some scalars $a_1, \ldots, a_p$ with $p < n$ and $a_p \neq 0$. Applying $(T - \mu_n I)$ to both sides of this equation we would obtain

$$0 = (T - \mu_n I) y_n = a_1(\mu_1 - \mu_n) y_1 + \cdots + a_p(\mu_p - \mu_n) y_p.$$

Since $(\mu_j - \mu_n) \neq 0$ for $j < n$ it follows that we could solve this equation to obtain an expansion for y_p in terms of $\{y_1, \ldots, y_{p-1}\}$, which contradicts the minimality of n.

Let F_n be the linear span of $\{y_1, \ldots, y_n\}$. By Lemma 35.5 there is u_n in F_n with $\|u_n\| = 1$ and $\|u_n - v\| > \frac{1}{2}$ for all v in F_{n-1}. The vector u_n has the form $b_1 y_1 + \cdots + b_n y_n$ so that $(T - \mu_n I) u_n \in F_{n-1}$. If $n > m$ the vector $w_{n,m} =$

$(u_n - \mu_n^{-1} Tu_n) + \mu_m^{-1} Tu_m$ is in F_{n-1} so that we have

$$\left\| T\left(\frac{y_n}{\mu_n}\right) - T\left(\frac{y_m}{\mu_m}\right) \right\| = \| u_n - z_{n,m} \| > \frac{1}{2}.$$

This implies that no subsequence of $\{T(u_n/\mu_n)\}$ converges even though $\| u_n/\mu_n \| \leqslant 1/\varepsilon$ for each n, contradicting the compactness of T. This completes the proof of the assertion that the sequence $\{\lambda_n\}$ must converge to 0.

The set of nonzero points in the spectrum of T is the union of the countable collection

$$\sigma_n = \left\{ \mu\text{: } \mu \text{ is in the spectrum of } T \quad \text{and} \quad \frac{1}{n} \leqslant |\mu| \leqslant n \right\}.$$

If any set σ_n is infinite it must contain a convergent subsequence by the Bolzano-Weierstrass Theorem. Since we have already ruled this out, each set σ_n must be finite so the entire spectrum must be countable and have only 0 as a possible limit point. This completes the proof of (a).

To prove (b) suppose for the sake of obtaining a contradiction that F_λ has infinite dimension for some λ in the spectrum of T. Then we can use Lemma 35.5 to obtain a sequence (x_n) in F_λ such that $\| x_n \| = 1$ for each n but no subsequence of (x_n) converges. But since T is compact and $x_n/\lambda = Tx_n$ for each n it follows that some subsequence of (x_n/λ) converges. This is a contradiction. ■

Suppose K is a continuous real valued function on a square $[a, b] \times [a, b]$. A *Fredholm integral equation* is one of the form

(F) $f(s) = \phi(s) - \lambda \int_a^b K(s, t)\phi(t)\, dt.$

Here f is a real valued continuous function on $[a, b]$, λ is a real number, and ϕ is a solution to be determined. The *homogeneous* Fredholm equation has the form

(H) $\psi(s) = \lambda \int_a^b K(s, t)\psi(t)\, dt.$

The Fredholm equation *dual* to (F) has the form

(D) $g(s) = \phi(s) - \lambda \int_a^b K(t, s)\phi(t)\, dt.$

and the homogeneous form of the dual equation has the form

(DH) $\chi(s) = \lambda \int_a^b K(t, s)\chi(t)\, dt.$

The following theorem proved in 1903 by I. Fredholm is a comprehensive description of the solutions of (F). The Fredholm theory provided the model for the theory of compact operators, now it is a striking application.

Theorem 35.7: Except for a discrete countable set $\{\lambda_n\}$ of real numbers the integral equation (F) has a unique continuous solution ϕ for each continuous function f and the dual equation (D) has a unique solution as well. For each λ_n, there is an integer $m = m(\lambda_n)$ such that the space F_n of solutions of the homogeneous equation (H) and the space G_n of solutions of the dual homogeneous equation both have dimension m. The equation (F) then has a solution if and only if the function f satisfies the condition

$$\int_a^b f(s)\chi(s)\,ds = 0$$

for each χ in G_n, and if ϕ is any solution then $\phi + \omega$ is a solution for every ψ in F_n.

Proof: The equation

$$T\phi(s) = \int_a^b K(s, t)\phi(t)\,dt$$

determines a continuous linear operator from $L^2[a, b]$ into itself. If B is a bounded subset of L^2 then the set $T(B) = \{Tf: f \in B\}$ is an equicontinuous subset of $C[a, b]$. To verify this let $\varepsilon > 0$ be given. Since K is uniformly continuous on $[a, b] \times [a, b]$ there is $\delta > 0$ such that $|K(s, t) - K(u, v)| < \varepsilon/(M(b-a)^{1/2})$ when $\|(s, t) - (u, v)\| < \delta$. Here $M = \sup\{\|f\|: f \in B\}$. If $\|f\| \leqslant 1$ and $|s_1 - s_2| < \delta$ we have

$$|Tf(s_1) - Tf(s_2)| \leqslant \int_a^b |K(s_1, t) - K(s_2, t)\|f(t)|\,dt$$

$$< \left\{\int_a^b \frac{\varepsilon^2}{M^2(b-a)\,dt}\right\}^{1/2} \left\{\int_a^b |f(t)|^2\,dt\right\}^{1/2} \leqslant \left(\frac{\varepsilon}{M}\right)M = \varepsilon.$$

If (f_n) is a bounded sequence in $L^2[a, b]$ then (Tf_n) contains a subsequence that converges in $C[a, b]$. Since the inclusion from $C[a, b]$ into $L^2[a, b]$ is continuous it follows that this subsequence converges in L^2 as well. This proves that T is a compact operator from $L^2[a, b]$ into itself.

The first assertion of the theorem now follows directly from Theorem 35.6 since $(I - \lambda T)\phi = f$ has a unique solution in L^2 if and only if λ^{-1} is not in the spectrum of T. Since f is continuous, and T takes members of L^2 into continuous functions, it follows that ϕ is continuous because $\phi = f + \lambda T\phi$. The sequence (λ_n) is discrete because the only possible limit point of $(1/\lambda_n)$ is 0.

Now suppose $\mu = 1/\lambda$ is in the spectrum of T. If f and g are in L^2 we have

$$\langle Tf, g\rangle = \int_a^b \left\{\int_a^b K(s, t)f(t)\,dt\right\} g(s)\,ds$$

$$= \int_a^b f(t) \left\{\int_a^b K(s, t)g(s)\,ds\right\} dt$$

so that we have

$$T^*g(t) = \int_a^b K(s, t)g(s)\,ds$$

for each g in L^2. This means that the dual problem (D) is represented by the operator conjugate to T. By Theorem 35.5 the range of $(\mu I - T)$ and $(\mu I - T^*)$ are closed for T a compact operator and the null space of at least one of the operators is nonzero. Let F denote the null space of $(uI - T)$, and let G denote the null space of $(\mu I - T^*)$. Since we have

$$\begin{aligned} F &= \{f\colon \langle(\mu I - T)f, g\rangle = 0 \quad \text{for all } g \text{ in } L^2\} \\ &= \{f\colon \langle f, (\mu I - T^*)g\rangle = 0 \quad \text{for all } g \text{ in } L^2\} \end{aligned}$$

it follows that

$$F = \mathscr{R}(\mu I - T^*)^{\perp}$$

where $\mathscr{R}$ denotes the range space. Since $\mathscr{R}(\mu I - T^*)$ is closed it follows that $\mathscr{R}(\mu I - T^*) = F^{\perp}$. Similarly, we have $\mathscr{R}(\mu I - T) = G^{\perp}$. Since L^2 is the direct sum of G and $\mathscr{R}(\mu I - T)$ it follows that each f in F has a unique expansion $f = (\mu I - T)u + v$ where v is in G. Define the linear operator Q on F by the formula $Qf = v$. To see that Q is one-to-one suppose $Qf = 0$. Then $f = (\mu I - T)u$ so that

$$\langle f, f\rangle = \langle(\mu I - T)u, f\rangle = \langle u, (I - T^*)f\rangle = 0,$$

which implies that $f = 0$. Therefore, F and G have the same dimension. This dimension is finite by Theorem 35.6.

There is ϕ such that $f = (I - \lambda T)\phi$ if and only if $\mu f = (\mu I - T)\phi$, which holds if and only if μf, hence f, is in the range of $(\mu I - T)$. Since $\mathscr{R}(\mu I - T) = G^{\perp}$ it follows that $f = (I - \lambda T)\phi$ has a solution if and only if $\langle f, \chi\rangle = 0$ for each χ in G. If ϕ is such a solution and ψ is in F then certainly $\phi + \psi$ is also a solution. ■

Exercises

A

1. Prove Theorem 35.1.
2. Show that a compact operator is always bounded.
3. Suppose S is compact from F into G, T is continuous from E into F, and R is continuous from G into H. Show ST is compact from E into G and RS is compact from F into H.
4. *(a)* Show that every linear operator from a finite dimensional space into a normed space is compact.
   (b) Show that every continuous linear operator from a normed space into a finite dimensional space is compact.
   (c) Why do we need to specify continuous in (b)?
5. Show that the linear operator S defined by $(Sf)(t) = \int_0^t f(s)\,ds$ is compact in $C[0, 1]$ and that the spectrum of S contains only 0.
6. Define T on c_0 by $T(x_1, x_2, \ldots) = (x_2, 0, x_4, 0, \ldots)$. Show T is continuous and nilpotent $(T^2 = 0)$ but not compact.
7. In the proof of Lemma 35.5 we use the fact that $1/(2 - r) > r$. Establish this inequality.

B

8. Use Lemma 35.5 to prove that a normed space E has finite dimension if and only if the identity operator on E is compact.

9. *(a)* Prove that if T_n is a compact linear operator from E into F for each n and (T_n) converges in $L(E, F)$ to T then T is also compact.

(b) Thus the set $C[E, F]$ of all compact operators from a normed space E into a Banach space F with the uniform operator norm forms a Banach space.

10. Suppose E is one of the BK-spaces, $l^p\, 1 \leqslant p \leqslant \infty$, or c_0.

(a) Show that a diagonal matrix mapping from E into E is compact if and only if the diagonal converges to 0.

(b) Show that the nonzero spectrum of the matrix consists of precisely the members of the diagonal.

(c) Describe the adjoint operator and its spectrum.

11. Let $(S, \mathscr{A}, \mu)$ denote a measure space, $1 < p < \infty$ and $1/p + 1/q = 1$. Suppose K is a scalar valued measurable function on $S \times S$ such that

$$\left\{\int_S \left\{\int_S |K(s, t)|^p \, d\mu(s)\right\}^{q/p} d\mu(t)\right\}^{1/q} = M < \infty.$$

If we define T by $(Tf)(s) = \int_S K(s, t) f(t)\, d\mu(t)$, show T is a compact operator on $L_p(S, \mathscr{A}, \mu)$.

12. Use Lemma 35.5 to prove that a normed space E has finite dimension if and only if the identity operator on E is compact.

13. Show that if a compact operator T in a normed space E has an inverse then E has finite dimension.

14. Prove that if T is a compact linear operator on a normed space E and $\lambda \neq 0$ then the space of all x in E such that $(\lambda I - T)x = 0$ has finite dimension.

15. For (s, t) in the square $[0, \pi] \times [0, \pi]$ define $K(s, t)$ by the formula

$$K(s, t) = \sum_{n=1}^{\infty} \frac{\sin ns \sin nt}{n}.$$

(a) Show that this series converges uniformly on the square to a function continuous in both variables.

(b) Show that the equation

$$f(s) = \lambda \phi(s) + \int K(s, t) \phi(t)\, dt$$

has a unique solution for each f if and only if λ is not equal to $2n\pi$, $n = 0, \pm 1, \pm 2, \ldots$.

C

16. If $p > r \geqq 1$ show every continuous linear operator from l^p into l^r is compact.

36

Ergodic Theory

Ergodic theory is concerned with the existence of time averages of functions defined on a mechanical system. The state of a mechanical system at time t is represented by a point $x(t)$ in a subset S of R^n. If the system is deterministic then the state $x(s)$ for $s > t$ is completely determined by $x(t)$. If Δt is a small unit of time then a *flow function* Ψ is determined from S into S by the formula $\Psi(x(t)) = x(t + \Delta t)$. This function Ψ has the property that $\Psi(x(n\Delta t)) = \Psi^n(w)$ for each n.

A numerical quantity determined by the momentary state of the system (for example, the force exerted on the walls of a vessel by gas molecules contained in it) is given by a real valued function f on S. If the time t is small the quantities $f(\Psi^n(x))$ may fluctuate erratically with n but the average quantity

$$\text{(AVG)} \quad A(N, x) = \frac{1}{N} \sum_0^N f(\Psi^n(x))$$

may be more predictable and measurable in a laboratory. Ergodic theory is concerned with determining conditions under which the limit $\lim_{N\to\infty} A(N, x)$ will exist. In our elementary treatment of ergodic theory we shall restrict ourselves to discrete time t whereas in the continuous model of a physical process one would allow t to go to zero and inquire about the existence of the limit

$$\lim_{T\to\infty} \frac{1}{T} \int_0^T f(\phi_t(x))\, dt.$$

There are two types of ergodic theorems: (1) the individual ergodic theorems that concern the convergence of $A(N, x)$ for almost all x in $(S, \mathscr{A}, \mu)$ where μ is some measure on S and (2) the mean ergodic theorems that concern the convergence of $A(N, x)$ as a function of x in some space $L^p(S, \mathscr{A}, \mu)$. Of central value in ergodic theory is a theorem of Liouville that asserts that under most conditions, transformations of the type Ψ are measure preserving. That Ψ is measure preserving is a frequent hypothesis in ergodic theorems.

Our first result is the *Individual Ergodic Theorem* due to G. D. Birkhoff. We begin with an easy combinatoral lemma. Let $\{a_1, a_2, \ldots, a_n\}$ be an indexed set of real numbers. A member a_i of this set is called *m-favorable* for some integer $m > 0$ if there is some $p < m$ such that $a_i + a_{i+1} + \cdots + a_{i+p} \geqslant 0$. If a_i is m-favorable and negative then $a_{i+1} + \cdots + a_{i+p} > 0$ so that a_{i+1} is m-favorable.

Lemma 36.1: If F_m is the set of all m-favorable members of the set $\{a_1, a_2, \ldots, a_n\}$ then the sum of the members of F_m is nonnegative.

Proof: The proof proceeds by induction on the number of members of F_m. If F_m has only one member, then it is nonnegative by the observation above. Suppose the lemma is true when F_m contains k members. Let F_m have $k + 1$ members and let a_i be the member with smallest subscript. Then there is $p < m$ with $a_i + a_{i+1} + \cdots + a_{i+p} \geqslant 0$. Assume p is minimal. By the induction hypothesis the set G_m of m-favorable numbers in the set $\{a_{i+p+1} a_{i+p+2}, \ldots, a_n\}$ has a nonnegative sum; but this set is $F_m \setminus \{a_i, a_{i+1}, \ldots, a_{i+p}\}$. Since p is minimal the number a_{i+p} is positive and thus m-favorable. By the observation preceding the theorem each a_j with $i<j<i+p$ is m-favorable. Combining these facts, we see that the sum of the members of F_m is

$$a_i + a_{i+1} + \cdots + a_{i+p} + \sum \{b \colon b \in G_m\} \geqslant 0. \ ■$$

Theorem 36.2: Let $(S, \mathscr{A}, \mu)$ be a finite measure space and let Ψ be a one-to-one function from S into S such that $\mu(\Psi(A)) = \mu(A)$ for every A in $\mathscr{A}$. Given f in $L^1(S, \mathscr{A}, \mu)$ let

$$A_n(x) = \frac{1}{n}\{f(x) + f(\Psi x) + \cdots + f(\Psi^{n-1} x)\}.$$

Then $\lim_{n\to\infty} A_n(x)$ converges for almost every x in S.

Proof: For every pair of rational numbers $a<b$ let $S(a, b)$ denote the set of all x in S for which

$$\liminf_n A_n(x) < a < b < \limsup_n A_n(x).$$

It suffices to prove that $\mu(S(a, b)) = 0$ for all $a<b$.

We first note that $\Psi x \in S(a, b)$ if and only if $x \in S(a, b)$ because of the way $S(a, b)$ is determined. Since $\liminf_n A_n(x) < a$, it follows that for each x in $S(a, b)$ there is m such that $A_m(x) < a$. Let S_m denote the set of all x in $S(a, b)$ such that $A_m(x) < a$.

Let $g(x) = a - f(x)$ so that $g(\Psi^n x) = a - f(\Psi^n x)$. Then we have

$$g(x) + g(\Psi x) + \cdots + g(\Psi^m x) = ma - mA_m(x) \geqslant 0.$$

For each given M, consider the indexed set

$$g(x), g(\Psi x), \ldots, g(\Psi^{M+m-1} x).$$

For $k \geqslant 0$ let T_k denote the set of x in $S(a, b)$ such that $g(\Psi^k x)$ is m-favorable. By Lemma 36.1 we have

$$\sum_{k=0}^{n+M-1} \int_{T_k} g(\Psi^k x)\, d\mu \geqslant 0. \tag{36–1}$$

For each $k < M$, we have $\Psi(T_k) = T_{k-1}$ and $T_0 = S_m$ so we have

$$M \int_{S_m} g\, d\mu + \sum_{k=M}^{n+M-1} \int_{T_k} g(\Psi^k x)\, d\mu \geqslant 0,$$

which yields

$$M \int_{S_m} g(x)\, d\mu + m \int_S |g|\, d\mu \geqslant 0$$

since $m \int_S |g|\, d\mu$ is clearly greater than $\sum_{k=M}^{M+m-1} \int_{T_k} g(\Psi^k x)\, d\mu$. Now M is arbitrary so we must have $\int_{S_m} g\, d\mu = \int_{S_m} a - f\, d\mu \geqslant 0$. Since $S(a, b) = \bigcup_m S_m$ we have $\int_{S(a,b)} (a - f(x))\, d\mu \geqslant 0$.

An argument similar to that in the preceding paragraph shows that $\int_{S(a,b)} (f(x) - b)\, d\mu \geqslant 0$. Adding, we see that $\int_{S(a,b)} (a-b)\, d\mu \geqslant 0$. But since $a<b$ it follows that $\mu(S(a, b)) = 0$. ■

Before presenting an example of a *Mean Ergodic Theorem* (Theorem 36.6) we first must derive three facts about Banach spaces.

Theorem 36.3: If the conjugate space E^* of a normed space E is separable so is E.

Proof: If $\{f_n\}$ is a countable dense subset of E^*, let x_n in E be such that $\|x_n\| \leqslant 1$ and $|f_n(x_n)| \geqslant \|f_n\|/2$. The set F of all finite linear combinations of members of $\{x_n\}$ that have rational coefficients is countable. Suppose f in E^* is 0 on F. If (f_{n_k}) is a subsequence of $\{f_n\}$ that converges to f then since

$$\|f-f_{n_k}\| \geqslant |(f-f_{n_k})(x_{n_k})| = |f_{n_k}(x_{n_k})| \geqslant \frac{\|f_{n_k}\|}{2}$$

it follows that $f_{n_k} \to 0$ so that $f=0$. By Corollary 20.4 on page 157 this implies that F is dense in E. ■

Theorem 36.4: A closed subspace F of a reflexive Banach space E is reflexive.

Proof: Let J_E denote the James map from E onto E^{**} and let J_F denote the James map from F into F^{**}. We shall prove that J_F is onto. Define T from E^* into F^* by $(Tf)(y) = f(y)$ for $y \in F$. Then T is onto because each continuous linear functional on F can be extended to all of E by the Hahn-Banach Theorem (20.2 on page 156). Suppose $\hat{y} \in F^{**}$. Then the functional defined on E^* by $\hat{x}(f) = \hat{y}(Tf)$ is linear because it is the composition of linear operators, and it is continuous because $|\hat{y}(Tf)| \leqslant \|\hat{y}\|\,\|f\|$ for f in E^*. Therefore, $\hat{x} \in E^{**}$ so there is x in E such that $J_E x = \hat{x}$. If x were not in F we could find a continuous linear functional g on E that is 0 on F but with $g(x) \neq 0$. Then we would have $g(x) = \hat{x}(g) = \hat{y}(Tg) = \hat{y}(0) = 0$, which is a contradiction. Therefore, x must be in F. Finally $J_F x = \hat{y}$ because T is onto and $\hat{y}(Tf) = \hat{x}(f) = f(x) = (Tf)(x)$ for all f in E^*. ■

Theorem 36.5: If $\{x_n\}$ is a bounded sequence in a reflexive Banach space E, then there is a subsequence $\{x_{n_k}\}$ and an element x in E such that for each f in E^* we have $\lim_k f(x_{n_k}) = f(x)$.

Proof: Let F denote the closed linear span of $\{x_n\}$. Then F is reflexive by Theorem 36.4. By Theorem 36.3, F^* is separable since its dual space is isometric to F. Let $\{f_n\}$ be a countable dense subset in F^*. Use a diagonal argument to find a subsequence $\{x_{n_k}\}$ of $\{x_n\}$ such that $\{f_m(x_{n_k})\}$ converges for each m. By the Banach-Steinhaus Theorem (21.3) there must be $\hat{x}$ in F^{**} such that $\lim_k f(x_{n_k}) = \hat{x}(f)$ for all f in F^*. But F is reflexive so there is x in F with $f(x) = \hat{x}(f)$ for all f in F^*. ■

Theorem 36.5 can be restated as follows: A subset of a reflexive Banach space is weakly sequentially compact if and only if it is bounded and weakly closed. We are finally prepared to prove our Mean Ergodic Theorem.

Theorem 36.6: Let $(S, \mathscr{A}, \mu)$ be a finite measure space and let Ψ be a function from S into S such that $\Psi^{-1}(B) \in \mathscr{A}$ for all $B \in \mathscr{A}$, and such that there is $K > 0$

for which $\mu(\Psi^{-n}B) \leqslant K\mu(B)$ for all n and for all B in $\mathscr{A}$. For each f in $L^p(S, \mathscr{A}, \mu)$ with $1 \leqslant p < \infty$ let A_n be defined by

$$(A_n f)(x) = \frac{1}{n}\{f(x) + f(\Psi x) + \cdots + f(\Psi^{n-1}x)\}.$$

Then $\{A_n\}$ is a sequence of continuous linear operators from L^p into L^p that converge pointwise to a continuous linear operator A. The operator A is a projection of norm $\leqslant K$ from L^p onto the space of all f in L^p such that $f \circ \Psi = f$.

Proof: The function $Tf = f \circ \Psi$ is a linear operator from L^p into L^p because $f \circ \Psi$ is measurable. Moreover $\| T^n \| \leqslant K^{1/p}$ for each n because for each simple function $f = \sum a_j \chi_{B_j}$ we have

$$\| T^n f \|^p = \int_S |f(\Psi^n x)|^p \, d\mu = \sum |a_j|^p \mu(\Psi^{-n}(B_j))$$

$$\leqslant K \sum |a_j|^p \mu(B_j) = K \int_S |f|^p \, d\mu = K \| f \|^p.$$

The set of simple functions is dense in L^p so that $\| T^n f \| \leqslant K^{1/p} \| f \|$ for all f in L^p. Since $A_n = (1/n)\{I + T + T^2 + \cdots + T^{n-1}\}$ it follows that A_n is linear and has norm at most $K^{1/p}$.

We continue the argument for the case when $1 < p < \infty$. In this case the spaces L^p are reflexive so that by Theorem 36.5, for each f in L^p there is a subsequence $(A_{n_k} f)$ of $(A_n f)$ and f in L^p such that $\lim_k F(A_{n_k} f) = F(f)$ for each F in $(L^p)^*$. We shall prove, in fact, that $(A_n f)$ is weakly convergent for each f in L^p. For otherwise there would be two subsequences $(A_{n_k} f)$ and $(A_{m_k} f)$ that converge to different functions g and h respectively. We observe that if the function k is in the range $R(I - T)$ of $(I - T)$ then $A_n k$ converges strongly to 0. This is true because if $k = (I - T)l$ we have

$$A_n k = A_n(I - T)l = \frac{1}{n}(I - T^n)l \to 0.$$

We also observe that $Tg = g$ because for each F in $(L^p)^*$ we have

$$F((I - T)g) = \lim_{n_k} F((I - T)A_{n_k} f) = \lim_{n_k} F\left((I - T^n)\frac{f}{n}\right) = 0;$$

and similarly we have $Th = h$. Since $Tg = g$ and $Th = h$ it follows that $A_n g = g$ and $A_n h = h$ for each n so that $\lim_n A_n(g - h) = g - h$. We have assumed that $g \neq h$ so there is F in $(L^p)^*$ such that $F(g - h) \neq 0$ and F is 0 on $R(I - T)$. For each n we have

$$F(f) = F(T^n f + (I - T^n)f) = F(T^n f) + F((I - T^n)f) = F(T^n f)$$

because $(I - T)$ is a factor of $(I - T^n)$. Consequently $F(A_n f) = F(f)$ for all n, which implies

$$F(g) = \lim_{n_k} F(A_{n_k} f) = F(f) = \lim_{m_k} F(A_{m_k} f) = F(h),$$

which is a contradiction. Therefore, we are forced to conclude that for each f in L^p there is Af in L^p such that the sequence $(A_n f)$ converges to Af.

We leave it to you, the reader to show that A is a linear operator from L^p into L^p. That A is bounded follows because for each F in $(L^p)^*$ and f in L^p we have

$$|F(Af)| = \lim_n |F(A_n f)| \leqslant K^{1/p} \|F\| \|f\|.$$

In the preceding paragraph we have noted that $T(Af) = Af$ for each f in L^p so that $A_n(Af) = Af$ for each n so that $A^2 f = Af$. This means that A is a projection and that h is in the range of A if and only if $Th = h$ or $h = h \circ \Psi$.

We have noted above that if g is in the range $R(I-T)$ of $I-T$ we have $Ag = 0$. This means that $\overline{R(I-T)} \subset N(A)$. We shall prove that $N(A)$ is the closure of $R(I-T)$. Suppose F is any member of $(L^p)^*$ that is 0 on $R(I-T)$. Then as above $F(f) = F(T^n f)$ for all n, from which it follows that $F(Af) = F(f)$. Thus if $f \in N(A)$ we must have $F(f) = 0$ so that $R(I-T)$ is dense in $N(A)$ by Corollary 20.4. Now if $f \in N(I-T)$ we have $\lim_n A_n f = f$, and if $f \in R(I-T)$ we have $\lim_n A_n f = 0$. Therefore, if $f \in \overline{R(I-T)}$ we have $\lim_n A_n f = 0$. Since each f in L^p has a decomposition $f = g + h$ with g in $N(I-T)$ and $h \in \overline{R(I-T)}$, it follows that $\lim_n A_n f$ converges to AF for all f in L^p. This establishes the theorem for $1 < p < \infty$.

For each B in $\mathscr{A}$, we have $\lim_n A_n \chi_B$ converges in L^2 by the above reasoning. Since the inclusion from L^2 into L^1 is continuous for μ a finite measure it follows that $\lim_n A_n \chi_B$ converges in L^1 to $A\chi_B$. The operators (A_n) are bounded and $\{\chi_B : B \in \mathscr{A}\}$ is a fundamental set in L^1 so by the Banach-Steinhaus Theorem (21.3 on page 163), it follows that $(A_n f)$ converges to AF for all f in L^p. The other details of the proof are like those for L^p with $p > 1$. ■

Exercises

A

1. Suppose $(S, \mathscr{A}, \mu)$ is a finite measure space and Ψ is a one-to-one function from S into S such that $\mu(\Psi(A)) = \mu(A)$ for all A in $\mathscr{A}$. Show that the function $Tf = f \circ \Psi$ is an isometry on $L^p(S, \mathscr{A}, \mu)$ for $1 \leqslant p < \infty$.
2. Let $S = [0, 1)$ and let m denote Lebesgue measure restricted to S.
   (a) Define u from S into S by
   $$u(t) = t \quad \text{if } 0 \leqslant t \leqslant \tfrac{1}{2}, \qquad u(t) = \tfrac{3}{2} - t \quad \text{if } \tfrac{1}{2} < t < 1.$$
   Define Ψ on the set $\mathscr{M}$ of Lebesgue measurable subsets of S by $\Psi(A) = u(A)$, the range of A under u. Then $\mu(\Psi(A)) = \mu(A)$ for all A in $\mathscr{M}$ even though u is not continuous, let alone absolutely continuous.
   (b) Describe other examples of functions v from S into S that preserve Lebesgue measure but are not continuous.
   (c) Prove that there are only two continuous measure-preserving functions from S into S.
3. Let S and Ψ be as in Exercise 2. Show that $\Psi(\Psi(x)) = x$ for all x in S so that if A_n is given as in Theorem 36.2 then
   $$\lim_{n \to \infty} A_n(x) = \frac{f(x) + f(\Psi(x)}{2}.$$

4. What is an example of a separable normed space whose conjugate space is not separable?
5. Suppose Ψ is a function from $\mathscr{A}$ into $\mathscr{A}$ for which there is r with $0 < r < 1$ such that $\mu(\Psi^{-1}(A)) \leqslant r\mu(A)$ for all A in $\mathscr{A}$. Show that $A_n f$ converges to 0 for each f in $L^p(S, \mathscr{A}, \mu)$ for $1 \leqslant p < \infty$. [*Hint*: First show it for f, a characteristic function of a set, and then apply the Banach-Steinhaus Theorem.]
6. Prove that if $\|T\| < 1$ then (T^n) converges to zero in the uniform operator topology so that $A_n = (1/n) \sum_{j=0}^{n} T^j$ also converges to 0.

B

7. Suppose T is a continuous linear operator on a Banach space E such that $A_n x = (1/n) \sum_{j=0}^{n} T^j x$ converges for each x in E. If $Px = \lim_n A_n x$ show P is a continuous projection from E into E such that the null space is the closure of the range of $(I - T)$ and the range space of P is the null space of $(I - T)$. Thus P is nonzero if and only if 1 is in the spectrum of T.
8. Suppose $I = [0, 1)$, $I_j = [j - 1/n, j/n)$ for $j = 1, 2, \ldots, n$, and w_{jk} is a linear function from I_j into I_k. If π is a permutation on the integers $\{1, 2, \ldots, n\}$ define Ψ on I by

$$\Psi(x) = w_{j\pi(j)}(x) \quad \text{if } x \in I_j.$$

   (a) Show Ψ satisfies the conditions of Theorem 36.4 where $\mathscr{A}$ denotes the Lebesgue measurable subsets of I and μ is Lebesgue measure m.
   (b) Describe Af for f in L^p.
   (c) How many possible functions Ψ can there be?

Index

The letter d after a page number means the theorem is stated or concept defined on that page. The letter f after a page number means that subsequent pages, usually the rest of the section, contain discussion of the topic.